I.E.Timchenko, E.M. Igumnova, I.I.Timchenko

ADAPTIVE BALANCE MODELS FOR ENVIRONMENTAL-ECONOMIC SYSTEMS

UDC 5513465.7:504.42

Adaptive Balance Models for Environmental-Economic Systems

I.E. Timchenko, E.M. Igumnova, I.I. Timchenko

The monograph presents a new method of Adaptive Balance of Causes (*ABC*-method), based on second-order negative feedback links between model's variables and their velocities of change. The method is applied to the problem of terrestrial and marine resources consumption with the purpose to assess profitability of their use in coastal zone areas (*CZA*s). Drawing on principal steps of systems thinking and modeling, conceptual models of *CZA*s industrial, biological, recreational and environmental resources were developed and studied in view of simultaneous dynamics of environmental management and economic development. The models containing large numbers of control parameters were formalized by the *ABC*-method, which allowed to develop prediction scenarios of ecological and economic processes in the *CZA*s. The *ABC AGENT* informational technology was applied with the purpose of using management agents in model equations to perform control of factors defining balance between economic profitability and environmental protection activities in *CZA*s. The principal advantage of this management technology and its capacity to perform simulation experiments were proven by a variety of prediction scenarios; the goals of *CZA*s sustainable development were satisfied.

CONTENTS

INTRODUCTION

Sustainable development of territories depends on rational use of their natural and economic resources. Intensive consumption of natural resources, which is essential for economic growth, inevitably leads to deterioration of the environment, if there is no balance between the economic benefits of resources consumption and the environmental expenditure. This book, grounding on examples of coastal zone areas, shows that the problem can be resolved by applying systematic approach to building informational technologies of the areas' nature and economy management.

Every management technology is based on a model, designed to forecast development scenarios. This is why, we focused our attention on a new method of systems' models construction, which was proposed by us in [], published in Russian. This method, whicn we called Adaptive Balance of Causes (ABC-method), can be regarded as further development of the well-known System Dynamics method (SD-method), which is used to create models of causality (cause - effect) of complex systems. The ABC method has several advantages due to its fundamental difference from the SD-metod: if models developed by SD method have stabilizing feedback of first order, which are common for all equations, the ABC method models have second order negative feedback in each equation. This helps significantly simplify models' construction and improves stability and convergence of numerical solutions of model equations.

Since the moment of its development, *ABC*-method was probated numerous times by construction of socio-economic and ecological-economic systems' models. A general model of economic production processes, which contained management agents (ABC AGENT-technology) was suggested. Applications of this technology were presented in our monograph [] devoted to management of educational socio-economic systems.

Considerable number of ABC-method applications deal with the problem of natural resources consumption in coastal zone areas, which is due to the work of the authors on the subject. The results of that research were published in a number of scientific articles and monographs, mostly in Russian. Some of these results were presented in [], which was translated into English by I.I.Timchenko and published in Kiev. This book is a revised edition of that translation.

In this monograph, we look into the problem of near-shore sea and ocean resources consumption, aiming to ensure sustainable development of seashore territories. Due to the multidisciplinary character of sustainable development, all natural processes of the off-shore marine environment, as well as the socio-economic processes taking place in the adjoining seashore territories should be studied together as a system which we will furtheron call a Coastal Zone Area (*CZA*).

The *CZA* resources management is a structural part of the overall integrated sustainable development management of such areas. The urgency of careful attitude towards natural processes developing in the Sea – Land interface was first highlighted by The World Congress "Earth Summit" in 1992, which adopted a so-called «Agenda-21» document [66]. Chapter 17 of that document was devoted to the tasks of marine environment protection. It provided recommendations to all countries on rational consumption of unrenewable *CZA* resources and on preserving of living resources in the seas and oceans.

In subsequent years, the UN Environment Program [142], the Intergovernmental Oceanographic Commission of UNESCO [61], and governments of several countries (the USA, the European Union, and others) developed and carried out numerous projects aimed at promotion of rational use of the *CZA* resources [187].

The experience obtained during implementation of those programs resulted in development of conception of Integrated Coastal Zone Management –*ICZM* [61]. The urgent necessity in a generalized, but scientifically grounded methodology of the *CZA* integrated management became evident [22].

The coastal zone areas all over the globe are multifunctional. Being the trade and transportation centers that bring in enormous returns (ports and harbours), they have unique economies. In the ecological-economic aspect, they represent the richest pantries of marine biological and mineral resources. Socially, they are the areas of seashore cities concentration, being especially attractive for their natural variety of landscapes and marine climate.

Combination of saline and fresh water in coastal estuaries creates the richest bioresources and the most productive ecosystems. In many regions, the millennia-old coastal landscapes attract millions of tourists, ensuring the growth of recreational infrastructure. Approximately half of the world's population is already concentrated along the narrow ribbons of earth which surround oceans, seas and large lakes. Two thirds of the world's largest cities are located along coastal lines; their populations grow much faster compared to that of the mainland cities.

High population density in the *CZAs* multiplies the antropogenic loading on these regions and threatens their valuable natural properties. Contamination of the off-shore waters considerably reduces production potential of fishing industry, causes degradation of mangroves and coral reefs, violates the balance of processes in island and off-shore ecosystems, which have been forming through ages. The uncontrolled construction and mining for mineral resources in the *CZAs* sharply reduces their attractiveness for tourism and recreation. In many countries of the world, rapid urbanization and economic activity in *CZAs* generates sharp conflicts between the consumption of natural resources and environmental protection.

Management of a whole *CZA* or its structural parts must be regarded as a flexible, continuous decision-making process, aimed at compromising between setting development goals and selecting social, ecological and economic criteria for estimation of attainability of these goals. Sustainable development management is hardly possible without evaluation of consequences that may follow the resources-related decisions. As the socio-economic development processes can only be examined in connection with each other, the complex systems approach to management of the *CZA* resources must be applied [22, 181].

This monograph studies systems methodology of resources management in coastal zone areas. Continuing the early researches of this subject [39, 171, 177, 181], we are determined to show that the most generalized methodology of the natural and socio-economic resources consumption must be grounded on systems principles, systems thinking and consequently, on systems modeling and systems management techniques. Systems methodology is aimed at development of informational technologies that support decision-making. The ultimate goal of systems methodology is to determine stages of development of such technologies.

In order to predict possible scenarios of development, mathematical models of the proper processes are needed. Two approaches to the design of natural marine ecosystems and coastal ecological-economic systems are developing presently: the differential approach which is based on partial derivative equations, and the integrated approach which uses ordinary differential equations.

The *CZA* resources management is related to development of each of these two classes of models. Presently, as the integral models are considerably simpler than the differential models, the *CZA* resources management is mainly directed at application of the integral models.

In the majority of marine ecosystem models developed by the integral approach, the analytical formulas are used. They link the modeling processes to each other and close the systems of model equations. These formulas are of approximate character and are obtained, as a rule, from the data of laboratory experiments. To verify the adequacy of such models, enormous volumes of field measurements' information are needed, which can only be obtained by means of development of specific geo-informational systems of observations and control.

However, there exists another method of building ecosystems' integral models; it is based on the analysis of mutual tendencies in variability of the ecosystem's state parameters. In the simplest case it means that all parameters of the state are the linear combinations of each other. Thus, the linear forms' coefficients which link them can be interpreted as levels of mutual influences between the processes. If we apply the methods of identification of these coefficients from observational data and follow the changes of coefficients in time, such ecosystem model will become considerably simplified. It will represent a fully competitive version of traditional integral models. The development of this very method of the ecological-economic systems modeling is the object of this monograph.

Chapter 1 of the book offers a short analysis of *CZA* systems modeling and management problems. The method of system dynamics is known as the most elaborate instrument of integrated development processes' modeling in complex social ecological-economic systems [52]. As an example of application of this method, a *CZA* model was built. It contains 3 system levels (the population's life standard, the ecological state of the natural environment, and the gross domestic product), and 13 informational levels of the system's state which describe cause-effect interactions between the processes of development. The advantages and limitations of the system dynamics method are discussed.

The method of Adaptive Balance of Causes (*ABC*-Method) is suggested as a perspective trend of the system dynamics, which is free of some limitations that the method of system dynamics has. The method of Adaptive Balance of Causes is discussed in detail below, in Chapter II.

The book offers fundamentals of systems methodology of the complex adaptive systems management. It describes the structure of the systems methodology, outlines philosophy of systems approach and provides practical steps to building informational technologies of management, which are attributed to the systems analysis. The systems approach is based on six basic concepts (also discussed in this chapter), which allow experts to link the goals of management with the choice of models that suggest actions to achieve these goals.

The systems analysis lays the basis for applied methods of construction of integral models to forecast dynamics of goal–seeking scenarios and their adaptation to observational data.

Chapter 2 describes a new method of complex systems modeling – the method of Adaptive Balance of Causes (*ABC*-method [181]), which is built in accordance with systems principles (as reviewed in Chapter 1). We provide a justification for a unified system of equations of this method, and show its difference from the Method of System Dynamics [50]. The accent is made on such properties of the *ABC*-method as the simplicity of models' construction and the analysis of intra-relationships between the simulated processes, even with no external influences on the modeling system.

Applications of the systems methodology and the *ABC*-method are illustrated with a number of examples. In the first example, the method was used to construct a model of environmental management. We looked at the process of making up a scheme of cause-effect relationships between the key processes of the study – the demand for natural resources, resources consumption level, and ecological state of the environment.

The second example illustrates the ways of analyzing scenarios of development processes in the socio-economic system of the *CZA*, in connection with the task of local budget distribution between industrial and social needs.

In the third example, a so-called Daisy World model [199] was constructed by the *ABC*-method. This model was proposed according to the *GAIA*-theory [100] to explain the reaction of the global system "The Atmosphere - The Earth's Surface" to the greenhouse gases content in the atmosphere. The notion of self-organization processes in the nature (introduced by this theory) was illustrated by construction of a generalized social ecological-economic adaptive model of global development, which connects 18 processes. This *ABC*- model allowed us to simulate self-organization scenarios in global natural and socio-economic development processes.

One of the serious problems in modeling of *CZA's* natural-economic systems is the choice of the model equations' coefficients, which should reflect cause – effect interactions between the processes subject to modeling. The closer are these coefficients to the reality, the more adequate is the model, and consequently, the more accurate are the forecasted scenarios which are obtained by the model.

The concluding paragraphs of the Chapter Two offer a way to solve this task by using the *ABC*-method. It describes an objective method of identifying coefficients for *ABC*-models. The method is grounded on the results of the optimal filtration theory which is applied to the observational data of the development processes.

The *ABC*-model for setting influence coefficients was proposed. It allows to implement variable coefficients calculated simultaneously with prediction of the development scenarios. An example of statistical identification of such coefficients using Kolmogorov's optimal interpolation method [90], was supplied.

Application of the optimal filtering theory helps to resolve such an important task of the development management as adaptation of predicted scenarios to the observational data. The solution can be based on assimilation of the observational data in the *ABC*-models of complex *CZA* systems by implementing Kalman's filter method [88]. We provided an example of predicted values corrections by means of data assimilation, which serves to clarify this method and to show its efficiency.

Chapter 3 deals with the task of modeling of the development processes in coastal zone marine ecosystems. The *ABC*-model of integrated chemical-biological processes developing in the upper layer of the north-west shelf zone of the Black sea (assumingly in the central part of the region) was built. The annual cycle of the sea surface temperature, the solar illumination, the intensity and duration of the wind stress, as well as the nutrients coming in from the river runoff, were taken as external climatic forcing and were applied to the model input. They formed the set of external climatic conditions determining the ecosystem's variability.

In the proposed model, the anchovy biomass was defined by the cold- and heat-loving species of zooplankton, as well as by the fish larvae and by the availability of nutrients. The Analytical Hierarchy Process method by Saaty [150] was used to determine coefficients of the model. Scenarios of integrated processes development were constructed, taking into account the dependence between plankton intra-dynamics on the sea surface, and the winter-time wind regime in this region.

The systems principle of adaptive balance of causes reflects the desire of living organisms of the marine ecosystem to adapt to changing environmental conditions. In each case, the ability to adapt is limited by livelihood resources, and that should be reflected in models of marine ecosystems. Chapter 3 deals with techniques of development processes' resources limitation in marine ecosystems. The agents of limitation were introduced here. They are logical operators which control the growth of aquatic organisms'

15

concentrations and their death rate, resulting from insufficient vital resources.

Limitation processes of development resources affect both, the living organisms in marine ecosystems, and the formation of new substances in course of chemical reactions. Within the integral description of these phenomena, we studied formation of nutrients during the oxidation of detritus, since the dying organisms produce detritus, while the living organisms need nutrients for their existence. During computational experiments with integral models of the ecosystem, the influence of homeostasis zones' boundaries on aquatic organisms was demonstrated by formation of complex scenarios of their concentrations.

The role of the management agents in formation of the ecosystem processes' scenarios is shown on an example of a generalized *ABC*-model of the marine ecosystem, which contained 17 interconnected chemical and biological processes and included three groups of management agents, which followed the environmental changes of the marine organisms' habitats.

The computational experiments on the model helped us to construct integrated scenarios of concentrations of phytoplankton, zooplankton, fish, oxygen, nutrients, detritus, carbon dioxide, nitrates and other elements. Scenarios induced by the decreasing oxygen concentration in the sea resulting from growing amounts of organic matter and nutrients from a river runoff, were considered. The results of applying systems management methodology to construction of models of ecosystems demonstrated that dynamic *ABC*-models of the *CZA* ecosystems can be built for prediction of integral development scenarios in the ecosystems. The models can be grounded on causal relationships between major processes taking place in the ecosystem, and can utilize the data obtained by previous studies.

Chapter 4 contains results of research on construction of the ecological-economic models, which include models of two subsystems: the marine ecosystem, as well as the ecological and economic subsystems of a production unit located on the sea shore.

The natural, ecological and other types of coastal zone resources can be used in different production technologies and services. Each local region, as well as the *CZA* marine areas, should be considered as sources of resources production, depending on economic benefits and environmental feasibility of consumption of these resources.

A convenient general characteristic of the local *CZA* resources' properties is the resource potential of these sites, which is conditional related to the alleged production technology [177].

Rational environmental management requires to keep dynamic balance between receiving economic benefit from consumption of natural resources and observing accepted norms of environment protection. The concept of conditional resource potential, presented in this chapter, makes it possible to characterize the ecological-economic potential of this environmental area by forecasting the profitability scenario of it's resources' utilization by means of a certain technology.

In order to evaluate profitability of a production process which involves consumption of *CZA* resources, an ecological-economic model is needed. This model should allow us to calculate the anticipated economic benefits, considering the cost of maintaining natural environment in a normal, well-balanced state.

In the model described in this chapter, the cost of the natural resources utilization appears in the form of a credit extended to a production company by administrative authorities of the *CZA* [181]. The size of the credit can not exceed the maximum allowable quantities controlled by the authorities. Thus, the society has the ability to manage resources consumption, taking into account the economic and environmental criteria simultaneously.

These considerations are illustrated with the example of a local area environment resources management, provided in this chapter. Taking a production industry ecological-economic model, we calculated development scenarios of economic processes for a technology which requires three types of natural resources on the local coastal zone area.

These scenarios allowed us to build time graphs of production profitability depending on a variety of factors (ecological state of the environment, changes in prices for resources, etc.). This way, the resource potential of a local part of the *CZA* was calculated, as a conditional value towards the given production technology. The predicted dynamics of the production profitability serves as an example of informational support to administrative decisions in the *CZA* resources management.

The economic model of production built by the *ABC*-method with the use of management agents, was named the *ABC-AGENT Informational Technology* [181, 183]. In further discussion, this technology is applied to the problem of finding environmental and economic balance of the *CZA's* resources consumption: in chapter five it is related to bio-resources, in chapter six it deals with recreational resources, in chapter seven it looks at ecological resources of the marine environment (in cases of industrial operations at the sea).

Chapter 5 looks at the problems of rational use of marine biological resources and introduces an averaged model of processes involved in formation of bio-resources' concentration. It contains 9 variables: phytoplankton and zooplankton, fish larvae, fish, oxygen, carbon dioxide, nutrients and detritus influenced by solar radiation, sea-surface wind, sea temperature and chemicals coming in from river flow. To describe homeostasis conditions in the marine ecosystem, logical operators or management agents are included in the model. They monitor the onset of limiting conditions and switch on operations which limit concentrations of living organisms.

The model linked the integral parameter of fish concentration in the sea with the amount of nutrients and detritus entering the sea with the river runoff. Further on, simulation of management over the chemical substances' inflow into the sea was performed by means of establishing connection between the amount of these substances and oxygen concentration in the marine environment. A logical operator included into the ecosystem's model was intended to ensure that the oxygen concentration did not fall below the minimum allowable value. In cases when this happened, the agent began to reduce the amount of leaching nutrients and detritus from the river inflow compared to the unregulated scenario. This way, imitation of the environmental activities aimed at preserving certain concentrations of bio-resources in the sea was performed.

The balance of consumption and reproduction of the marine resources in the "sea - land" system requires that the land production subsystem which performs harvesting and processing of bio-resources must allocate a part of its profit to environmental objectives which maintain normal state of the marine ecosystem. Therefore, the model of bio-resources' consumption management should contain a mechanism of continuous evaluation of the ecosystem's state, and should be able to control the expenditure allocated for the purpose of protection of the nature.

We described an integrated ecological-economic model of environmental management, built on the basis of the *ABC-AGENT* technology. Indexes of biological diversity and pollution/contamination of the marine environment were used as composite indicators of the marine ecosystem's ecology. The average pollution level in the considered *CZA* was presented with an integrally weighted summary concentration of the major contaminants, specific for this particular zone.

A general ecological-economic balance management scheme of the marine resources consumption was suggested. A computational algorithm to estimate profitability of utilizing these resources for marine industry production was built.

In course of the simulation experiments, scenarios of economic processes in the "sea - land" system were obtained to illustrate various options of financing environmental protection activities, with funds being subtracted from the profit of the land economic subsystem that consumes the marine resources. It became obvious from the experimental results that the informational technology of environmental management in the coastal zone enables us to provide evident prediction scenarios of development processes which play decisive role in administrative decision-making related to the use of the *CZA's* biological resources.

Chapter 6 is devoted to construction of an informational technology of management for the recreational resources use in *CZAs*. It contains a dynamic model of the sea coastal zone factors, which form its recreational attractiveness. The following factors were selected as the core resources for recreational services- the properties of natural environment, the economic infrastructure, the quality of recreational services, and the general state of socio-economic system in the *CZA* region. The recreational attractiveness generates a demand for recreational services, and thereby it creates prerequisites for successful economic activity.

At the same time, the recreational services are strongly related to the cost of consumption and reproduction of environmental resources, which include the nature-protection activities, as well. The informational technology of management is designed to forecast profitability of recreational services under administrative control over the volume of consumed resources.

We proposed a dynamic model of recreational attractiveness, which was based on the method of adaptive balance of causes (*ABC*-method) with the inclusion of management agents in the right parts of the model equations. To assess its coefficients, the method of analytical hierarchy process mentioned above was applied. The model was constructed to analyze possible scenarios for consumption of the recreational resources under control of environmental pollution levels. It was assumed that administrative bodies that manage the recreation and control environmental state of the local coastal zone area, can choose policies of financing environmental actions that could ensure economic viability of the resort and recreational resources resumption.

To determine the capacity of the model, some computational experiments were carried out, where the model's reaction to external conditions was studied. Initially, we selected a scenario, where the recreational economy system could increase volumes of the recreational services up to a certain level, which depended only on the value of the credit gained by the system. To achieve this level, the system was able to acquire loans for consumption of the recreational resources.

The level of environment pollution was considered to be proportional to the volume of the offered recreational services; it had no direct influence on their intensity. As the volume of the services grew, the amount of funds invested by the economic system of recreation into the consumed resources, also increased. Therefore, the assumption was made that the restriction of total investment into acquisition of the resources, or a so-called accumulated credit, should provoke the economic system of recreation to reduce their economic activity and, consequently, the level of environmental pollution will go down. In addition to the loan-return on each step of the economic system's turnover, 20% of funds were deducted as net profit.

Simulation experiments showed that profitability of the recreational services was totally dependent on the demand for these services. The demand's variability closely followed the seasonal factors of recreational attractiveness. However, the scheme of the system's limited crediting proved to be ineffective from the environmental point of view, as the maximum accepted value of accumulated loan failed to limit the intensity of the recreational services and to reduce pollution of the environment. Therefore, a conclusion about taking more stringent economic sanctions was made in order to limit commercial activity in the system of recreation, by means of exemption of a part of income for the environmental purposes.

Calculations conducted in these conditions, proved that the method allows to achieve balance between economic viability of the system and the necessary environment-protection activity of the local government. The model was able to provide management scenarios, where economic subsystem obtains maximum profit, while the funding assigned by it for the environmental activity is sufficient to prevent the level of pollution from going below the critical value. These results confirmed possibility to manage scenarios of the *CZA* recreational resources by selecting appropriate steps of external management of the system.

One of the most significant problems of *CZA's* sustainable development relates to supplying these areas with energy, which inevitably affects the ecology of the surrounding environment. As recreational attractiveness of CZAs is growing, the energy consumption tends to increase. In most cases the energy is obtained by one of the traditional ways: by burning fossil fuels in boilers, and using thermal power. We know how bad is the pollution caused with air emissions of greenhouse gases that accompany traditional methods of energy supply. We also know that alternative energy sources – solar panels, wind generators, heat pumps and others are considerably inferior to traditional sources of energy in cost, and hence, in profitability of production.

The problem of traditional energy supply for *CZA* potentially limits the use of its recreational potential. Using alternative energy sources seems to be more beneficial for the local CZA sites with high recreational potential, despite its higher cost. To make decisions related to development of recreational establishments, having an informational technology becomes crucial, as it can help to carry out comparative assessment of economic benefits of traditional and/or alternative methods of energy conservation in view of the main parameters of ecological state of the environment

In section 6.7 we considered an adaptive transformation model of demand for energy supply of recreational services in conditions of management of balance between consumption of traditional and alternative energy. As the main incentive for the transition from traditional to alternative energy suppliers, we took the following objective of development management: reduction of hazardous caused by traditional energy production, through application of economic sanctions over any violations of the established requirements. As a further incentive, we considered a factor of growing demand for development of environmental awareness among people who visit CZAs for leisure and travel, as they must be aware of the risk caused to their health in association with traditional energy utilization .

Aggregated demand for energy is distributed between traditional and alternative energy suppliers under the influence of price control mechanisms and ecological state of surrounding environment. Constantly updated information about the level of pollution, provided by a system of territorial monitoring, is the basis for management of traditional energy costs, which can be done by means of introduction of differentiated environmental taxes. The tax rate for «traditional» energy should significantly increase and take the form of environmental penalty when its cumulative impact on the environment grows higher than the maximum allowable level of pollution.

Transformation of aggregated demand leads to redistribution of energy production and to achieving dynamic balance between traditional and alternative energy supply in the region. The idea of achieving and maintaining dynamic balance within a system is the essence of systems modeling concept. Therefore, to build a simulation model of transformation of the regional power economy, we used systems method of adaptive balance of causes (*ABC*-method) [181].

We developed a structure of dynamic equations for economic blocs of traditional and alternative energy supply. By using a differentiated environmental tax and environmental fines, the cost of traditional energy sources could be increased substantially, when the state of environment is worsening compared with the average long-term statistics, or when pollution levels exceed maximum permissible values. The growing cost, in its turn, should reduce the demand for traditional forms of energy, and thus the demand for alternative types of energy will be increased.

In calculation experiments performed the redistribution of energy demand was obtained: the share of traditional energy decreased considerably and the share of renewable energy in total energy demand increased by almost two times. After applying the environmental tax, the level of environmental pollution decreased significantly.

One of computational experiments studied the role of environmental awareness of CZA' population in the transformation of demand from using traditional towards utilizing alternative sources of energy. The public awareness of the necessity to shift towards environmentally friendly methods of energy production, provided roughly the same effect as the cost (tax) limitation of production volumes of traditional energy. The proposed model allowed us to consider not only economic, but also social methods of control over the state of natural environment in the area of recreation.

Chapter 7 discusses possible ways to manage marine production operations, such as cargo transportation, disposal and industrial waste burying, extraction of hydrocarbons on the shelf, etc. which affect the condition of the marine environment. All these production operations, later referred to as "maritime operations", should be regarded as another important type of the *CZA* resources, along with ecological, biological and recreational resources. All types of the maritime operations present potential danger for the marine environment, and therefore, they conflict with the objectives of conservation and preservation of the natural *CZA* ecosystems.

The problem of bringing economic benefits of maritime operations and ecological state of the sea into balance is no less important than the issues of marine biological and recreational resources consumption. The deterioration of ecological state of marine environment in the areas of maritime operations should be regarded as consumption of natural marine environmental resources by the production system. Therefore, each maritime operation in the sea must be accompanied with a degree of profit, as compensation for the damage caused to the marine ecosystem. In other words, the economic system of production must acquire a certain amount of the marine environmental resources for implementation of maritime production operations.

Proceeding from these considerations, the *ABC-AGENT* informational technology and the method of adaptive balance of causes were used to build an integrated model of maritime operations in a coastal zone area. As an example of a typical maritime operation in *CZA*, a series of cargo transportation / shipment operations carried out by an average tonnage vessel, was studied. It was assumed that the volume of operations must be restricted in cases when it can cause danger of irreversible degradation to the marine ecosystem, due to lack of environment protection measures. Similar restrictions should take place in cases of insufficient funding to ensure risks of natural disasters and manmade emergencies. Thus, we came to a generalized conclusion that maritime operations should be conducted under the balanced use of three kinds of resources providing appropriate industrial, environmental and proactive interventions. The existing working capital of the economic system should be distributed proportionally for acquisition of economic, environmental and insurance-covered types of resources.

Taking these assumptions into account, we constructed a model of the maritime operations management. It provides forecasts of the three main scenar Taking these assumptions into account, we constructed a model of the maritime operations management. It provides forecasts of the three main scenarios of development processes in natural and economic systems of a coastal zone: the level of economic cost-effectiveness of the maritime operations, the index of marine biodiversity in the area and the level of marine pollution resulting from production operations and/or from emergency situations. Computational experiments provided evaluations of profitability of the maritime operations conducted under pollution control, as well as evaluations of ecological condition of the marine environment.

The general model of the coastal zone maritime operations management was applied to the case study of the Kerch Strait area. This area represents a complex ecological-economic system where intensive cargo transportation/shipment operations conflict with the objectives of the marine environment protection. More than seven million tons of oil and about three million tons of bulk cargo (sulfur, coal, grain, fertilizer) are reloaded in the Strait annually. Extreme weather conditions can significantly affect the maritime operations in the Strait. Thus, during a violent storm of November 11, 2007, four ships were crashed in the Kerch Strait. This resulted in contamination of the sea with nearly 7 tons of sulfur and up to 2 tons of petroleum products. As the result of oil and chemical pollution, the fishing industry resources of the Azov-Black Sea basin suffered considerable damage.

One of the tasks of the resources management in the Kerch Strait area is to provide balance between the highest possible profitability of maritime operations, the best possible ecological state of marine environment, and the minimal risk of adverse effects caused by potential emergencies. In this formulation of the problem, the industrial profitability depends on the costs of the maritime operations in the Strait, on the expense of environmental protection measures and on the cost of anti-emergency measures.

The model of maritime operations management in the strait was based on the *ABC-AGENT* informational technology. It consists of three main blocks: the maritime operations, the resources supply and the integrated assessment of the marine ecosystem's state. The model introduced a number of management agents which control parameters of the ecosystem and conduct the intensity of maritime operations, based on ecological criteria of marine environment. The management of environmental economic scenarios was carried out by formation of cost and profitability of marine operations, in accordance with biodiversity and pollution indexes in the Strait area.

Depending on the ecological state of the marine ecosystem in the strait, two modes of ecological-economic balance formation for resources consumption in the Strait area were examined. In the first case, the growth of biodiversity index was achieved by cutting down investments (loans) for marine operations. In the second case, the costs of ecological resources and of the emergency situations prevention increased with the decrease of biodiversity index and with the growth of marine environment pollution index. Dynamic equations of an *ABC*-model applicable for all types of resources involved into maritime operations were developed.

The chapter offers results of computational experiments on ecological-economic balance management of industrial operations in the Strait. The production cost of the operations was considered proportional to deviation values of biodiversity index and pollution level index from their mean annual values. For the purpose of comparison, simulation calculations were carried out for scenarios of marine operations profitability and ecological state of the Strait. It was done prior to and after the performance of environmental actions, which were paid for by the profits of the production system. It was shown that cooperative planning of production operations volume and of environmental actions' expenditures can help to achieve a reasonable balance between them.

To assess the effect of weather factors on profitability of the maritime operations, we utilized archive data about the average monthly temperatures in the upper sea layer, the intensity of light flux, and the sea surface wind velocity modulus in the Kerch Strait. These data were applied to simulation of weather conditions on a certain year, taken in the form of certain parameters' random deviations from their averaged long-term variability.

Initially, inclusion of the weather factors had resulted in some decrease of biodiversity level. This can be explained by the fact that the imitated temperatures and light inter-annual scenarios differed significantly from the average long-term norms. Simulation of the environmental actions on bio-resources supposed a slow increase of the biodiversity index to the average long-term level. However, the taken time period of the environmental actions appeared to be insufficient. Thus, the model scenarios of simulated processes suggested the direction in which the parameters of environmental management actions should be changed.

In the subsequent series of simulation experiments, the most effective options of funding environmental protection actions were suggested in order to preserve the ecosystem in the Strait area. These experiments confirmed the possibility to plan the volumes of necessary environmental actions.

By choosing anticipated maritime operations' demand curves at the model's input, by forecasting the weather factors and by proposing average resources' prices, it is possible to obtain prognostic scenarios of production costs and operations' profitability. It is also possible to plan necessary actions to support the biodiversity index value at the appropriate level.

Summing up this brief discussion of the monograph, it is necessary to mention its principal objectives once more. The authors had set the task to expound the basic overall systems method, which is used in general to achieve the development goal, and to provide examples of its use in integrated management of coastal zone resources of the sea. The systems management methodology of the *CZA* development which is described and applied in the book, declares the need for informational technologies to support administrative decisions when we choose development goals and use natural resources to achieve these goals.

Such informational technologies should provide the *CZA* administrative authorities with forecasts of basic development scenarios, which provide an idea of anticipated impact of various administrative decisions. Therefore, forecasting and suggesting scenarios for various ways to use the resources has become one of the most important tasks of the *CZA* resources management. From the economic point of view, such models allow us to suggest management actions and ensure cost-effective consumption of all kinds of resources, which is especially important when planning investments into the infrastructure and development of *CZAs*. To assist the administrators of environment protection activities, such models can provide control of the resources consumption and ensure that their levels are balanced in accordance with current ecological standards.

We proceeded from understanding that management of spatially distributed *CZA* resources requires more sophisticated differential models of ecosystems. The economic models of natural resources consumption suggested in this book, are waiting for further improvement, as well. However, in our opinion, the development of the *CZA* management systems must begin with construction of informational technologies, where the forecast development scenarios are based on observational data (rather than on sophisticated theoretical models), which feed relatively simple models. Therefore, the focus of attention was given to the *ABC-*method of complex systems modeling, which is relatively simple and provides good results in describing integrated development processes.

CHAPTER I. SYSTEMS APPROACH TO SUSTAINABLE DEVELOPMENT OF COASTAL ZONE AREAS

1.1. Coastal Zone Areas Sustainable Development

Coastal zone areas (CZA) of the sea represent an organic union of two natural systems, the land and the sea. This enriches them with unique properties, and brings up a requirement of setting specific goals in order to manage their sustainable development. The plentiful marine resources of coastal zones, as well as traditional use of sea routes for transportation and trade, have empowered *CZAs* with a special role in the development of mankind. Today, these areas retain their higher growth rates compared to the continental areas of the Earth. The productivity of *CZA* waters plays crucial role in providing food supplies to many nations. The *CZAs* perform important function of recreation, contribute to overall economic growth and improvement of people's life standards. Coastal waters are the most productive areas of the seas with the greatest biodiversity. About 90% of the global fish catch comes from the coastal waters. Most of the coastal lands are heavily populated. Nearly 60% of the world's population today lives within 200 km of the coast zone.

The coastal zones of seas and oceans can be identified by five distinguishing kinds of local areas: the areas remote from the sea (they affect oceans mainly via rivers and via particular sources of pollution); the land lying in the immediate vicinity to the shore (meadows, marshes and lowlands, where human activity causes direct impact on the marine environment); the coastal waters (estuaries, lagoons and small bays, where anthropogenic influence is strong); the off-shore waters (extending within the limits of national jurisdiction, which is 200 nautical miles to the open sea); and the open sea (which is beyond the limits of national jurisdiction) [22].

Significance of the Coastal Zone Area. A *CZA* is a zone of active interaction of three physical environments: the atmosphere, the ocean and the land. Their effects on people become particularly evident in CZAs: tsunami, hurricanes, algae blooms and eutrophication of water. Many of the effects of global climate change show up in coastal areas, as well: on-elevated sea levels, coastal flooding, coastal erosion, changing weather cycles, etc. In addition to this, high concentration of economic activity and sharp competition for *CZA* resources, cause frequent conflicts in fishing industry, tourism and recreation, industrial construction, use of mineral resources (oil, gas, corals), aquaculture etc.

Along with population growth, industrialization of coastal areas accelerates. All this brings up the need in development management of economic, environmental and social processes in these areas. The mankind's dependence on *CZAs* is increasing, as these zones experience constant growth of economic activity associated with the development of housing, recreation, expansion of ports and their infrastructure, construction of roads and strengthening of shore lines. The territories adjacent to the *CZA*, are responsible for inflow of people, goods and energy into the CZAs, but they also flow there masses of polluted water, industrial and agricultural waste, and domestic waste of towns and villages. The rapid increase of anthropogenic loading in the *CZAs* greatly exacerbates the task of finding balance between economic interests of people and ecological health of the biota.

Currently, there exists a well-known paradox related to consumption of natural resources, when the desire of society to preserve and restore natural ecosystems interferes with the commitment to provide complete and stable satisfaction of people's needs in the natural resources. Science should develop a framework for interaction between the groups in society which represent different approaches to environmental management, and propose a compromising solution to overcome the conflict. The scientific principles of sustainable development are based on understanding that sustainable biosphere is beneficial to society, not only environmentally, but also economically.

This raises awareness of existence of a so-called *human factor* in the consumption of natural resources. As the scientific approach to sustainable development has not been formulated clearly enough, there is no unanimous perception of challenges of sustainable development in the society. Moreover, there are contradictions in opinions even among those who support the general doctrine of sustainable development. Realization of the social nature of the environmental problems, brings up the idea of taking cooperative measures in determining environmental policy in the *CZAs*. The problem should be solved by united efforts of administrative bodies, private users of the CZA resources and public organizations. Using the human factor means looking for decisions which would make sure that majority of society members can influence, evaluate and utilize the natural resources of the *CZA*. It is essential to understand the social causes of changes in the environment, to be able to find appropriate ways of coping with the contradictions.

Systems Analysis of Sustainable Development. The complex social eco-economic system of the *CZA* is connected with the dynamic processes occurring in the natural environment via bilateral interaction. The processes developing in it depend on critically important contradictions between labor and capital, between the growing demand for natural resources and decreasing amounts of these resources. The UN World Commission on Environment and Development formulates the problem of sustainable development as follows: "Sustainable development responds to the needs of the present generation of people and is not inconsistent with the possibility of future generations to meet their own needs" [74]. This formulation intends to improve the living conditions of people throughout the world by taking into account the limited resources that the natural environment has today.

Sustainable development is a continuous process of finding and implementing current and long-term development goals, which should be directed at global changes in the socio-economic life and at protection of environment by human activity. In today's rapidly changing world the search of current challenges of development must begin from analysis of the natural, economic, environmental and political conditions. It should be based on experience and active position of current generation of people. Movement to the goals of development must be accompanied with their refinement in accordance with currently changing conditions and circumstances.

Following this statement, the complex problem of sustainable development must be seen from the standpoint of systems approach to this problem [181]. The systems approach is grounded on the concept of inseparable unity between goals of the society and those objects and connections that create a system which can help it to achieve the set goals. Traffic management of the goals by means of rational use of resources resulting from continuous adaptation (or at list clarification) of the goals, is the subject of systems analysis [171].

In several studies cited below the systems approach to the problems of social ecological-economic systems' sustainable development has been considered [39, 82, 181, 183]. These studies suggest that analysis of the problem of sustainable development must begin from definition of its basic concepts. A special place in the systems analysis is given to definition of the term *system*. Several definitions have been suggested, from axiomatic definitions provided by mathematical theory of systems [87, 112], to philosophic ideas discussing integrity of sets of interrelated objects [67].

Development means any change of state. Progressive development means changes of state of the society, which intend to bring people closer to achievement of their goals. The notion of *development* is conditional, both in relation to the future goals of the society, and in relation to a set of parameters which characterize the *state* of the society. The *states*, taken as functions of time, are traditionally called *scenarios*.

Grounding on these basic concepts, sustainable development should be considered as consequent movement of the society towards achievement of its goals. If a society undergoes sustainable development, its current state tends to continuously approach the anticipated state that brings satisfaction to the society members and encourages further actions to enhance the progressive development tendency. The actions that pursue accelerated movement of the society towards its goals, are called *management of sustainable development* [181]. In cases when the systems approach is applied to sustainable development management, continuous adaptation of control operations takes place, based on comparison between the anticipated goals of development and current ecological-economic conditions of the natural environment and the socio-economic system of the *CZA*.

In application to coastal zones' management, sustainable development means making continuous decisions directed at reproduction and rational utilization of the *CZA's* natural and economic resources, which ensure growth of production and improvement of social well-being, simultaneously with taking measures for protection of environment. Management of natural, economic, labor and intellectual resources of coastal areas involves obtaining current scenarios of development and selecting the best of them. However, certain intellectual and financial resources must be assigned for decision-making, analysis and adjustment of the development goals.

Consequent adoption of correct decisions is hardly possible without informational (computer-based) support. Proper management of resources should simultaneously satisfy several criteria of quality management, which frequently fall into contradiction with each other. For the administrators who take decisions, it is critically important to have prognostic scenarios of possible impact of their decisions. Therefore, management of the *CZA* sustainable development, (as well as the management of any social ecological-economic system), brings up an urgent need in an informational technology to support such decisions.

In what follows we will consider its application to justify the construction phases of informational technologies designed specifically for decision-making support. We will look into the whole process, starting from defining the management problems to obtaining forecasted development scenarios and analysing them. Fundamental importance must be given to the *CZA* systems modeling methods and, in particular, to the models of marine ecosystems. Therefore, in the following discussion we will focus on the principal simulation techniques, which are commonly used in the systems analysis of sustainable development [181].

Sustainable development management can be defined as a controllable movement of social ecological-economic systems to their short-term and long-term development goals, which have to be currently adapted to available resources (or to limitations in their use). They must satisfy the criteria of economic efficiency, social equity, rational environment protection and civil society. In the cases of sustainable development, the current state of a system tends to approach the anticipated state, which satisfies members of the society and encourages them to take further actions towards enhancement of positive development tendency.

Everything that we mentioned above about management of socio-economic development is true for marine ecological and economic systems. In application to such systems, sustainable development means consequent decision-making process with the purpose of multiplication and utilization of natural and economic resources, to ensure simultaneous production growth and environment protection in the coastal areas. United management of natural, economic, labor and intellectual resources of coastal areas involves obtaining current scenarios and choosing the best one among them. Again, certain intellectual and financial resources should be applied to the selection, justification and adjustment of the development goals.

Let us take a look at a general diagram of sustainable development management of *CZA*, which contains several stages. The most prioritized stages are- making a current state diagnosis of natural and social ecological-economic system of "sea-land" type; forecasting possible scenarios of its development; choosing the best possible scenarios and execution of operations for carrying out the chosen scenarios. The choice of scenarios is based on certain selection criteria. The diagnosis of current state makes it possible to monitor the deviations of actual scenarios from the previously chosen (planned) ones. Comparison of these scenarios provides important information for the further management of sustainable development. The difference between a planned scenario and the actual one, allows us to introduce corrections not only into the control system, but also to revise the previously established development goals in terms of their feasibility.

Thus, systems analysis of a sustainable development problem leads to creation of an informational technology of development management, which includes basic operations of collecting, processing, organizing and storing of information about the state of a socio-economic system, as well as about attainable goals of its development. The informational technology provides processes of setting the goals, suggesting possible development scenarios and developing selection criteria for the most rational, attainable goals. The management operations are performed consistently, they lead to achievement of the goals and support adjustment of the development scenarios.

A general diagram of sustainable development management is presented in Fig. 1.1. Each individual block in it presents a structural element of the informational technology of systems analysis, as discussed below. Each block reflects information on certain results, obtained during the process of development. The arrows indicate the sequence of obtaining the results. They can be interpreted as informational flows between the blocks, or as cause-effect relationships. Each arrow represents a direct informational link.

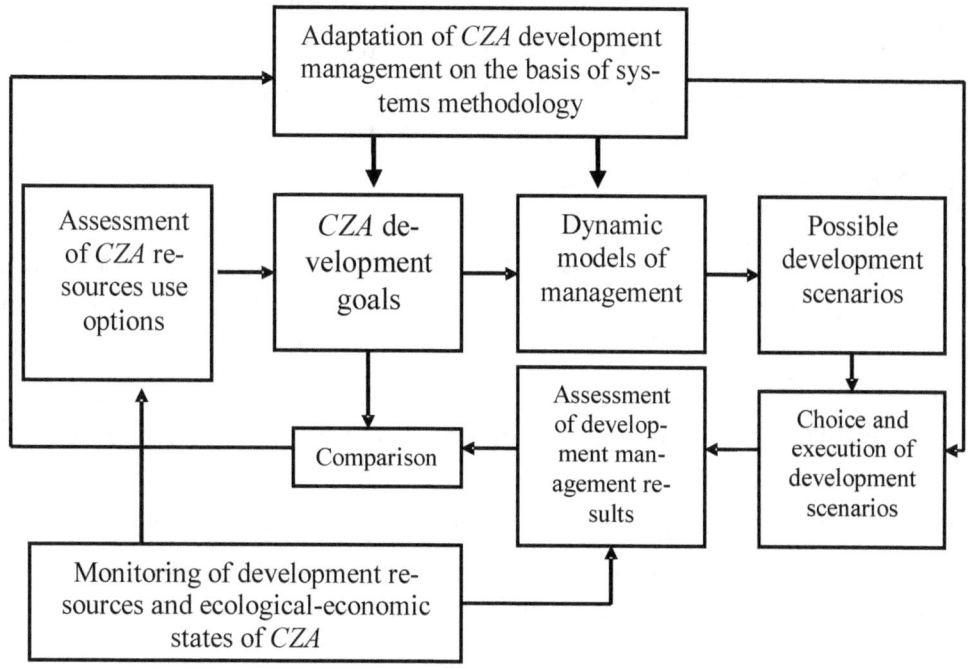

Fig. 1.1. General diagram of sustainable development management.

If several blocks connected with arrows form a complete, closed cycle, a feedback connection takes place. The feedback returns the transformed information into the block from which it had previously come out of. It is not difficult to note that there are several feedbacks in the Fig. 1.1. They provide continuous adjustment of management operations and adapt development goals, and by doing this, they approach the actual development scenario to the anticipated (planned) one.

Marine Ecological-Economic System as a Part of Natural-Industrial Complex of a *CZA* Territory. Sustainable development of coastal zone areas should be based on the rational use of resources of the natural-industrial complex "Sea-Land". Creating models of problem-oriented systems represents the least developed segment of information management technology for such complexes. We are talking about building a numerical dynamic model of ecological-economic system, which would link together the economic problems of industrial production cycles utilizing natural resources, and the environmental issues of biodiversity conservation. Putting it generally, creation of such models should consist of two phases: at the first phase a conceptual model of cause-effect relationships in an ecological-economic system is built; at the second phase it should be turned into a formal model.

Let us consider an example of consumption of marine biological resources (biodiversity). Let us denote its concentration in the marine environment as X_1. We will assume that the bio-resources distributed about the sea, serve as a raw material for some seafood production. Concentration of the bio-resources depends on parameter X_2, which determines the quality of marine environment, i.e. the living conditions of biological objects constituting the bio-resources. It also depends on the concentration of pollutants in the sea X_4 and on the volume of seafood production X_5 using these bio-resources.

Profit Y is a stimulating factor for the seafood production, which depends on the value of seafood in the market X_3, on the production cost X_7, as well as on the amount of fine X_6 withdrawn for pollution (contamination) of the sea area. In its turn, the price of seafood depends on the demand for it in the market X_8. The demand depends on the resource quality of the marine environment X_2 and on the price X_3. Pollution level X_4 affects the quality of the environment X_2, which determines the production cost X_7.

These causal relationships provide the basis to construct a conceptual model of the marine ecological-economic system, shown in Fig. 1.2. The main challenge of managing the system is to maintain a balance of bio-resources concentration in the level exceeding some limit value norm X_{1L}. The closer the bio-resource value to this norm, the higher is the production cost. When the bio-resource concentration falls down to level X_{1L}, the administrative sanctions Z begin to limit the consumption of bio-resources.

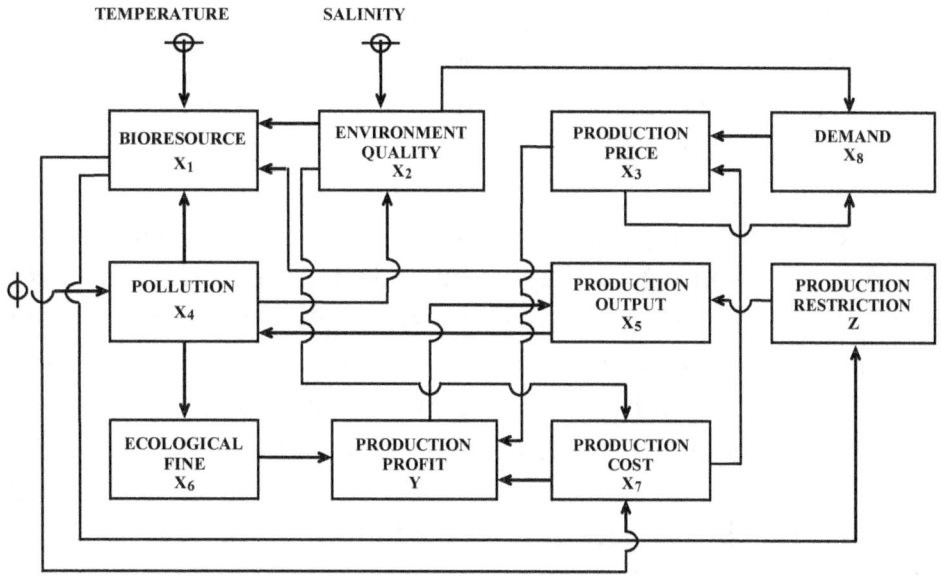

Fig. 1.2. Cause-effect relationships in a marine ecological-economic system

The conceptual model of ecological and economic subsystems of the "land-sea" complex system should be formalized and converted into a numerical dynamic model for prediction of complex scenarios for all variables of states X_1 ,..., X_{1L}. The effectiveness of management depends essentially on the choice of method of constructing these scenarios. Thus, it is critically important to create a system basis for modeling and management of the *CZA*.

1.2. Problems of the Coastal Zone Resources Management

Sustainable development of the coastal zone area is associated with solution of a large number of social, economic, environmental and institutional issues. The foremost need in sustainable development of the *CZA* is predominated by the necessity to eliminate controversies and search for compromises between conflicting demands for use of natural resources. It is well-known that creation of new environmental technologies facilitates sustainability of development on the one hand, but it complicates the development management economically and politically, on the other. In many regions revenues associated with the use of the *CZA* resources continue to increase dramatically. Unfortunately, it happens simultaneously with rapid increase in consumption of energy and other natural resources, which adversely affects the natural environment [22].

The administrative control bodies of *CZAs* seek balance between centralized resources management based on environmental criteria and local control over the resources, which is performed by private ownership groups. The complexity of this task lies in presence of several criteria, which cannot be satisfied simultaneously. There is no regulatory framework for environmental management which would be applicable to this situation. To succeed in sustainable development management, it is necessary to predict ways of secure use of the natural resources for the future generations.

Some Environmental Problems in *CZA*. The coastal landscapes and the coastal marine areas are influenced by human economic activity both, in the *CZAs* and in the basins of the rivers flowing into them. Changes in the upper reaches of the rivers cause noticeable changes in the variability of water flows, sediments, nutrients and pollutants, which include:
  – water discharge of the river systems;
  – diversion of water for the operation of urban, industrial and rural systems;
  – flows of energy within the *CZA;*
  – regional decrease in sedimentation due to deposition of rocks in tanks;
  – regional increase in the supply of sediment and nutrients at the expense of increasing soil erosion; and
  – changes in water flows in estuaries due to irrigation and performance of earth works.

In addition to the changes in the upper reaches of the rivers, many other human activities can affect the state of the *CZA* ecosystems:
  – strengthening of river banks, ports and urban development;
  – consumption of marine resources;
  – increasing use of sea areas;

– increasing pollution of the environment, due to atmospheric and coastal sources, resulting from urbanization and industrial production;

– modification of types and volumes of surface and subsurface water discharge;

– elimination of flood plains due to the use of their land for other purposes, and

– appearance of new biological objects.

These changes have far-reaching implications: they affect viability, productiveness and biodiversity of ecosystems, they weaken their stability and morphology. Transformation of particulate and dissolved substances in coastal and offshore waters is subject to changes. All this reduces the potential of sustainable development and makes the *CZA* management even more complicated.

At present time, the *CZA* resources management issues are studied and solved separately, as the general policy of the natural resources use has not been developed yet. For example, coastal waters zoning and establishment of fishing quotas is done without consideration of spatial-temporal processes in marine ecosystems, such as migration and reproduction of fish stocks. Search for compromising decisions in resources consumption is becoming a priority for sustainable development of the *CZAs*. The following circumstances must be taken into account [22].

(1) The complexity of natural systems forces users to take reductive approach to the resources management, considering the consequences of their use separately from the overall objectives of sustainable development. In addition, experiments with resources consumption at local *CZA* stations are practically ineffective for the large CZAs, because they fail to suggest right decisions about large-scale resources use.

(2) With growing economic benefits, political and public pressure on the environment is increasing, as it is often used to promote new projects of further unlimited exploitation of natural resources.

(3) Traditional demographics and economics do not devote sufficient attention to environmentally justified principles of the nature use. Moreover, ecologists often ignore the anthropogenic influence on ecosystems by focusing solely on the scientific side of the ecosystems' dynamics and functioning.

The dynamics of natural processes, including tidal cycles, the variability of weather and mechanisms of sediment transport, the relationship between the watershed of rivers and their estuaries within *CZAs*, are the major factors to determine the intensity of economic activity. This is confirmed by the increasing number of emergency situations and the growing costs of eliminating effects of the sea-level rise due to climate warming or subsidence of the soil. In many areas the economic activity exacerbates these problems. These are the main risk areas that may cause disbalance in the ecosystems and destruction of the *CZA* resources.

Growing population of *CZAs* (both, local and transient) increases the conflicts of interest in the use of marine and coastal shoreline resources. The intensive use of natural resources often provides short-term positive effect, but it leads to serious consequences in the long term perspective. It sharply reduces the resource potential of *CZA* and weakens the resistance of ecosystems. The increasing number of residing and visiting population causes disbalance in use of the *CZAs* natural resources. This frequently leads to degradation of natural systems and economic instability.

These problems have been causing serious concern in the European Union lately, due to complex interaction between the marine and the land systems. The EU program on the CZA integral management outlines the following areas of concern [66]:

Unauthorized construction. Natural and social degradation rapidly exhausts the productive capacity of the *CZAs*. It causes pollution over its territory, destroys natural landscapes, deteriorates life standards of its residents. The unplanned construction mostly affects the resources base of *CZA's* economic system. The most affected coastal zones are the Mediterranean Sea off-coast area of

Spain (Andalusia), as well as the countries of Central and Eastern Europe.

Changes in traditional sectors of economy associated with natural environment. They lead to unemployment, people's migration, and social instability. When due to some reasons, the traditional ways of earning (eg. coastal fishing) become unprofitable, unemployment rate and related social problems begin to rise. Recently, Gironde inhabitants (France) became victims of the process. Their traditional fishing method could not compete with the new high-tech fishing techniques.

Coastal erosion. It destroys the natural habitation of living organisms and human settlements, slows down economic activity and threatens people's lives. The sea level rise that occurs as a result of climate change compounds the erosion. The East coast of England suffers from massive erosion and floods, despite the large investments which are assigned for construction of dams.

Lack of necessary infrastructure of roads and communications. The remote location of some *CZAs* explains the gap in their cultural and economic development. Examples are some Greek islands, which are deprived of visitors in winter months.

The coastal zone of most countries has abundant natural resources: coastal and offshore stocks of fish and shrimp, mangroves and other forests, land, cattle, salt, minerals, renewable energy sources, including wind and solar energy. Medium- and long-term *CZA* management policy should guarantee sustainable development of both, biological and economic resources. This complex problem requires study and should be addressed to some specific problems of the *CZA*, which can be classified as follows:

Rational use of land. Scientifically-grounded planning of land use should prevent unauthorized and irrational distribution of enterprises and infrastructure facilities.

Water consumption. Prevention of salinization of coastal lakes and estuaries. Collection and storage of fresh water.

Fishery and aquaculture. Rational balance of consumption and reproduction of biological resources.

Problems in agro-culture. Prevention of pollution of coastal waters by pesticides and agricultural waste.

Energy production. Evaluation of potential to use the energy of waves and tides to generate electricity, particularly on the coastal islands, where these sources of energy can be much cheaper than delivery of energy from the shore.

Conservation of critical ecosystems. Special attention in the preservation of ecosystems must be given to the areas of spawning fish and marine reserves. The sea coast guard can provide significant assistance in conservation of such ecosystems.

Pollution control. Terms of zoning areas and coastal territories must be observed during construction of new facilities. All production sites must be equipped with necessary means of protection to prevent hazardous waste from entering the *CZA* environment. The most common marine environment pollutants contain the following chemicals and microorganizms, affecting the flora and fauna of the coastal waters [22, 34].

1. *Hydrogen carbonates of oil*. Its main source is burning of diesel fuel at thermal power stations, industrial plants and transport. Discharge of manufacturing and transport waste containing diesel fuel and lube oil, as well as oil spills and urban sewage are harmful to the living objects of the sea. They spoil beaches and harm the inhabitants of flood plains and wetlands. Issues of sanitation can be toxic to marine organisms, causing their disease, reproduction problems and death.

2. *Chlorine*. It enters the coastal waters from sewage and runoff as a consequence of the of fresh water utilization in residential and

urban areas. It causes ruining impact on all living organisms in the sea.

3. *Nutrients*. They enter the marine environment from numerous land-based sources: liquid waste from timber harvesting, urban and agricultural runoff, industrial and marine release of waste water, waste processing plants and packaging enterprises, places of cattle-growing, urban sewage, construction of canals. The flushing of ship tanks directly in the sea is also a source of nutrients. The nutrient-enriched coastal waters cause the algae blooms, which can destroy food chains and cause sharp reduction in oxygen concentration (eutrophication). Resulting from eutrophication, massive death and diseases of fish can happen.

4. *Bacteria and viruses*. Their sources are the same onshore facilities and production technologies as for the nutrients, as well as the waste pits located on porous soil and having leakages. Penetration of bacteria and viruses into the offshore water can cause diseases of the offshore fish. Infection of the upper layer of the sea can cause human illnesses of swimmers through common infection.

5. *Heavy metals*. Their sources are- automobile and motor boats' exhaust, industrial emission and waste, waste processing plants, dumps of garbage, city sewage systems, spills and the spread of waste. They accumulate in the tissues of fish and can pass into humans. Coming into a human body through drinking water, they can cause brain damage, defects in infants, abortion and infant mortality.

6. *Synthetic organic substances*. Among their sources are liquid waste timber, urban and agricultural runoff, industrial and household waste, floods. They can cause cancer, birth defects and chronic diseases after the use of contaminated drinking water or seafood.

Less typical pollution include an abnormally high intake of fresh water in estuaries, which can cause changes in salinity and, consequently, slow down reproduction or cause death of living organisms. Some damage to estuary ecosystems can be made by construction of canals and drainage of wetlands. Soil erosion and sedimentation of dust often lead to obstruction of marine waters and harm marine organisms. Wind and waves may not only destroy the seacoast, but also affect the bottom sediments, especially in shallow coastal waters.

Distribution of Resources Between Short-Term and Long-Term Development. The primary task of the *CZA* management is to acquire rational balance between social, economic and environmental components of the complex system of this zone. It is necessary to differentiate between development scenarios caused by natural processes in coastal ecosystems, or climate variations and the scenarios which reflect the ecosystem's response to human interference into the environment. One must learn to establish social and environmental priorities in the use of *CZA* resources, to prevent their excessive economic exploitation.

Therefore, oceanographers and experts representing related disciplines, as well as social organizations, government, industry, and society should cooperate to manage these resources and achieve balance between healthy environment in the *CZA*, improved quality of life for its population, and beneficial consumption of its resources. This approach to management is called The Integrated Coastal Zone Management (ICZM). It has been approved by the United Nations and reaffirmed by global international community [142, 187]. The concept of sustainable development originates from people's desire for economic well-being, social justice in society and solution of environmental problems, as all of these goals are inextricably linked in the future perspective. Sustainable *CZA* development intends to provide equal opportunities for people to use its resources, without creating obstacles to future generations. Economic, social and environmental objectives of this development must foresee a certain "threshold level" of non-renewable resources consumption.

International organizations tend to pay special attention to development of the strategy of Integrated Management of the Coastal Zone Area. The strategic planning should follow the steps: policies – strategy – investment priorities – management program. To support it, the following principles have been formulated [34, 66].

A. The strategy of the *CZA* Sustainable Development should be developed and pursued by combined effort of all related ministries and bodies, by means of effective dialog.

B. All materials on the strategy of sustainable development, containing guidelines to action, must be prepared and distributed among all program participants.

C. The implementation of the program should be organized to encourage the responsible departments and agencies to cooperate with local governments, businesses, nongovernmental organizations and civil community.

D. In a longer perspective, international standardization of plans and development programs of the *CZA* will be needed. It will take into account the accumulated experience of their resources management.

In addition to monitoring the state of the socio-economic system, information about the outside processes developing in the environment, is need. Some of these processes will have significant impact on internal processes and therefore, they must be studied for successful management.

Let us take a look at the ways to describe a controlled system. There are three most common ways to describe systems: mental (verbal), conceptual (graphical, in the form of diagrams) and formal (in the form of equations and formulas). Each of these descriptions results in creation of an appropriate model of the system. A conceptual model of resources monitoring and their allocation is presented in fig. 1.3.

The administrative body of sustainable development should be in charge of taking current decisions about distribution of limited development resources between short-term (S) and long-term (L) goals of development. The principle of subordination suggests that the long-term development program should be based on an L –

scenario, provided by a control system which occupies the top position in the hierarchy of the two systems, L and S. The L – scenario model will be in higher position in the hierarchy and it will play the role of external influence on the S – model, located below.

The monitoring data should serve as the basis for development of S and L scenarios, for which a dynamic model of the *CZA* resources is needed. The long-term scenario which we chose for implementation, will serve to describe the external (and majorly restrictive) conditions for the development of operational short-term scenarios of sustainable development. Facing the task of rational allocation of the available resources, we need to undertake adaptive planning of the short-term and the long-term development scenarios of the system.

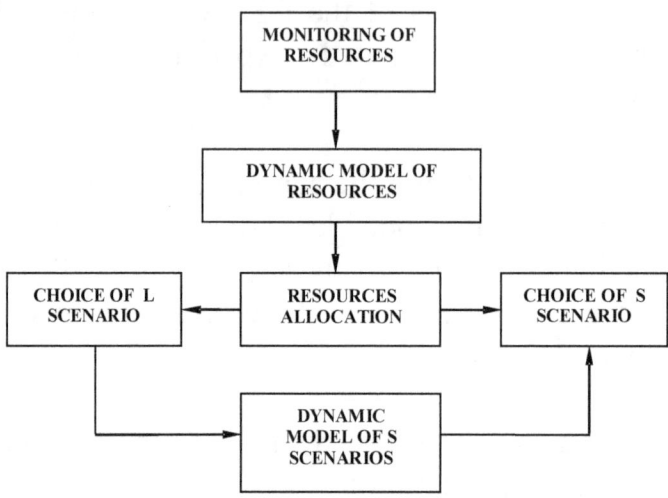

Fig. 1.3. Monitoring of resources and their allocation between long-term   (*L*) and short-term (*S*) development scenarios

Model predictions of S and L – scenarios can be developed from realistic assessment of resources availability, with consideration of some restrictions which may be necessary to bring short-term and long-term goals into concordance. These goals should be continuously adapted, based on incoming flow of information about current results of the development management. An important role must be given to the sustainability criteria of short-term and long-term development, provided by the administrative authorities of the *CZA*.

Main Types of the *CZA* Development Resources. The uniqueness of coastal zone areas originates from close relationship between two natural objects: the land and the sea. The land adjoining to the sea possesses all kinds of natural and socio-economic resources, which characterize an inhabited territory:
- the natural resources;
- the infrastructure and capitalization of the economy;
- the labor and intellectual resources;
- the technological level of production;
- social and environmental consciousness of the population;
- socio-political system of governance; and
- some level of development of civil society.

Adaptive planning of short-term and long-term development scenarios of the *CZA* can be represented by the scheme shown in Fig. 1.4.

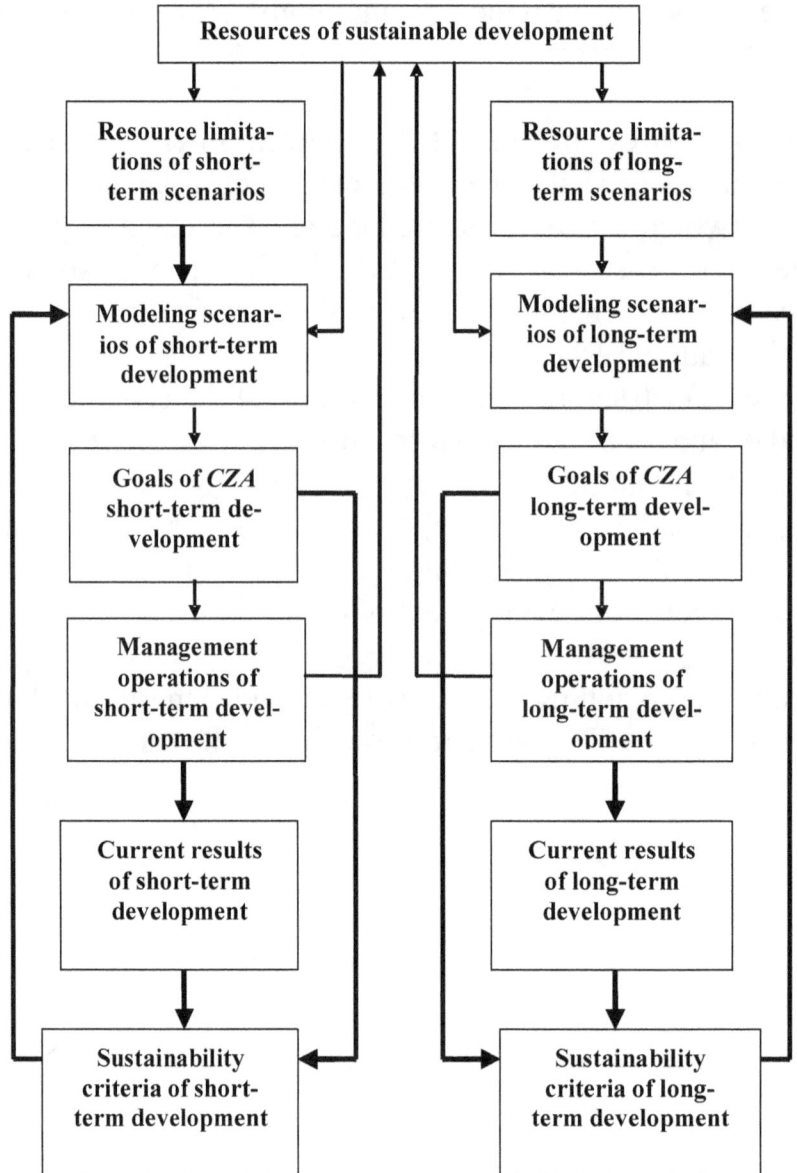

Fig. 1.4. Adaptive planning of short-term and long-term development
scenarios of the *CZA*

The land adjacent to the sea in the coastal zone areas has the types of marine resources which are traditionally utilized by socio-economic systems of all regions of the Earth. The marine resources possess a few categories of the following useful properties of the marine environment:

1. Food resources: the biota of the marine environment, used for the seafood supply;

2. Recreational resources: marine climate, beaches, coastal landscapes, sailing, diving, etc.;

3. Use of water areas for economic purposes: assimilation of sewage into the sea, marine transportation and cargo shipment, oil and gas offshore production and transportation via undersea pipelines, production of construction materials, etc.

The present study will focus on the use of marine resources. The systems approach allows us to formulate environmental and economic criteria for sustainable development of the *CZA*, which provide the ability to manage both, immediate and long-term development scenarios, based on preserving rational balance of consumption and reproduction of renewable resources.

1.3. Coastal Zone Models Construction by the Method of System Dynamics

Fundamentals of System Dynamics Method. The method of system dynamics was proposed and developed by professor Jay Forrester of Massachusetts Institute of Technology [50]. The method is intended for quantitative analysis of complex systems with multiple feedback loops which describe causal relationships between elements of a system. The main processes in the simulated system must be represented by a flow of material substance or information flows that come into the system from outside, pass through it and go beyond it. On its way, each flow goes through some store of the flow substance, called a *level*. The substance of a flow stored in a level characterizes current value of the process, being modeled by the level. This value changes constantly, as the velocities of incoming and departing flows vary. These velocities are called *temps* in system dynamics method. If a time instant is fixed, then all processes in the system will stop and the flows will disappear. However, the levels which have accumulated substance or information will maintain their values. These values represent the *state* of the system at any given time.

Forrester suggested to relate the value of the level to the temp of flow departing from it so that each flow passing through the level, can be controlled by the other levels. For graphic representation of levels and temps of incoming and outgoing flows, there is a special module in system dynamics method, as shown on Fig. 1.5. Level X_1 is formed by the temp U_1 of an entering flow and the temp of outgoing flow V_1. Management of the flows is marked on the flows' arrows with symbols resembling valves. Rates of the incoming stream U_1 are controlled with external influences f_1, attached to this module. This way, the impact on this process from other processes developing within the system or in the surrounding external environment, are taken into account. The temp of the outgoing flow is controlled with a negative feedback, which is shown in fig.1.5 as a chain $x_1 \to a_{11} \to V_1$. The purpose of this feedback is to force the level to summarize all external influences applied to this system module.

Let us assume that the temps of incoming and outgoing flows are equal. Then, the value of the level X_1 is constant and corresponds to the external influence f_1 applied to the system module. If we further assume that the external influence has increased then, consequently, the rate of the incoming flow U_1 has increased. Now, the value of the level will also begin to increase, and information about this will come through the negative feedback loop a_{11} to the "valve" which conducts the temp management of the outgoing flow V_1. The valve will increase the rate of the stream leaving the level, just to a degree which is needed to balance the rates of the incoming and outgoing flows. The balance of flows will be restored, but the level will take a new constant value, corresponding to the changed value of external influence f_1. As the result of continuous changes of external influences, the level will continuously monitor these changes. The ability to quickly adapt to total external influences is the first feature of the system module shown in Fig. 1.5.

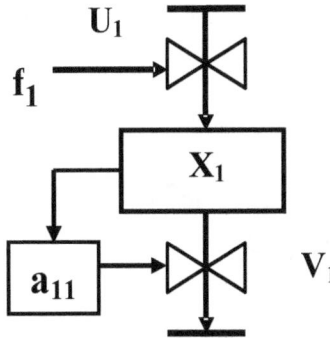

Fig.1.5. Systems diagram of the standard module used in the system
dynamics method

Another significant feature of the method is that a formal presentation of the module's adaptation to external influences can be expressed with a simple differential equation

$$\frac{dX_1}{dt} = U_1 - V_1 = f_1 - a_{11}X_1, \qquad (1.1)$$

in which parameter a_{11} serves to agree dimensions of the flow rate and the level value and can be put equal to unity in the transition to dimensionless variables.

Consider three consecutive moments of time: $t_j < t_k < t_l$, separated by equal intervals of time $\square$. We assume that the rates of flow during these time intervals remain constant. Then, the value of the level at time t_k will be equal to its value at time t_l plus an addition equal to the product of difference between the rates of incoming and outgoing flows at the time interval $\square$:

$$X_{1k} = X_{1j} + \tau(U_1 - V_1)_{jk} = X_{1j} - \tau(f_{1j} - a_{11}X_{1j}) \qquad (1.2)$$

This expression provides a forecast of the level values at a time t_k, if we specify the initial condition X_{1j}. Using the resulting value X_{1j} as a new initial condition, we can now get a forecast for the next moment in time t_l. For any initial value of X_{1j} each iteration process will direct the value of the level to the number

$$X_1 = a_{11}^{-1} f_1.$$

Here are some general notions of system dynamics method, formulated by Forrester [50].

1. Any complex system is a combination of feedback links forming its internal structure. All dynamic characteristics, representing the processes which cause interest for management of a system, should be included in this structure.

2. Every decision that controls behavior of the system is developed within a feedback loop. Making decisions changes the values of the system's levels, which in its turn influences the subsequent decisions.

3. Each feedback loop is a combination of levels (states of the system) and temps (actions). The levels integrate flows in the system and can not change instantly. The values of the levels depend only on the temps of flows.

4. The temps of the flows can not be determined (measured) immediately. In system diagrams, they appear as values averaged over a certain time interval. A temp of a flow should only depend on the levels.

5. The levels describe the state of a system completely. Temps are calculated depending on the values of the levels. In system diagrams of models, temps and levels must be alternated between each other.

6. Temps of a flow include following operations:
   - Definition of local objectives at which a decision should be directed
   - Comparison of this goal with the observed state of the system

– Identifying the fact that the state of the system differs from its goal

– Development of a control action, which reduces this difference

All these operations are included in the temps' equations.

7. The structure of any system presents a conjunction of following basic elements and limitations:
– A closed border of the system
– A structure of feedback chains within the boundaries
– Substructures of levels and temps within the feedback chains
– Components of internal substructure of temps: local goals, mismatching goals and observations, decision-making

These notions can not be regarded as rules for constructing models. They represent the conditions to which the structure of a model must be subordinated. The process of a model construction contains three main steps:
– Development of a conceptual model which reflects major cause-effect relationships between all elements of the system
– Construction of a system diagram (in accordance with the abovestated conditions) which represents conceptual model as a system of levels and temps, combined with feedback chains
– Formalization of the model, i.e. obtaining the explicit dynamic equations for levels and temps

In models of complex systems, the modules of the type shown in Fig. 1.5 (which are linked with subsidiary chains of information transformation) form a very complex system diagram. Information about the levels influences the temps of the adjacent levels' incoming flows through the chains of transformations.

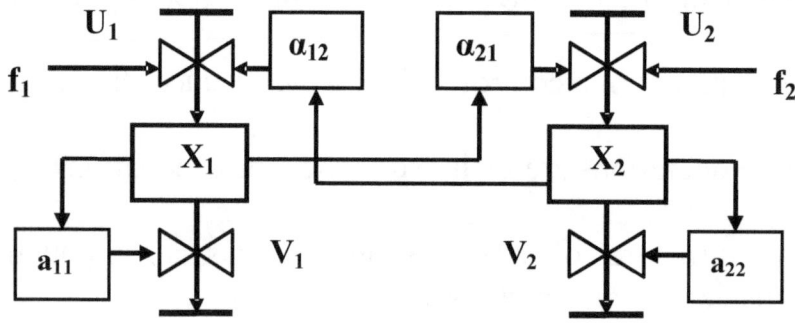

Fig. 1.6. Connection between two modules in the system dynamics method

The abovementioned basics of system dynamics method can be illustrated with connections between two modules of a system shown in Fig. 1.6. Each of the two modules performs its main function of adjusting to a summary of influences coming to its input. The first module, along with the external influence f_1 is affected by the second module, which results from conversion of the system's second level value X_2 by some function α_{12}. Since the level X_2 is influenced by external forcing f_2, the impact will spread through the function $\alpha_{12}(X_2)$ on the level X_1. Behavior of the second module is interpreted similarly. Eventually, the two modules adjust to each other and to the system of external influences.

By following the above rules of the formal submission of the temps of flow for the conceptual model in Fig 1.6, we obtain the following system of dynamic equations

$$\frac{dX_1}{dt} = f_1 + a_{12}(X_2) - a_{11}X_1 \tag{1.3}$$

$$\frac{dX_2}{dt} = f_2 + a_{21}(X_1) - a_{22}X_2$$

The system dynamics method has gained wide popularity worldwide and has become one of the modern instruments of complex systems management [109]. One of its advantages is the ability to transform verbal descriptions of complex systems into computer models relatively easily. This enables us to conduct simulation experiments on these models in order to find practical decisions of systems management. Because of their complexity, systems can generate scenarios which are hardly predictable intuitively. This counterintuitive behavior of complex systems manifests itself in unexpected and unforeseen scenarios of development processes, which could be however predicted by models of system dynamics. Therefore, the model scenarios are able to generate new knowledge about possible behavior of a system.

Scientific literature provides some examples of using the method of system dynamics to describe processes in the coastal zone areas [44, 45]. As a rule, relevant scientific publications do not contain formal models developed from system diagrams of this method. This can be explained by availability of the ready-for-application software packages that implement the method of system dynamics: POWERSIM [141], VENSIM [192], etc. In order to explain possibilities and some limitations of the method, we are going to build a formal model of a social ecological-economic system of a coastal zone area.

Model of Social Ecological-Economic System of a Coastal Zone Area. As for the basic processes that characterize states of a social ecological-economic system of a coastal zone area, we will choose the following: its population's standards of living X_1, the state of its natural environment X_2, and the gross domestic product of the *CZA's* economic system X_3. Each of the selected processes should correspond to a stream flowing through the system, and to a level representing scenario of this process. For a more complete description of the *CZA,* it is necessary to use different factors influencing development processes in each of the three areas: social, environmental and economic. Expert analysis of mutual influences between system levels and informative factors of development allows us to construct a scheme of causal relationships that characterize state of the *CZA*, which is shown in Fig. 1.7.

Among the social parameters of the system we will consider the following: public health α_1; public awareness of risks associated with possible decrease of environmental and social conditions of life α_2; educational level of the society α_3; public pressure on legislators (parliament) to increase funding for environmental and social programs α_4; part of the *CZA*'s budget invested in social needs α_5.

The ecological parameters in our system are: the integral index of environment pollution β_1 and index β_2 of environment protection efficiency.

In the sphere of economy we'll look at the factors: employment rate γ_1; the number of new (effective and resource-saving) technologies γ_2; economic efficiency of production γ_3; investment into new technologies and capital investment γ_4; pressure of corporations in order to move the *CZA's* budget investments from social needs to production projects γ_5; development budget of the *CZA* γ_6. We'll use the factors introduced above (α – social sphere, β – environment, γ – production) to construct a system diagram of the *CZA* model, as shown in fig.1.8. The levels control their incoming flows through the chains of functions that convert the information about these levels, according to the informational factors of development. Each transforming function must be verified by experts.

Gross domestic product of the *CZA* forms its development budget, which must be distributed by the Parliament or other administrative authority of the *CZA*, between social and industrial spheres. The increase of expenditure for social and environmental needs inevitably results in reduction of investment into new production technologies. Therefore, it is essential to balance investments within the *CZA* system to ensure the most favorable conditions for development of both: production processes and social sector.

Investments into social sphere increase general level of health protection and improve people's health. A part of the assigned funds is directed at nature conservation, which is ultimately beneficial to people's lives. In its turn, the improvement in living conditions of the *CZA* population creates opportunities for growth of their educational level. As a consequence, the employment increases, production efficiency grows up and new advanced technologies of consumption of natural resources are created. Thus, investment into the social sphere have indirect positive effect on the production sphere.

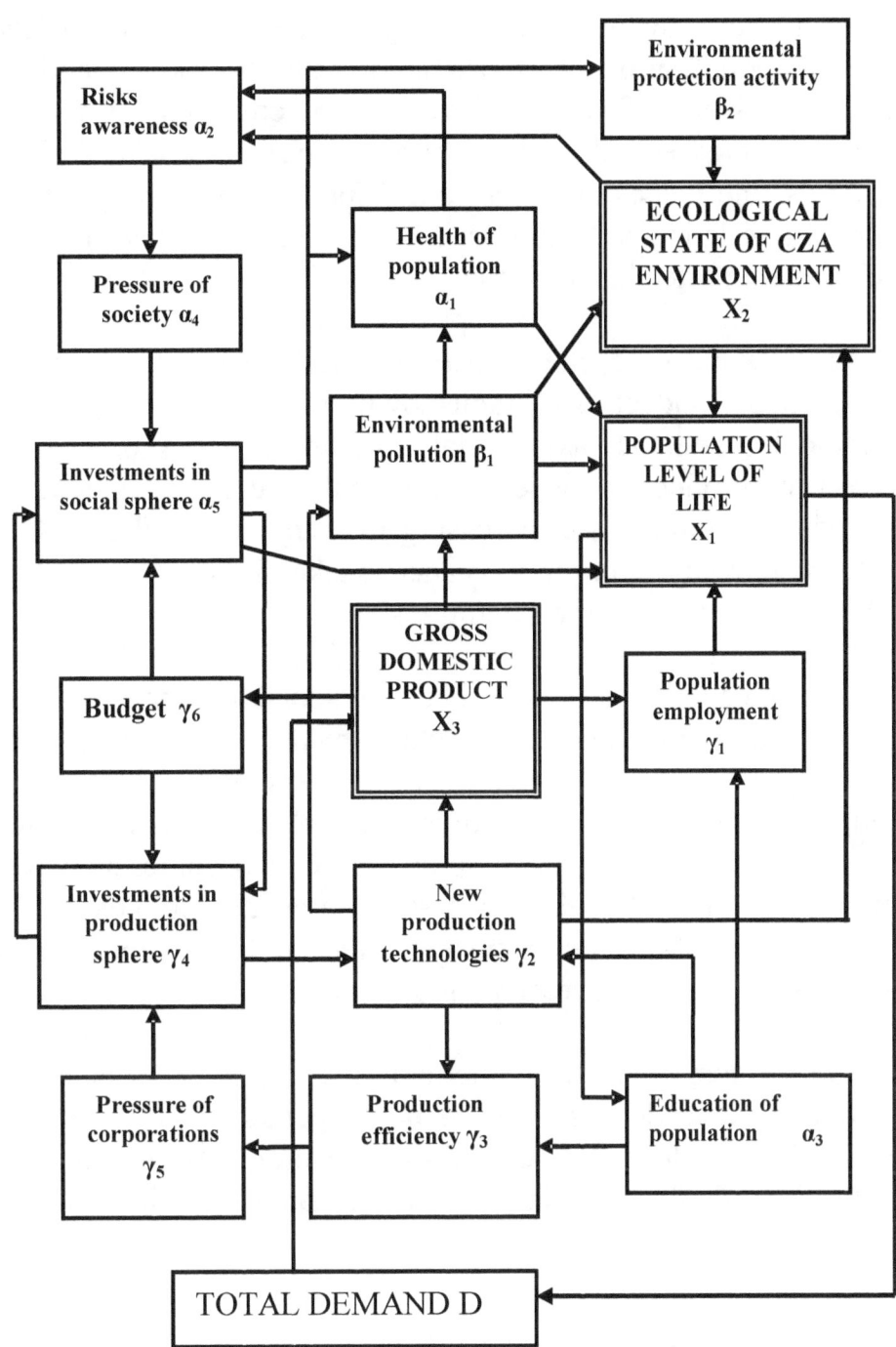

Fig. 1.7. Cause-effect relationships in a social ecological-economic
system of the *CZA*

As a rule, growth in the *CZA's* gross domestic product leads to increased pollution of the environment with by-products and wastes which negatively affect overall state of the environment and human health. It increases public awareness of possible risk to their living standards, and makes the *CZA* population to apply stronger pressure on local legislative and administrative authorities to invest more into environment protection activities.

At the same time, the industrial sphere seeks to use the *CZA's* development budget for their own purposes, to create new technologies and increase capital investment into production. The interest in production development makes enterprises and whole industries to compete for increasing investment into their sectors in the *CZA* economy. This circumstance is taken into account in Fig. 1.7 in the block "Pressure of corporations".

Natural conflict of interests between social and industrial spheres of life should be resolved by reaching balance via allocation of the CZA development resources. As a reasonable balance must simultaneously satisfy social, environmental and economic criteria at once, the management of development process becomes an extremely difficult task. Therefore, it is necessary to create complex models of social ecological-economic systems and to build computer-based scenarios of possible development options.

The system dynamics method uses expert evaluations of functional dependences that bind together many factors of development. A formal model of social ecological-economic system of the *CZA* built by using system dynamics method, takes the following general form

$$\frac{dX_1}{dt} = F_{11}(X_1) + F_{12}(X_2) + F_{13}(X_3) - a_{11}X_1,$$

$$\frac{dX_2}{dt} = F_{21}(X_1) + F_{22}(X_2) + F_{23}(X_3) - a_{22}X_2,$$

$$\frac{dX_3}{dt} = F_{31}(X_1) + F_{32}(X_2) + F_{33}(X_3) - a_{33}X_3 \qquad (1.4)$$

where, F_{ij}, $(i, j = 1, 2, 3)$ – the complex chains of successive transformations of information that have a visible exercise blocks, shown in Fig.1.8, in the process of transferring this information from level X_j to level X_i.

For instance, let us study the influence of the environment's ecological state X_2 on living standards of population X_1. In accordance with the scheme of causal relationships in fig.1.7 there are two chains of information transformations leading from the system level X_2 to the system level X_1: $F_{12}^{(1)}(X_2)$ and $F_{12}^{(2)}(X_2)$, and each of them represents successive transformations

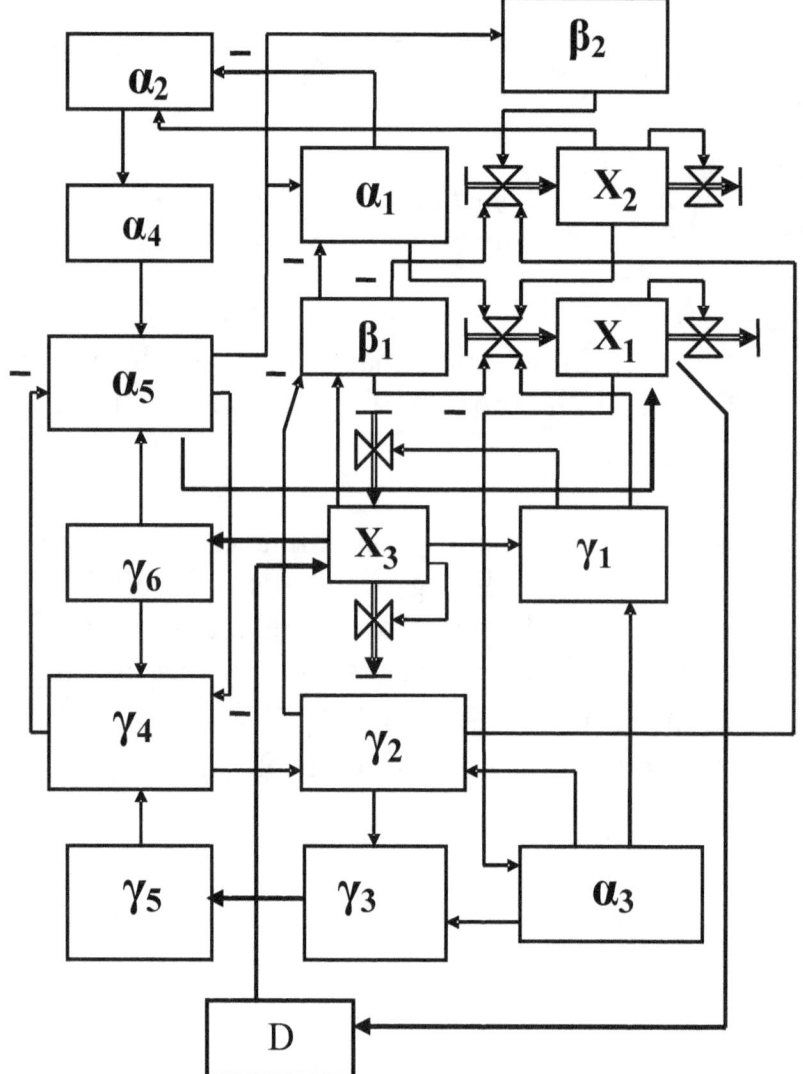

Fig. 1.8. System diagram of the social ecological-economic system of the *CZA*

inside the chain which form complex functions connecting both of the levels. Using the same characters that have marked the blocks in figure 1.8, the influence of environmental conditions X_2 on living standards X_1 can be represented by the following complex functions

$$F_{12}(X_2) = F_{12}^{(1)}(X_2) + F_{12}^{(2)}(X_2),$$

$$F_{12}^{(1)}(X_2) = \alpha_1(\alpha_5(\alpha_4(\alpha_2(X_2)))), \qquad\qquad (1.5)$$

$$F_{12}^{(2)}(X_2) = \beta_1(\gamma_2(\gamma_4(\alpha_1(\alpha_5(\alpha_4(\alpha_2(X_2))))))).$$

If we introduce a simplified assumption that all functional dependencies of the converters of information are linear, $\alpha_i(X_j) = \alpha_i X_j$, the complex functions of the form (1.5) will be replaced by the products of coefficients

$$F_{12}^{(1)}(X_2) = \alpha_1\alpha_5\alpha_4\alpha_2 X_2, \quad F_{12}^{(2)}(X_2) = \beta_1\gamma_2\gamma_4\alpha_1\alpha_5\alpha_4\alpha_2 X_2.$$

However, in this case, one can not replace the products of coefficients by a single coefficient, as each of them can be parametrically dependent on time.

Some Advantages and Limitations of System Dynamics Method. System dynamics model (1.4) reflects complex relationships that exist among major development processes in the coastal zone area. Each set of functional relationships between informational factors of development corresponds to definite equilibrium state of the system. The system should seek one equilibrium state when setting arbitrary initial conditions. The example of the model (1.4) outlines some general properties of the system dynamics method.

In models of system dynamics, scenarios are formed under the influence of external management f_1, f_2 and f_3 of levels which should be added to the right parts of proper equations of the model, as it was done in the model (1.3). The second possible way to manage the system (1.4) is to change functional relationships between informational factors of development. Management of these functional relationships must be grounded on expert evaluations of experimental data analysis. In practice, this procedure is considerably complicated by the fact that the right parts of the model equations

(1.4) represent complex functions of many variables, which in their turn depend on two or more variables each.

Development scenarios, derived as solutions of equations (1.4), can develop in quite unexpected ways, which cannot be foreseen even by the most experienced experts. This reflects the Forrester's effect of so-called "counterintuitive" behavior of complex systems [51, 52]. The ability to simulate and predict possible adverse scenarios (and thus to prevent them) is a very important feature of models in system dynamics method.

However, the task of complex, multi-step transformations of information coming from one level to the temps of incoming flows of the other levels, is one of the main practical limitations of the system dynamics method. Apart from the technical difficulties in computer implementation of such informational transformations, a number of purely mathematical problems arise, connected with the numerical solution of systems of nonlinear differential equations (1.4). The main problem among them is to ensure the stability of solutions which can not be guaranteed without some simplification of the model.

The method of system dynamics does not set a task to adapt model scenarios to observational data and assimilate the data in forecasted scenarios, although the use of observational data in calculations should lead to the updated forecast-benefiting development scenarios. The objective assessment of the links between informational factors of development by statistical analysis of observation data archives was not suggested. An objective evaluation of informational communication factors allows us to eliminate some errors in the subjective expert opinions. Therefore, there are ways to simplify and to ensure further development to the method of system dynamics which we will discuss below.

1.4. Integrated Management of Coastal Zone Ecological-Economic Processes

Integrated management involves utilization of information on the averaged (aggregated) development processes for decision making, leading to achieve its overall target-represented plans. The coastal-zone natural-industrial complexes represent complex ecological-economic systems where the marine environment and the shore land area are brought together. The integrated management of coastal zone systems which aims to address common priorities, should be considered as a necessary start-up management of the *CZA's* sustainable development.

The principle of management of such systems lays in finding and maintaining the necessary balance between economic benefits of the marine environmental resources' consumption and conservation of its natural biological and physical-chemical properties. For practical realization of this principle it is necessary to combine results of modeling of the processes in marine ecosystems with economic technology of production management based on marine resources' consumption.

Structure of the Model of Integrated Management. According to the mission of a system, it should contain a few basic blocks in its structure: dynamic models of marine ecosystem, a model of industrial production unit, a model of simulation of external conditions and the block of integrated control. Therefore, the main task is to create an imitational model which would be capable of reproducing dynamic balance conditions of environmental and economic processes in the coastal natural-industrial complex. An overall diagram of an integrated management model is shown in Fig. 1.9.

The structure of an integrated management model has certain symmetry which reflects the organic link between the *CZA* marine ecosystem shown in the left part of Fig. 1.9, and the production system, shown in the right part. The integrated management of the system is based on data of monitoring that reflect the state of natural environment, and on condition of production system that uses marine resources; the data should be obtained from observational systems. Observational data are compared with predictions of the environmental ecological state and with the planned volume of production in the block of the integrated management. Economic subsystem of the management model plans the level of its production, based on the information about profitability of industrial production and services, depending on market situation, as well as on the environmental actions that are necessary for observing proper ecological state of the environment.

Production plans determine consumption volumes of the *CZA's* natural resources, which are available in the model of the ecosystem. This model can be used to forecast changes in ecological state of the environment after a planned volume of natural resources has been used by the production system. Forecasting the environmental situation closes a feedback loop in the system of integrated management, since it provides rational basis for making decisions about necessary nature-conservation measures. Evaluation of the necessary environmental activity serves as an information management model of production dynamics which also controls the state of the marine ecosystem.

It is assumed that any industrial activity in the sea affects the state of the marine environment and results in dramatic reduction of environmental resources, as it tends to bring them down to critical zero point. Therefore, the economic subsystem of integrated management should cover the expense of the environmental activities. If this value falls down below the prescribed limits, then any economic activity in the sea must be stopped. The economic subsystem must be charged compensation payments.

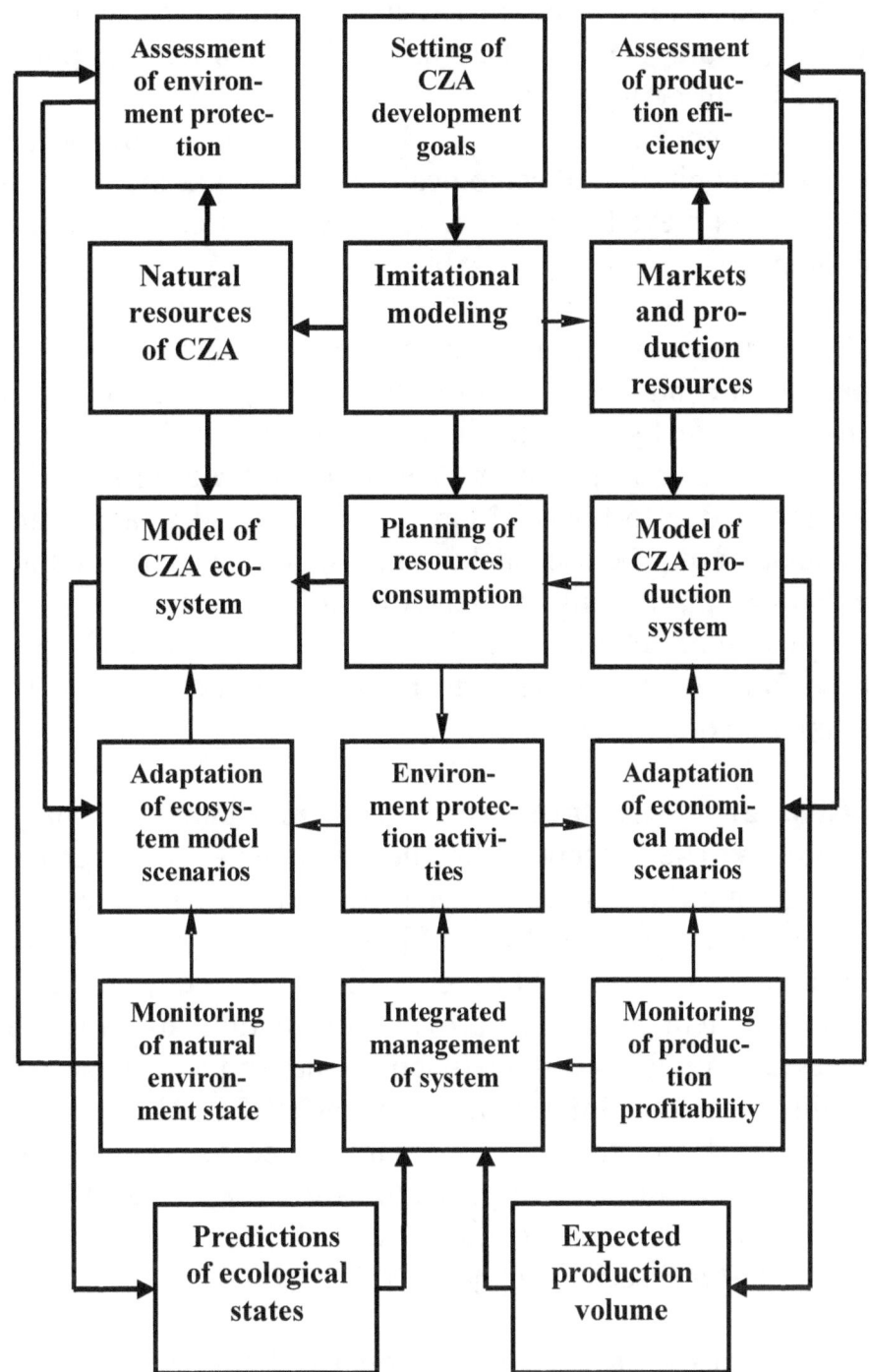

Fig. 1.9. Structure of a model of *CZA* integrated management

These charges can be divided into several categories:

– Resource rent – the cost (per unit) of the consumed marine resources, multiplied by the current level of consumption (e.g., during 24-hour period),

– Environmental taxes – the revenues taken from the profit to invest into monitoring of ecological state of the marine environment,

– Environmental fines for damage caused to the natural environment.

The structure of integrated management model for a coastal zone system must contain a mechanism for tracking current compensation payments made by the economic system of the land. Commonly, these governance operations are performed by control agents. Delays of payments due to financial circumstances, lead to appearance and accumulation of producer's debt to the society, which is "de jure" the owner of the marine environmental resources. There exists a limit to accumulation of debt; in case if the debt reaches its limit, the society will be forced to terminate further private industrial production and consumption of marine resources.

General Structure of Informational Management Technology. The concept of "intellectual agents" (Intelligent agents) is widely used in modern scientific literature on the theory of complex control systems [87, 168]. Such agents can be determined by a number of ways. With their help, we can describe parts of a complex system that can provide evaluation of environmental changes and perform certain functions depending on the situation. We will discuss the informational management technology in detail below, based on the use of marine ecosystems' models and production models which contain specialized agents in their management structure. In earlier studies, this technology was called the *ABC-AGENT* Technology [181, 182].

A general diagram of informational technology *ABC-AGENT* is shown in Figure 1.10. The technology should create a basis for making decisions about the use of natural and socio-economic development resources of the *CZA*. The objectives of management of social and economic resources are formed under the influence of selected scenarios of socio-economic development and by means of monitoring of the environment's ecological state. Therefore, there is an inverse relationship shown in Fig. 1.10, which establishes the environmental-economic balance of natural resources' consumption.

To obtain simulated development scenarios, a formal *ABC*-model is constructed. It contains agents in its structure. Reanalysis of ongoing observations of processes in a controlled system along with modeling of development scenarios, provide important information for improvement of the entire simulation technology and its adaptation to reality.

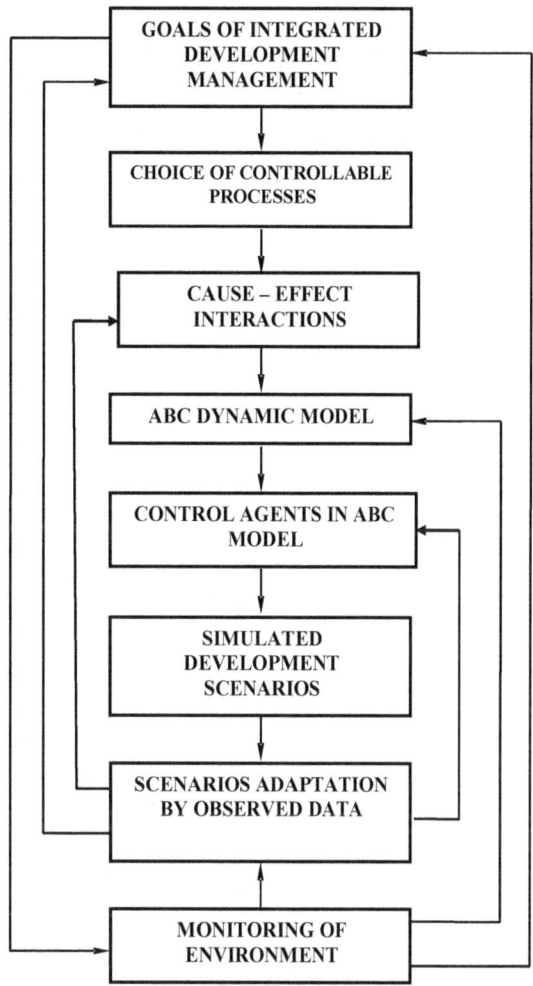

Fig.1.10. General diagram of informational technology *ABC-AGENT*

Based on the definition of agents described above, simulation technology of the development resources' management should contain several relatively independent models, each of which is composed of its own agents. These models express the essence of ecological economics and should be combined with each other, as shown in Fig. 1.11.

Any production is based on consumption of natural resources and is located between the two markets: the market of natural and the market of other socio-economic resources, as well as the market for their products. Therefore, along with the natural environment, it makes up a whole socio-economic complex. From the standpoint of systems analysis, this complex can be regarded as a substantially open system in a state of dynamic balance with the environment. Managing sustainable development is possible only

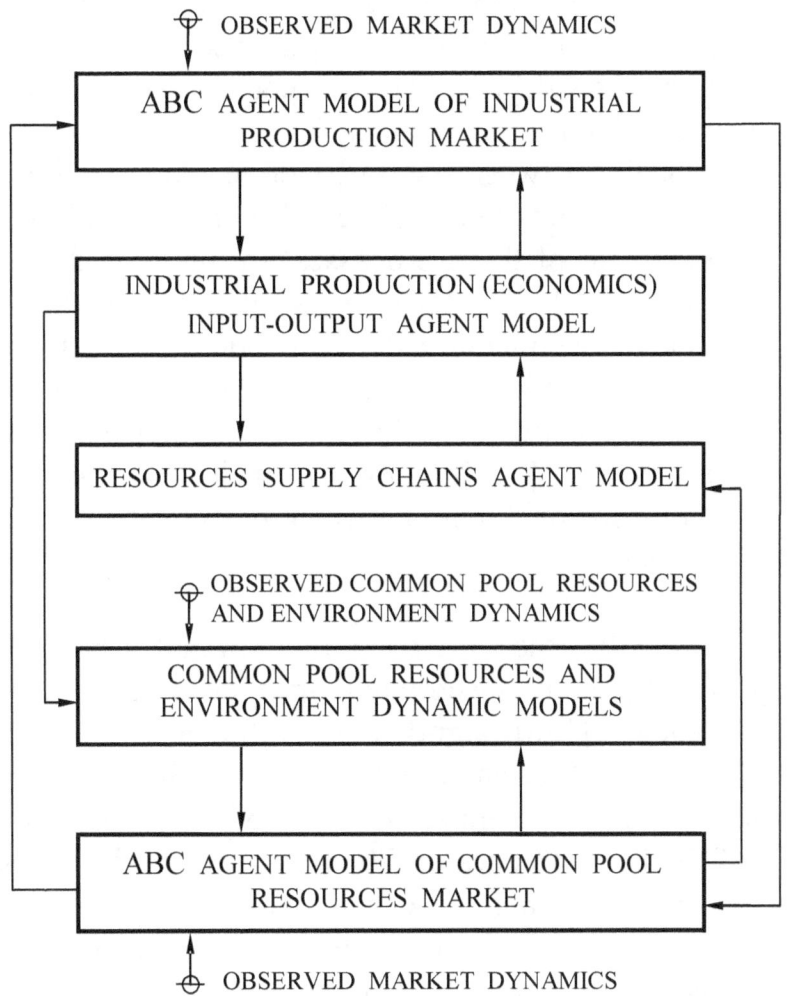

Fig. 1.11. Models used in the *ABC-AGENT* information technology of a *CZA* integrated management

via observations of environmental resources' dynamics and changes in market conditions. These considerations relating to the model structure of production technology, provide an opportunity to present it in the form of finite-difference algorithm for construction of a dynamic model of the economical production system.

1.5. Systems Methodology of Development Management

Some Reasons for Systems Study of Marine Environment. The need for a comprehensive systematic study of phenomena taking place in the ocean and its coastal areas has become apparent by the

1970's. Along with the rapid development of research and satellite observations over processes in the ocean, attention of scientists was attracted with fast degradation of coastal ecosystems due to anthropogenic influence. To create systematic description of complex causal relationships between socio-economic and natural processes, it was necessary to formulate principles of modeling and prediction of further development of these processes. In this regard, the research conducted in the Marine Hydrophysical Institute of NASU in Sevastopol should be mentioned. It intended to lay foundation to systems management methodology for development of marine natural-industrial complexes [120,170, 171]. In papers of Belyaev and Timchenko [11, 12], an attempt was made to formulate a systems principle of informational unity between theoretical and experimental study of the ocean.

In 1965-1970, the method of optimal interpolation of Kolmogorov [90] and its spectral analogue, developed by D. Petersen and D. Middleton [138], significantly influenced the formation of System Concepts of oceanographic research in the MHI NASU. These methods were applied for creation of algorithms for objective analysis of fields in the ocean [170, 171]. Along with the search for rational ways of mapping the ocean fields, the method of optimal interpolation provided the methodological basis for the beginning of systems research. In particular, rational methods for the planning of observations in the ocean were developed, which were directed at assimilation of observational data in theoretical models of the ocean [120].

During those years, the task of modeling dynamic balance between the research phenomena and the environment was set up. The researchers then intended to adapt these models to varying external influences. With this purpose in mind, the first works on the application of the Kalman's adaptive filtering method were carried out in oceanography [169], which later opened a very fruitful way to solve the problem of data assimilation in models of ocean dynamics [120, 170, 145].

The following concepts were identified as the basic systems principles of investigation and control of the environment:

– *Uncertainty about the goals* – the principle which ensures continuous adjustment of goals of the natural environment study;

– *Macrodeterminizm* – the principle of conditional averaging (conditional in relation to observations) of the equations describing models of the ocean dynamics;

– *Decomposition and synthesis* – the principle establishing reasonable balance between goal-setting for study of a phenomenon, and the complexity of the model used for this purpose.

Using the principle of conditional averaging for making up equations of the ocean dynamics laid foundation to development of a new direction in oceanography called dynamic-stochastic modeling of the ocean [170]. Soon after this, methods of systems four-dimensional analysis of oceanographic fields started to develop [55, 89, 171].

Recent trends in development of marine science reveal increasing interest of scientists towards the problems of ecology and rational use of environmental resources. A number of suggestions have been made in order to formulate systems principles of sustainable development of marine natural-economic complexes. Among the most recent systems conceptions it's necessary to mention a few recent attempts to justify management structure of sustainable development for such complexes. In [181] six basic systems principles were summarized, which we are going to consider in our further discussion as the basic principles of systems methodology of goal-setting decisions. These principles (of uncertainty of goals, integrity, causality, subordination, adaptive balance of causes and informational unity of theory and experiment) will be discussed in detail in Chapter II, because of their importance to justify management structure of sustainable development of natural-economic complexes *CZA* and to create an informational technology to manage them.

System principles are critically important for creating dynamic models of development scenarios, in process of defining stages of collection and compilation of information about the modeled system. The role of dynamic models of controllable systems in the overall scheme of sustainable development management, shown in Fig. 1.1, is illustrated in Fig. 1.12

Dynamic models perform the most important mission in management of development: it ensures predictions of possible scenarios that provide a basis for making decisions about the choice of development options. Only a formal model can provide scenarios' forecasts for a controllable system, as it supposes solution of dynamic equations of the model, contained in it. Creation of formal dynamic model of a controlled system is preceded by construction of its mental and conceptual models. A mental model represents synthesis of the original information about the managed system and contains verbal description of the problem of sustainable development, as well as a set of goals which should be reached by the controllable system at the end of the planned phase of development. We will use this information later, to develop a conceptual model (or a scheme) of causal relationships between the processes of development. Therefore, the process of creating development scenarios consists of a sequence of all three forms of description of the controlled system. This is shown in Fig. 1.12, where the block "information about the controllable system" refers to its verbal description.

Of great importance is the adequacy of dynamic model to real processes, taking place in the system. To verify the adequacy it is need to compare scenarios, predicted by the model with observational data of the actual scenarios. In addition, these observations should be assimilated into the model. They should be used (wherever possible) as initial conditions for forecasts. The model's adaptation process, as well as assimilation of observational data into it, are presented in Fig. 1.13.

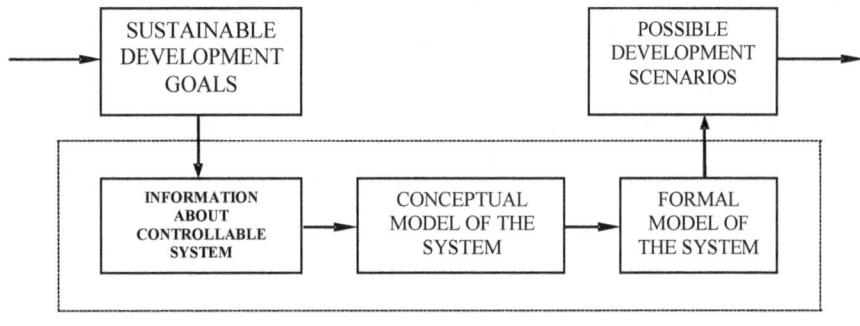

DYNAMIC   MODEL   OF CONTROLLABLE   SYSTEM

Fig.1.12. Dynamic model of controllable system in overall scheme of
sustainable development management

It is assumed that management of processes in ecological-economic system is accompanied with continuous collection and processing of information about the state of the system. These operations are carried out in a block named "monitoring of the state". Then, observational data enter the block of "data assimilation", where – with the delay of the time period between model prediction and the actual observations – the data are turned into a model forecast of possible development scenarios.

The data assimilation allows to use the concept of informational unity which was mentioned above, and to obtain the most realistic diagnosis of the system's state. It consists of a weighted average of the model prediction and direct observations of the system's state parameters. Evaluation of accuracy of the model scenarios by observational data makes it possible to vary the coefficients of the dynamic model, which helps minimize the prediction error. This way we can carry out adaptation of possible development scenarios to reality, as well as structural adaptation of a controlled dynamic system. This is the essence of systems simulation.

In addition to monitoring of socio-economic system's state, information about the processes developing outside of the system, i.e. in its environment, is needed. Some of them will have a significant impact on internal processes, and thus, successful management is hardly possible without thorough monitoring of these processes. Introduction of system principles of modeling has led us to a series of simulation experiments with integrated models of marine ecological-economic systems [38, 82]. Examples of this approach are provided below.

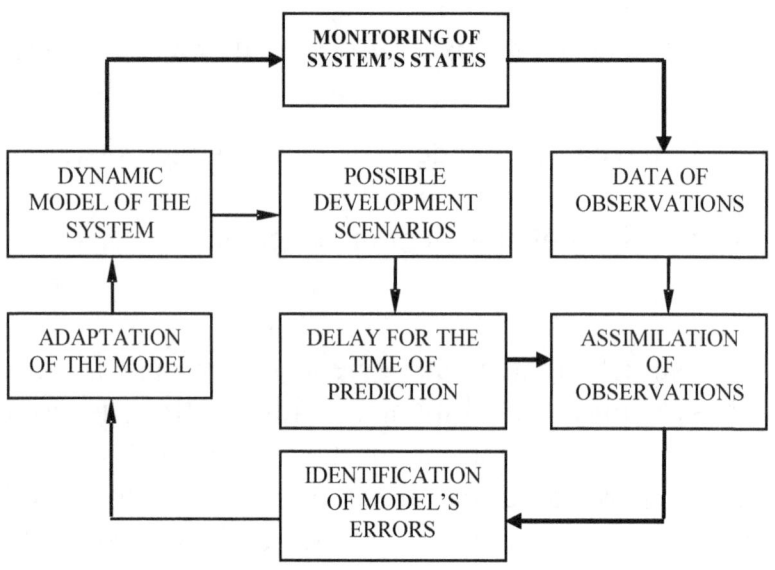

Fig. 1.13. Adaptation of dynamic model of a controllable system

Formation of Systems Methodology Foundations. Goal-oriented actions lay foundation for all development processes. Movement towards a set goal relies on a model which describes the reality and allows us to predict the results of actions that can bring the subject of modeling to its goal (or move it away from its goal). The management of development means implementation of a decision-making technology and execution of actions which should lead the subject to achieving its goal. Thus, applying scientific methodology to development management becomes a task of primary importance, and the systems approach will play the role of scientific methodology in our research.

The systems approach as a discipline appeared at the junction of a few previously existing sciences: the control theory, the theory of decision-making, the theory of operations analysis, and others [112]. This approach brings into focus organization of objects and processes into a system which ensures their adaptation to the processes in the surrounding environment and brings the system's current state closer to the desired one, i.e. approaches it to its development goals. It is essential that the systems approach allows to change goals at any stage of the process. Therefore, among a few specific features of scientific methodology of the systems approach, we can highlight the complexity of researched systems, mutual dependence between their elements and their self-organization, and their ability to adapt to changing goals and surrounding environment.

Initially, the systems approach developed in the framework of general systems theory. Significant contribution into methodology of the systems theory was made by scientists who applied systems theory methods in biology, chemistry, economics and social sciences. H. Odum proposed a way of describing complex systems from the standpoint of energy flows [128]. J. Forrester developed the method of system dynamics, which allows to represent complex systems as a result of interaction of flows of various substances, passing through them [52].

The systems approach today is based on new ideas about characteristics and behavior of complex systems, which were obtained due to the use of advanced computer technologies and due to related development of discrete mathematics. A complex adaptive system can be viewed as a large number of interrelated non-linear processes which interact within the system and with the processes in surrounding environment. These systems have proven to be capable of producing scenarios, which cannot be predicted by experts (so-called counterintuitive scenarios). Computer is the only adequate tool to study and manage such systems. Many new areas of complex adaptive systems theory have been recently developed since the beginning of the computer age.

Bearing in mind the idea of applying the systems approach to problems of *CZA* sustainable development, we will consider a relatively simple definition of the systems methodology of management that is based on several concepts of the general management theory and the theory of complex adaptive systems [34, 87]. Let us briefly examine these concepts.

Among classical methods of complex systems' study it is necessary to prioritize reductionism – the concept of complex objects' decomposition into simpler structural elements, with successive study of these parts, and making general conclusions concerning the system on the whole. Newton and Descartes suggested this principle as a method of synthesis of knowledge about the behavior of complex objects of research.

Development of the systems theory has led researchers to a conclusion that, along with reductionism, an opposite concept, called holism, has a significant value. A complex system may have new (emergent) properties which prove to be new for all of its parts, and which can not be detected through study of the system's separate parts. This property of integrity of systems is one of the most important concepts of systems methodology. Relying on it, L. Bertalanfy provided the following, often quoted definition of the concept of system: "System is an entity that maintains its existence through the interactions of its parts" [13]. A system can meet its goal only when its structure will include all the most important elements and processes relevant to this goal.

Interaction of elements and processes in structure of a system allows us to select closed chains of interaction in it. Influence of an element on itself is transmitted along these chains. This way, another important concept called feedback, can be introduced. Positive feedback connections reinforce processes within the system and lead to increase in their values. The negative feedback drive values of the processes to zero.

The knowledge gained in course of development of scientific theories and during numerous experiments, lead us to the conclusion that all objects of animate and inanimate nature tend to acquire a state of rest, when these processes are balanced, but the object itself consumes and releases minimum of power. A special case of this important concept of balance is a state of static equilibrium of a mechanical system. In complex systems, the state of equilibrium can be achieved through a balance of positive and negative feedback connections. This is one of the concepts of the systems theory [112].

The concept of balance of positive and negative feedback connections within the system brought up the emergence of several expert-analytical methods of complex systems' modeling. Among them, the method of system dynamics developed by J. Forrester [50], which became famous especially in solving practical problems of management in various areas: strategy planning, social economy, ecology, and business.

Management of goal-seeking movement should use both, the model predictions of development scenarios, and the observational data of the actual development processes which take place within the systems and in the surrounding environment. Implementation (assimilation) of observational data in calculations of projected scenarios has also acquired a certain degree of importance. A great significance in this area of knowledge is given to the theory of optimal interpolation of stationary random functions, developed by A. Kolmogorov [90], and to the method of optimal filtering by R. Kalman [88]. Both methods use the concept of conditional (in relation to observations) averaging of dynamic equations of the model. In accordance with this concept, the observations generate conditional probability distributions of the measured processes in the vicinity of points of observation. The best estimates (predictions) of the process are the conditional means of these probability distributions.

The model predictions which utilize observational data, provide informational support to administrators, who are in charge of making management decisions. However, constructing models of modern social ecological-economic systems is a very complex task. The management must respond to multiple criteria of quality, take into

79

account changes in the environment and assess the available resources to achieve goals. As a rule, development of management technology begins with creation of a general scheme that links goals of development with many processes occurring in the system and in its environment. This scheme represents the initial mental model of management.

The problems of building adequate mental models of management are being actively studied and developed by the theory of learning organizations, proposed by P. Senge [155]. The concept of systems thinking lies in the foundation of this theory, which evolved as a result of application of J. Forrester's system dynamics method for managing complex systems [166]. Systems thinking involves special methods of data collection, analysis and information sharing in a team of co-workers who are seeking to achieve a common goal; systems thinking is principal for management of the movement towards this goal.

The Constituent Elements of Systems Methodology. Systems management methodology has not become an established and widely accepted discipline yet. The difficulties in its advancement can be explained with very deep content of its core concepts of "system" and "systems thinking", which have not been universally defined and accepted [183]. The alternative approach suggested here is related to constructing systems methodology of management of goal-seeking movement. It should be regarded as one of possible ways of organizing management of development processes in complex systems. The following definitions can be selected as key statements of the suggested methodology:

1. The mission of systems methodology is to set up stages for building informational technology to lay foundation for achievement of development goals.

2. Systems methodology includes theoretical methods (philosophy) of informational technology construction (which we will further call systems approach) and applied aspects of development of these technologies (which we will further call systems analysis).

3. Systems approach is based on four key elements: principles, thinking, modeling and management.

4. Systems analysis is aimed at rational use of all available information to achieve the set goals.

Based on these statements, let us define systems management methodology and the techniques which it uses [183]:

Systems methodology is a set of scientifically grounded steps intended for construction of informational technology of decision-making support, which are taken in course of movement to established goals and ensure adaptation of the goals to the available development resources and conditions of surrounding environment.

Systems methods are definition of principles, organization of thinking, models construction and goal-oriented management which constitute the basis of systems methodology of goals achievement.

It is also appropriate to introduce the four structural components of the systems approach, which we define as follows:

Systems Principles – the axiomatics of creation of informational technologies for setting up and achievement of the goals which correspond to the rational use of the available resources.

Systems thinking – the application of systems principles for construction of a conceptual model of two systems' hierarchy: a controlled system (introduced for setting up and achievement of the goals), and a managing system (which represents the processes taking place in the surrounding environment).

Systems modeling – the use of systems thinking and the concept of adaptive balance of causes for construction of a model that provides forecasts of possible goal-achievement scenarios based on management of processes taking place in the environment.

Systems management – finding and implementing of rational scenarios of goal-seeking movement with use of all available

81

information and continuous adaptation of the goals to the resources available for development.

Informational technology for decision making support – a computer-based adaptive model for prediction of possible scenarios of goals achieving, selection of the most desirable one, and adjustment of the selected scenario and the development goals to observations of the development processes and their environments. Some illustration of interrelation between the notions introduced above can be seen on the diagram presented in Fig.1.14.

As shown on the diagram, the systems management methodology begins with design of principles of the systems approach. Therefore, we pay special attention to these principles. Systems principles are necessary for building informational technology to support decisions taken in the process of achieving goals, which are ensured with the available resources. Based on these principles, the systems thinking allows us to analyze and summarize the available information in constructing the initial mental model for movement to the goal. The cause-effect scheme between the elements and processes of the mental model serves to build chains of positive and negative feedback connections, which form the conceptual model.

Systems thinking allows us to build a conceptual model of the management system and to provide foundation for scenarios forecasting. It introduces a model of environment, which constitutes a hierarchical system of movement to the goal. In addition, systems thinking helps to bring up suggestions about the processes which should be added to the conceptual model of the system in order to close the feedback loops in it and to ensure its equilibrium state.

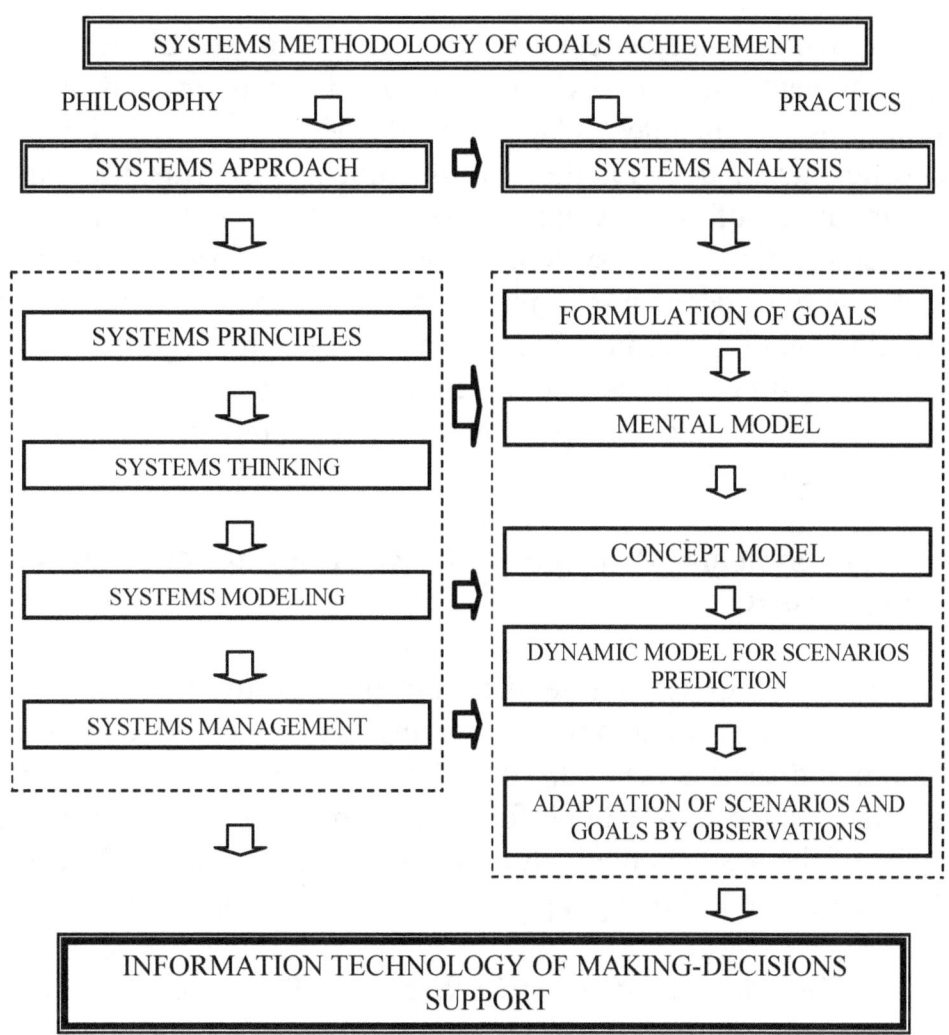

Fig.1.14. Diagram of systems management methodology [181]

Systems modeling opens a possibility to formalize the conceptual model, by entering it into computer and receiving forecast scenarios of the development processes in the system. These processes are reactions of the system to the processes taking place in its environment; they reflect the system's ability to adapt to external influences accompanying movement to the goal. The peculiarity of systems modeling lies in application of universal (standard) equation to describe all the processes of development in a complex system.

Systems management is to adapt the model forecasts of development to the evidence-observed scenarios of predicted processes. It includes adaptation of the model by adjusting its coefficients, and adaptation of the model scenarios to reality by assimilation of observational data in the model. Along with implementation of systems management, the informational technology of decision-making support includes the following steps: development of criteria to assess the quality of management, analysis and forecasting processes in the environment, planning observations, systematization and storage of observed and predicted scenarios, and other functions.

1.6. Systems Principles of the Coastal Zone Resources Management

A wide variety of scientific and popular literature is available today to represent attempts of applying systems approach to study of nature and society [38, 68, 103, 113, 183]. Definitions of principles of systems approach can be found in many papers, as they made the basis of the systems methodology. For example, in [67] F. Heyligen formulates the following 6 principles of systems analysis.

1. *The Principle of Selective Retention*: The stable configurations are retained, the unstable ones are eliminated. The principle can be interpreted as stating basic distinction between stable configurations and the configurations undergoing variation. The word "configuration" denotes any phenomenon that can be distinguished. It includes everything that is called feature, property, state, pattern, structure or system. This principle sets out principal differences between objects, which change slowly over time, and those that are experiencing rapid variations. The more stable configuration is more difficult to remove. In evolution biology, this principle manifests itself in self-preservation of species and organisms in the struggle for survival.

2. *The Principle of Autocatalytic Growth*: Stable configurations that facilitate appearance of configurations similar to themselves,

tend to grow in numbers. The principle characterises development and growth. In a simplified explanation, the principle defines major differences between the slowly changing configurations and the ones that are subject to fast changes. In biology, such configurations are said to have high fitness and that gives them a selective advantage over configurations with lower fitness. Self-reproduction of species in biology or self-organization and growth of crystals in the crystallography are typical examples to illustrate the principle

3. *The Principle of Asymmetric Transitions: entropy and energy.* Transition from an unstable configuration to a stable one is possible, while the opposite transition is not. This principle implies fundamental asymmetry in evolution: one direction of change (from unstable to stable) is more likely than the opposite direction. The generalized, "continuous" version of the principle is as follows: the probability of transition from a less stable configuration A to a more stable one B is larger than the probability for the inverse transition.

F. Heyligen noted that a similar principle was proposed by W.R. Ashby in his Principles of the Self-Organizing System [5] "We start with the fact that systems in general go to equilibrium. Now most of a system's states are non-equilibrial ... So in going from any state to one of the equilibria, the system is going from a larger number of states to a smaller. In this way, it is performing a selection, in the purely objective sense that it rejects some states, by leaving them, and retains some other state, by sticking to it ". This reduction in the number of reachable states signifies that the variety, and hence the statistical entropy, of the system diminishes. This is a fundamental property of evolving systems. Any open system dissipates the most energy comes from outside, tending to a stable state of balance with external forces and, consequently, reducing their entropy as a measure of the degree of variability.

4. *The Principle of Blind Variation*: At the most fundamental level the variation processes "do not know" which of the options produced by them will turn out to be selected. The blindness of variation is obvious in biological evolution, based on random mutations and recombinations. Yet even perfectly deterministic

dynamic systems can be called blind, in the sense that if the system is complex enough it is impossible to predict whether the system will reach a particular attractor (select a stable configuration of states) without explicitly tracing its sequence of state transitions [67]. The meaning of this principle is that a complex system that has many possible stable states, not necessarily rush to any particular one. The principle is important for the process of thinking and for formation of knowledge about the subject of study. Knowing something is the result of experience, e.g. process of trial and error, engraved in the memory.

The memory selectively retains successful and unsuccessful experiments, and inductive reasoning makes it possible to visualize predicted results of such experiments. Therefore, knowledge could be interpreted as an associative guess, based on a combination of blind variation and selective preservation (blind variation and selective retention, or BVSR [67]). This formula summarizes the first three principles, since the configurations produced by blind variation can make transitions to the selective preservation.

5. *The Principle of Selective Variety*: The larger the variety of configurations which a system undergoes, the stronger the probability that at least one of these configurations will be selectively retained. If under a new selective regime configurations lose their stability, a large initial variety will make it possible that at least some configurations will retain their stability. To illustrate this principle, F. Heyligen presents a classic example of the danger of monoculture with genetically similar or identical plants: a single disease or parasite invasion can be sufficient to destroy all crops. On the other hand, if there is a variety, there will always be some crops that'll survive the invasion.

6. *The Principle of Recursive Systems Construction*: The *BVSR* processes recursively construct stable systems by means of recombination of stable construction blocks. The stable configurations resulting from *BVSR* processes can be seen as primitive elements: their stability distinguishes them from their variable background, and this distinction, defining a "boundary", is

itself stable. The relations between these elements, extending outside the boundaries, will initially undergo variation. A change of these relations can be interpreted as a recombination of the elements. Of all the different combinations of elements, some will be more stable, and hence will be selectively retained. Such combination will be a connection of elements of a higher order, and it can be called a system. The system is more than the simple sum of its parts. This is reflected in the so-called "emergent property" of systems [13].

The abovementioned principles, like many other attempts to define the essence of the systems approach, available in the literature [113, 188], are general in their nature. Therefore, they are closer to the philosophy of systems analysis, than to the practice of systems modeling and management. To ground the systems methodology of sustainable development, we have formulated such principles that can introduce the concept of "system" and to build informational technologies of goal-achievement management [181, 183]. These principles were used for creation of the new *ABC*-method of complex systems modeling, which will be discussed hereinafter. We will proceed to consider them in further discussion.

First of all, let us define the meaning which we are going to inlay into the concept of "system". We consider system as a convenient universal concept that brings together all the information that contributes to achievement of a goal. The system acquires meaning and significance only in connection with the problem of movement to a particular goal. To illustrate this, we will study the process of constructing a system step by step, assuming that each step contributes a portion of information about the problem into the system.

As we did above, we will consider a set of development goals as initial information for choosing parameters characterizing the movement towards these goals. An aggregate of the selected parameters will be called a "state vector". Each of the parameters, being a component of the state vector, is an element of the system corresponding to the development goals.

In case of management of a marine ecological-economic system, its elements, as we saw above, are the processes developing in the marine environment and in the economic system of marine resources consumption. Thus, the goals of development through the state vector selected for their representation define a core set of elements (processes) of which the system shall be composed. This amount of information enters the system at the first step of its formation.

The second step in construction of a system is taking into account the links between its elements. Uniting the system's elements together with the help of additional information about the connections existing between them (just as it was done with the blocks in Figure 1.7), creates the structure of the system. In addition to the components of the system's state vector, the structure of the system may include many other elements. While they are not critical for assessment of the system's state, these elements provide a more detailed information about intersystem communications, as well as the system's links with the surrounding "outside world."

The third step is definition of the system's boundaries. The border of a system is characterized with appearance of connections that influence its elements without encountering a return reaction from the system itself. These one-side links are directed into the inside part of the system and affect its states. If the system's diagram contains blocks with no arrows directed at them, but only arrows directed out of them these blocks must not be included into the system's structure. They lie outside its borders and belong to external influences. Information about external influences plays an important role in the behavior of the system. External influence applied to any of the internal elements of the system is transmitted to other elements and eventually forms the scenario of processes in the system.

The complexity and diversity of the processes occurring in the real world makes us look at whole hierarchies of systems. A hierarchy consists of several systems which are connected with a one-way influence. The upper system – the one which occupies a higher position in the hierarchy, affects the lower system through its boundary, without experiencing any influence in return from the bottom system. Study of a hierarchy of systems can be regarded as the next (the fourth) step in systems description of the problem. For example, the global socio-economic system influences the development of processes in the regional systems. Global political and financial institutions (the UNO, the World Bank and others) have a significant impact on the socio-economic structures of individual countries. Thus, looking into a hierarchy of systems is the most common method of describing development processes in our complex and rapidly changing world.

Systems management methodology was designed to lay the ground for development of informational decision-making technologies, with the purpose to ensure rational use of natural resources and to meet social, environmental and economic criteria of sustainable development. Rational environmental management means consumption and reproduction of natural resources in accordance with general criteria of controlled balance between social, environmental and economic factors of development which satisfy current demands of society and take into account the needs of future generations.

Systems methods ensure simultaneous optimization of use of natural resources for social, environmental and economic criteria. They have no alternative in dealing with management of sustainable development of natural-economic systems. Due to existence of the six principles of the systems approach, the methods acquire system character. The principles are depicted in Fig. 1.15 in the sequence which ensures successful construction of informational technology to support decision-making operations in management.

As it was noted above, the goals of sustainable development of coastal zone area represent the continuously adapted tasks of consumption, reproduction and storage of natural, social, environmental, and other types of resources. At every single moment in time the goals of the *CZA* inevitably contain some degree of uncertainty that exists due to incomplete information about available resources and continuously changing development environments. Recognition of the uncertainty or of relative character of the development goals means inclusion of a negative feedback into the system of the *CZA* resources management, which is going to set corrections on both, the movement to the goals, and the goals themselves.

It is natural to assume that in general cases, the adjustment of development goals does not mean their radical revision. It should be a partial change of development objectives and retention of the majority of the goals. Therefore, the decision-support system for the *CZA* resources use should be directed at achieving the goals which are not causing any degree of doubt. While keeping focus on these goals, it is necessary to identify the processes which most clearly indicate social, economic and ecological state of the *CZA*.

The processes selected in accordance with the principle of integrity of the system, form the foundation of its structure. It should be noted that selection of these processes is an informal process of subjective thinking, in which a person (or a group of persons) work out a system of the *CZA* development management, by using all of the available information, and create a mental model of resources' use which responds to the principal goals at some selected step of development. The principle of integrity requires that none of the most important processes of development can be omitted, and all environment changes affecting the processes of achieving the goals, must be taken into account.

Along with the principle of integrity, the causality principle is the second important source of information for those who are building the system of goals achievement. This principle introduces links (influences) taking place between the processes which were selected to compose the structure of the system. Thanks to these links, the processes are tied together into a common conceptual model, which acquires new (emergent) properties. Therefore, using the principle of causality after utilizing the first two systems principles (as depicted in Fig. 1.15), completes the construction of conceptual model of the system of movement to development goals.

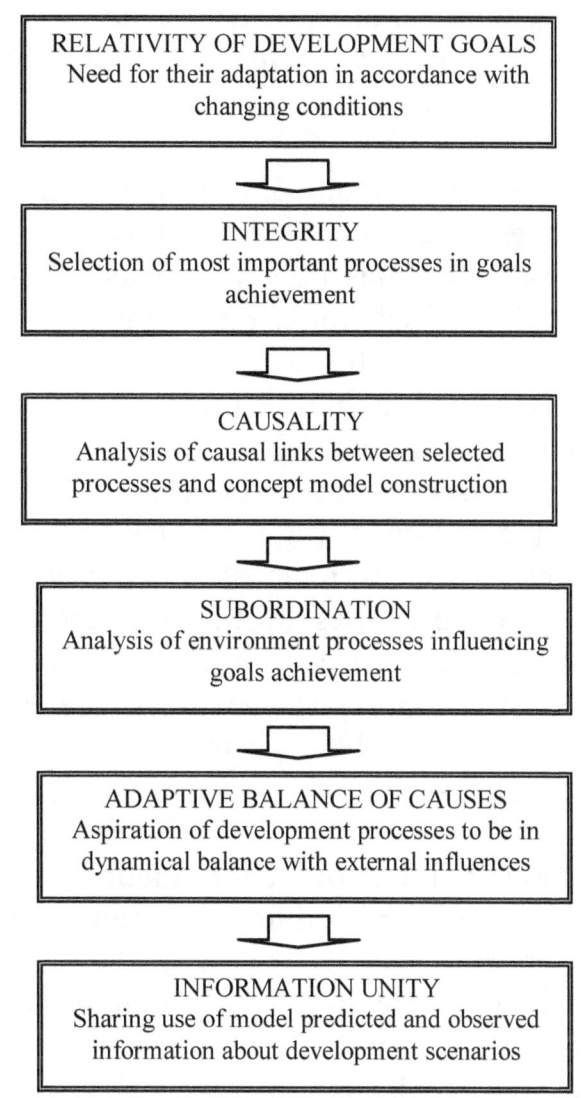

Fig.1.15. Essence of systems principles and sequence of their use

The principle of subordination allows us to specify the boundaries of our system, which we will call the "development system" for simplicity. The boundaries of the development system can be found in places where the development processes can still react to outside influences, but may no longer have any reverse affect on external sources of influence. The development system is in a subordinate position in relation to external influences, which represent the processes occurring in the system's environment. Thus, the principle of subordination introduces a hierarchical bundle of two systems, the development system and the system of their environment.

In relation to systems of *CZA* development, the role of the environment system is given to a social ecological-economic system of the region containing this *CZA* within its structure. Taking into account the principle of subordination, all the regional processes that are affected by counter-influence from a given *CZA,* should be excluded from the system of the region and introduced into the structure of the *CZA* system. The concept of systems thinking, considered above, assumes that the external influences directed from the system of the region towards the *CZA* development system, should represent a part of regional development processes. The regional development processes must be integrated into the system of the region with application of the same systems principles as for the *CZA* system construction. Thus, the use of the first four systems principles allows us to build a conceptual model of the hierarchy of two systems: "Region – *CZA*".

The next step in building of the informational technology of *CZA* management should be transition from a qualitative description of the hierarchy of systems: "region - *CZA*" to their quantitative formal modeling. It is about creating dynamic models of development processes and the processes in the system's environment. The systems principle of adaptive balance of causes postulates the characteristic tendency of development processes to reach dynamic balance with external influences. Practical realization of this principle leads us to the methods of systems modeling which will be considered in our further discussion.

Having a formal model of *CZA* development system at hand, the *CZA* administrative bodies acquire a tool for simulating scenarios of various resources use options. Conducting simulation experiments with such models has become an operational instrument for resources management worldwide, since they open a possibility to obtain and analyze forecasts of the *CZA* development, to take into account possible changes in external conditions, and to justify selection of specific management decisions.

However, the model predicted development scenarios will inevitably differ from the reality due to incomplete adequacy of the model processes to the ones that occur in reality. These differences will grow rapidly as the time of the forecasts increases. This phenomenon is peculiar to all complex systems, having nonlinear causal relationships between development processes in their composition. The system of atmospheric processes can serve as a good example to illustrate this. The predictability of these processes, which is needed for improvement of the weather forecasts' accuracy, is limited due to nonlinear interactions that are presented in dynamic models of atmospheric phenomena [99, 114].

While the experience in modeling dynamics of *CZA* is still insufficient to draw conclusions such as weather forecasting, a certain analogy to them can be driven as follows. The complexity of the *CZA* development models can be explained by the fact that mutual influences between processes in these models may change over time. For example, depletion of natural resources, or a change in the market of goods and services, can dramatically affect the profitability of production, and this would lead us to revision of the values of influence factors in the *CZA* development models. This is the reason for the nonlinearity of these models and, consequently, the possibility of non-reliable forecasts for the periods of time, comparable with the variability of reciprocal influences.

The principle of informational unity emphasizes the need to obtain all available information about development processes and the processes in environment. Except for model scenarios, an important source of information are direct observations (or measurements) of the processes. As a rule, the number of direct observations is insufficient to ground the *CZA* development management solely on the objective observational data. It is therefore necessary to create a combination of observations and model estimates. The principle of informational unity offers a similar united combination, and emphasizes on equal importance of both options for development

93

goals management- creating model predicted scenarios and providing objective assessments of development resulting from real time observations.

Practical techniques of combining information suppose evaluation of the model coefficients by observational data, then the use of processes observed in the environment as external influences on the development model, and then assimilation of observations about the development processes in this model directly. The assimilation of observational data in scenarios' calculations for *CZA* models can be achieved by variational method or probabilistic method [89, 170]. Below we consider the probabilistic method of observations assimilation as the most common way to adapt model predictions to reality. Note that it converts the *CZA* development model into a special kind of so-called dynamic-stochastic model [170].

Computer realization of a dynamic-stochastic model of the *CZA* development supplemented with management algorithms and supporting subsystems, data collection, analysis and processing of information, may serve as an example of informational technology of the *CZA* development management. The forecasted scenarios, adapted to the observations data for previous time periods allow us to establish achievable goals and to plan management operations in the system. Fig. 1.16 shows a scheme of systems principles application in the pursuit of development goals.

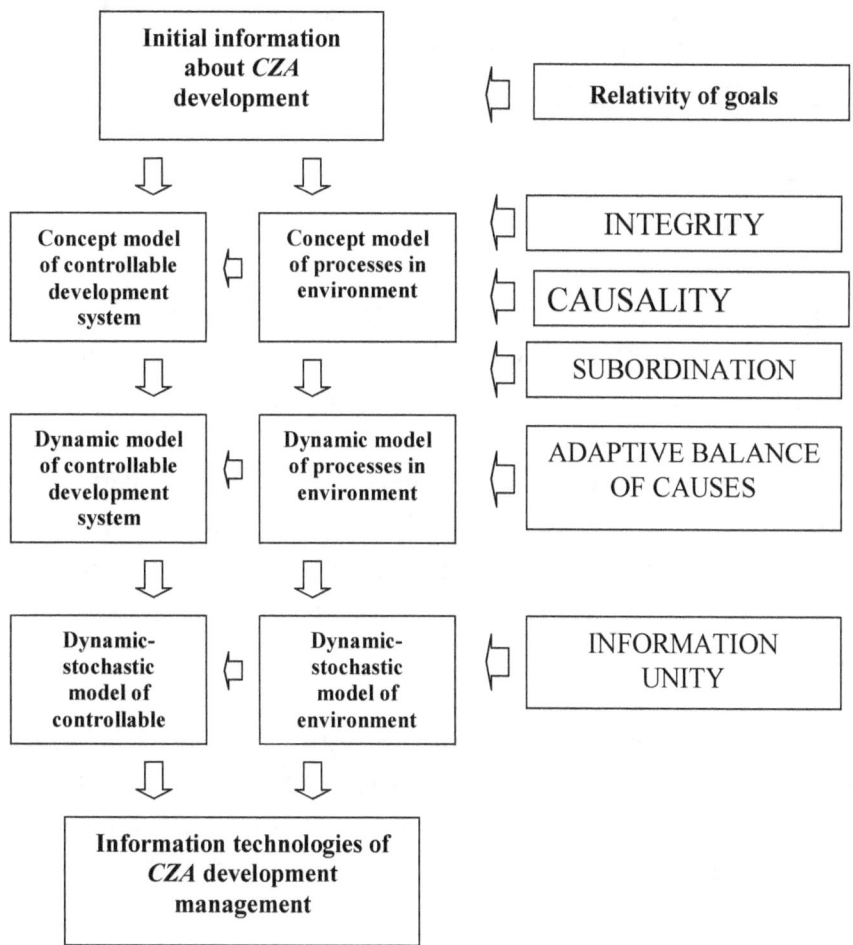

Fig. 1.16. Implementation of six systems principles for creation of

informational technologies of *CZA* development management

In general case, in accordance with systems thinking conceptions [183], and in parallel with the construction of models of development the models of environment can be created. Thus, informational technologies for decision- making support regarding the use of *CZA* development resources should take into account forecasts of processes in the region, which includes both, this *CZA* and the observational data which refine these forecasts.

CHAPTER II. METHOD OF ADAPTIVE BALANCE OF CAUSES (*ABC*-METHOD) IN COASTAL ZONE DEVELOPMENT MODELNG

2.1. Basic Equation of the Method of Adaptive Balance of Causes (*ABC*-Method)

The method of adaptive balance of causes was developed with the purpose to upgrade the method of system dynamics described in section 1.3. The developers of the method set the following tasks:

1. Develop a universal modular equation which represents all the processes in complex systems, like the equation (1.1) of the system dynamics' method. The equation must have the property of rapid convergence to a steady state solution in absence of external influences.

2. Suggest a principle of combining the modular equations into a system of model equations forming a complex system, to ensure rapid adaptation of the processes to each other, which would place a system experiencing external influences into a state of dynamic balance.

3. Suggest an objective method of analyzing available information about the simulated processes, in order to determine model coefficients of a complex system, which would ensure the adequacy of this model to real processes of development.

4. Develop assimilation techniques to utilize current observational data characterizing development processes that take place in the dynamic model of development management.

In this section, we will provide a brief introduction into the method of adaptive balance of causes, which we will further consider in detail as a general approach to formalization of conceptual models of *CZA* complex systems. We will then proceed to constructing a *CZA* development model, where the six basic systems principles introduced above will be applied. In accordance with the first four principles, we'll need to create a conceptual model in the hierarchy "*CZA* – Coastal Region", which correspond to *CZAs*' development goals. At the next step of modeling, this conceptual model should be formalized. Finding the right method of formalization is of principal importance, as it must be a universal one due to the fact that it must satisfy the principles of systems modeling.

A new method of complex systems' modeling, which corresponds to the four requirements mentioned above, was proposed in [181] and was named by the authors the Adaptive Balance of Causes (or *ABC-method*). In the next paragraph, we will look at the essence of this method.

Derivation of Basic Equation of the *ABC*-Method. A universal rule for constructing formal models of processes formulated by Newton [121] states that instantaneous rate of change of any process $x(t)$ is a function of the process itself and the impacts of external influences on it $f(t)$. Therefore, the dynamic equation for the process should be as follows:

$$\frac{dx}{dt} = F\left(x, f\right) \qquad (2.1)$$

The art of modeling lies in the ability to find an explicit form of function $F(x, f)$. The principles of systems modeling require that this function must ensure implementation of two conditions: rapid convergence of solutions of equation (2.1) to an equilibrium state in absence of external influences, and achieving the state of dynamic balance in the process $x(t)$ with the external influence function $f(t)$, when it exists.

First of all, let us consider the case when there is no external influence. If we convert the process x in dimensionless form on the interval (0, 1) and choose the simplest dependence $F(x) = x$, then the solution of the equation

$$\frac{dx}{dt} = x \qquad (2.2)$$

will be an exponentially increasing function

$$x = Ce^t \qquad (2.3)$$

Such a scenario is explained by existence of positive feedback between the change rate of the process and the process itself. Since they are equal to each other, as it can be seen from the equation (2.2), the increase of change rate will cause growth of the value of the process, which in its turn will further increase its rate, etc.

If we choose the dependence $F(x) = -x$ in the right side of the equation (2.1), this equation gives an example of a negative feedback. With increasing rate of the process its value will decrease, and with time, its scenario will take a form of unlimited decreasing function

$$x = Ce^{-t} \qquad (2.4)$$

One of the systems principles of adaptive balance of causes postulates existence of equilibrium in any dynamic system, under the condition that it must be achieved by finding balance between positive and negative feedback connections which take place within the system. When applying this principle, the authors of the *ABC*-method [181], simultaneously introduced in the right-hand side of the equation (2.1) both dependencies on x and $-x$ with some coefficients, which were marked as $F^{(-)}(x)$ and $F^{(+)}(x)$, and named as "basic functions of influence." Then the equation of the process took the form

$$\frac{dx}{dt} = F^{(-)}(x)x - F^{(+)}(x)x \qquad (2.5)$$

Basic functions were introduced to restrain both, growth and decay of x, by directing its scenario to a sustainable constant value. To ensure this, it is enough to require that, with growth of x, the basic function $F^{(-)}(x)$ must decrease monotonously, and the basic function

$F^{(+)}(x)$ must monotonously grow. To achieve overall balance, it is also necessary to introduce one more condition of normalization for the basic influence functions

$$F^{(-)}(x) + F^{(+)}(x) = 1 \tag{2.6}$$

Under this condition, the universal equation of the systems model for the process x takes the following form

$$\frac{dx}{dt} = x\left[1 - 2F^{(+)}(x)\right], \tag{2.7}$$

where $F^{(+)}(x)$ is any monotonously increasing function.

In general, when constructing models with equation (2.1), the selection of its right side is essentially based on the known properties of the modeled process. If we use equation (2.7) for systems modeling, the form of its right side is determined specifically with the choice of monotonously increasing basic influence function $F^{(+)}(x)$. Therefore, equation (2.7) has a universal character, which can be explained by its systems nature.

It is interesting to use the simplest basic function of influence again, by selecting it in the form $F^{(+)}(x) = x$. Then, equation (2.7) can be rewritten as follows

$$\frac{dx}{dt} = x\left[1 - 2x\right] \tag{2.8}$$

In this form it represents a special case of a nonlinear logistic equation [163] or Bernoulli's equation [118]

$$\frac{dx}{dt} + p(t)x + q(t)x^n = 0$$

if $p = -1$, $q = 2$, $n = 2$. By substitution $y = x^{-1}$ Bernoulli's equation reduces to the linear non-homogeneous equation

$$\frac{dy}{dt} + y - 2 = 0,$$

which can be solved by the method of arbitrary constant variation. The general solution of this equation is obtained in the form

$$y = \left[2e^t + C\right]e^{-t},$$

where C is an integration constant. A corresponding solution of the dynamic balance equation (2.8) takes the following form

$$x = \frac{x_0}{\left[2x_0 + (1 - 2x_0)e^{-t}\right]}, \qquad (2.9)$$

where x_0 indicates initial condition at $t = 0$.

The formula (2.9) implies that, for any positive x_0 the solution tends to reach equilibrium constant value of $x = 0.5$. This proves that the dynamic balance equation in the form of equation (2.8) provides a continuous time scenario $x = 0.5$ dimensionless units. Thus, the system module, represented by the equation (2.8), comes into the state of equilibrium and remains in it until it experiences an external influence.

Now, let us consider the case when a constant external influence is applied to a process x. Following the principle of adaptive balance of causes, it must be assumed that the response of x to that effect should lead to a new steady state x^*, supported by the external influence on the process x. To carry out this condition, the *ABC*-method proposes to add the external influence with the opposite sign to the basic function argument $F^{(+)}(x)$, which will ensure the increase of x for c > 0 and its decrease when c < 0 [181]. Then, equations (2.7) and (2.8) look as follows

100

$$\frac{dx}{dt} = x\left[1 - 2F^{(+)}(x - c)\right],$$

$$\frac{dx}{dt} = x\left[1 - 2(x - c)\right] \tag{2.10}$$

If we present the latter equation in finite differences by the Euler scheme, we find

$$x(n) = 2x(n-1)\{1 - [x(n-1) - c]\}. \tag{2.11}$$

The results of solutions convergence of equation (2.11) under the affect of constant external influences are shown in Fig. 2.1.

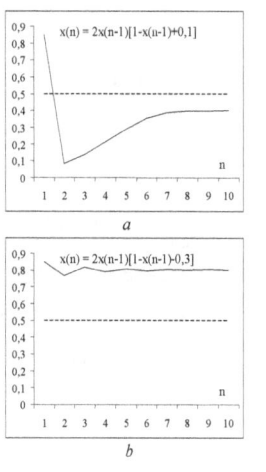

a

b

Fig. 2.1. Convergence of solutions of equation (2.11) to a stationary value at constant external influences: (a) temporal scenario with $c = -0,1$; (b) temporal scenario with $c = 0,3$

Rapid convergence to equilibrium allows us to suggest that the process x, represented by the equation of the *ABC*-method, would react similarly to variable external influence $f(t)$, and will continuously change its steady-state solution in accordance with current values of the influence. This way, the systems principle of adaptive adjustment to variable external influences is implemented in the system module.

The modular equation with variable external influences takes the form

$$\frac{dx}{dt} = x\{1 - 2[x - f(t)]\}$$

(2.12)

As an example, Fig. 2.2 shows the process of solutions' adaptation of finite difference version of equation (2.12) to external influence that has the form $f(t) = 0,25\sin[0,3t - \cos(0,6t)]$. The equation tracks the external influence, mapping it with the range of the process $(0,1)$.

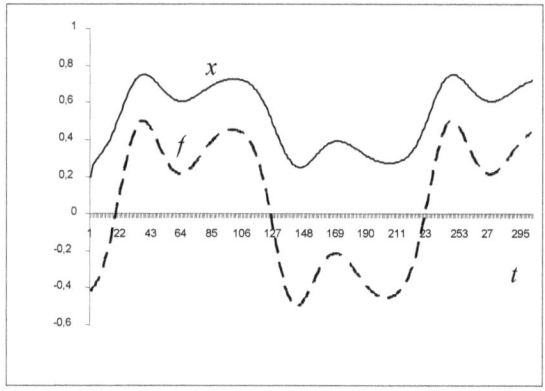

Fig. 2.2. Adaptation of solutions of (2.12) (curve x) to an external influence

(curve *f*)

2.2. General System of Equations of Development Processes in *ABC*-models

In the *ABC*-method, uniting various processes that influence each other into a general system is carried out by means of supplying variables representing the influence processes to the argument of the basic influence function. The principle of constructing systems of equations in *ABC*-models requires that the argument of the basic influence function $F^{(+)}(x)$ must be presented in the form of a weighted sum of all processes related to x with mutual influences.

***ABC*-Model of Two Interrelated Processes**. Let us take an example of two interrelated development processes, united into a system as shown in Fig. 2.3.

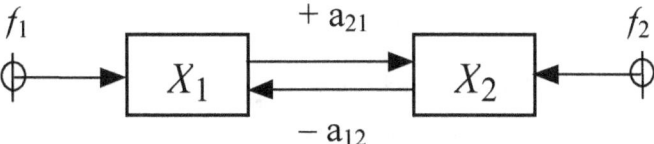

Fig. 2.3. Conceptual model of a simple system, formed by two interrelated processes x_1 and x_2

The weighting coefficients of functions x_1 and x_2 should reflect the degree of influence made by each of the processes on the other. Therefore, each process must be presented in the system by its module having an equation of the form (2.12) which, along with the function of external influence $f(t)$ will be a weighted sum of all other processes that influence it.

It is necessary to keep in mind that the signs of influences are to be reversed compared with Fig. 2.3, in order to preserve positive and negative nature of these influences. Taking that the basic influence functions are equal to their arguments, we'll obtain

$$\frac{dx_1}{dt} = x_1\left[1 - 2\left(x_1 + a_{12}x_2 - f_1\right)\right],$$

$$\frac{dx_2}{dt} = x_2\left[1 - 2\left(x_2 - a_{21}x_1 - f_2\right)\right]. \tag{2.13}$$

Combining the processes into the system (2.13) results in emergent effect which ensures adaptation of the processes to each other and to external influences f_1 and f_2. The system of equations (2.13) establishes a new equilibrium state for adapted processes x_1 and x_2.

To illustrate the processes' adaptation to each other, let us build scenarios of x_1 and x_2 with absent external influences, i.e. when $f_1 = f_2 = 0$. We can define degrees of the processes' influences on each other by choosing coefficients of the model: $a_{12} = 0{,}8$ and $a_{21} = -0{,}3$. The system of equations (2.13) represented in finite differences, takes the form

$$x_{1k} = 2x_{1j}\left[1 - \left(x_{1j} + 0{,}8x_{2j}\right)\right],$$

$$x_{2k} = 2x_{2j}\left[1 - \left(x_{2j} - 0{,}3x_{1j}\right)\right], \tag{2.14}$$
$$(k = j+1), \quad (j = 0,1,2,3...)$$

Let us supply external influences functions in the right sides of the model equations

$$f_1 = 0{,}5\sin[0{,}3t - \cos(0{,}6t)],$$

$$f_2 = \cos[0{,}5t + \sin(0{,}7t)],$$

with coefficients $-0{,}4\,f_1$ and $0{,}25\,f_2$. Then equations (2.14) take the form

$$x_{1k} = 2x_{1j}\left[1 - \left(x_{1j} + 0{,}8x_{2j} + 0{,}2\sin[0{,}3t_j - \cos(0{,}6t_j)]\right)\right],$$
$$t_k - t_j = 1, (k = j+1), \tag{2.15}$$

$$x_{2k} = 2x_{2j}\left[1 - \left(x_{2j} - 0,3x_{1j} - 0,25\cos[0,5t_j + \sin(0,7t_j)]\right)\right],$$
$$(j = 0,1,2,3...)$$

Graphs of the external influence functions f_1 and f_2. were shown in fig. 2.4, a. The resulting scenarios of processes presented in fig. 2.4, b.

In this version of the model, the processes x_1 and x_2 must simultaneously adapt their values to each other and to the external influences f_1 and f_2. If the processes do not affect each other, then one would expect that their scenarios will be similar to the scenarios of external influences attached to them, just as it happened in the experiment with equation (2.12), illustrated by Fig. 2.2. However, mutual influences between processes x_1 and x_2 lead to the fact that the external influence f_1 does not only affect the process x_1, but also influences the process x_2, and the external influence f_2. in its turn, affects the process x_1. Therefore, the resulting scenarios of the processes are significantly different from the external influences applied to them. It is expected that this difference will increase when the number of interrelated processes combined in the structure of the system, and the number attached to them outside influences, will grow.

General System of Equations in *ABC*-Method. As described above, basic properties of *ABC*-method allow us to outline general system of equations for n interrelated processes. In this general case, the system of *ABC*-model's dynamic equations and its finite-difference approximation are to take the following form

$$\frac{dx_i}{dt} = x_i\left[1 - 2\left(x_i - \sum_{j=1}^{n-1} a_{ij}x_j - f_i\right)\right],$$
$$(i,j = 1,2,...,n),\ (i \neq j). \tag{2.16}$$

$$x_{ik} = 2x_{ij}\left[1 - \left(x_{ij} - \sum_{m=1}^{n-1} a_{im}x_{mj} - f_{ij}\right)\right],$$

$$(i, m = 1, 2, ..., n), \ (i \neq m),$$
$$(k = j+1), \ (j = 0, 1, 2, 3 ...), \ t_k - t_j = 1 \qquad (2.17)$$

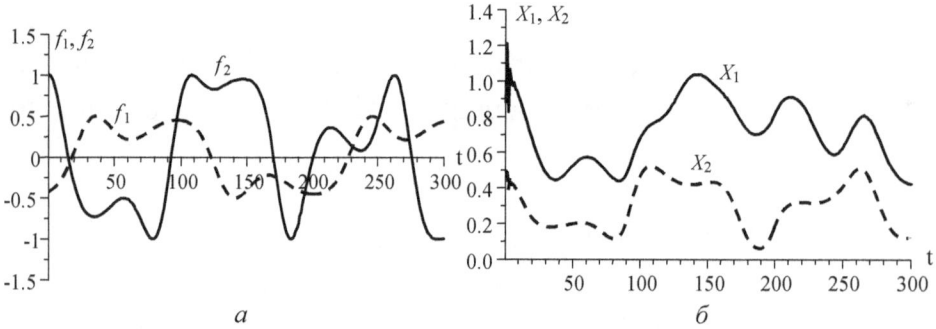

Fig. 2.4. Simultaneous adaptation of the processes x_1 and x_2 to each other and to external influences f_1 and f_2: (a) external influences; (b) results of adaptation

As it was proven by different researches [39, 82, 181, 183], when the system of equations of form (2.17) is provided with correct choice of influence coefficients, it has an ability of fast convergence to steady solution. The ease of transition from conceptual model of a system to its formal description is an indisputable advantage of *ABC*-models. In addition to this, there exists a number of expert evaluation methods of influence coefficients; these methods are utilized to construct models of specific systems [151]. The most objective methods of the coefficients' evaluation are the statistical ones, which will be discussed in what follows.

2.3. Comparison of the ABC-Method with the Method of System Dynamics

It is interesting to compare two approaches to the adaptive models' construction for interrelated processes: the *ABC*-Method [181] and the *SD*-Method [50] (the method of system dynamics, considered in Chapter I). In both methods the universal systems module for each of the processes in a complex system is used. These modules share a common property to adapt simulated process to external influences applied to the module, due to the negative feedback between the simulated processes and the external influences. Both of the methods define general rules of connecting individual modules in the structure of the model. These properties find their expression in right-side parts of the equations (1.1) and (2.8). The difference between them can be found in the ways of implementing the negative feedback entered in the right-side parts.

In system dynamics Method, the negative feedback of a simulated process x_1 is carried out by incorporating the term of form $- a_{11}x_1$ in the right side of the modular equation

$$\frac{dx_1}{dt} = -a_{11}x_1 .$$ (2.18)

Such feedback is of the first order (in terms of degree of x_1 in the right side of the equation), and in the absence of external influences it leads the process to converge to zero.

Unlike the other, the *ABC*-Method applies negative feedback of the second order, which is characteristic of the logistic equation (2.8)

$$\frac{dx_1}{dt} = x_1(1 - 2a_{11}x_1) = x_1 - 2a_{11}x_1^2 .$$ (2.19)

So, in case of absence of external influences, the solution of modular equation of *SD*-method (1.18) tends to zero, while the solution of modular equation of *ABC*-method (2.19) tends to the equilibrium value $x_1^* = 0,5a_{11}$. We can show that this distinction remains valid, when modular equations are united into a system of model equations, i.e. when in right sides of equations for both methods (which affect functions of the other processes) are presented.

To continue with our task, let us consider again the connections between the two modules, as shown in Fig. 2.3. We'll eliminate the external influences, setting $f_1 = f_2 = 0$, and choose influence coefficients' values, for example, as $a_{12} = -0,3$, $a_{21} = 0,3$. The system of equations of the *SD*-method takes the following form

$$\frac{dx_1}{dt} = -a_{11}x_1 - a_{12}x_2,$$

$$\frac{dx_1}{dt} = a_{21}x_1 - a_{22}x_2,$$

or in the finite-differences

$$x_{1k} = (1-a_{11})x_{1j} - 0,3x_{2j},$$

$$x_{2k} = 0,3x_{1j} + (1-a_{22})x_{2j}. \tag{2.20}$$

The system of equations of the *ABC*-method can be written as

$$\frac{dx_1}{dt} = x_1[1 - 2(a_{11}x_1 + a_{12}x_2)],$$

$$\frac{dx_2}{dt} = x_2[1 - 2(a_{22}x_2 - a_{21}x_1)],$$

or in the finite-differences

$$x_{1k} = 2x_{1j}[1 - (a_{11}x_{1j} + 0,3x_{2j})],$$

$$x_{2k} = 2x_{2j}[1 - (a_{22}x_{2j} - 0,3x_{1j})]. \tag{2.21}$$

Results of calculations by equations (2.20) and (2.21) are shown in Fig. 2.5.

As we can see from this Figure, in absence of external influences, there is no mutual adaptation between processes in the Method of system dynamics. The values of all processes tend to zero and the zero state vector characterizes steady condition of the system. Under the same terms, the model of interaction of the two processes constructed by the *ABC*-Method demonstrates mutual adaptation of the processes to each other in steady state. The state vector of the *ABC*-model is characterized with stationary values of the processes x_1* and x_2*, which are solutions of difference equations (2.21) converging to them.

Difference in approaches to description of complex systems between *SD*-method and *ABC*-method can be interpreted by physical analogies.

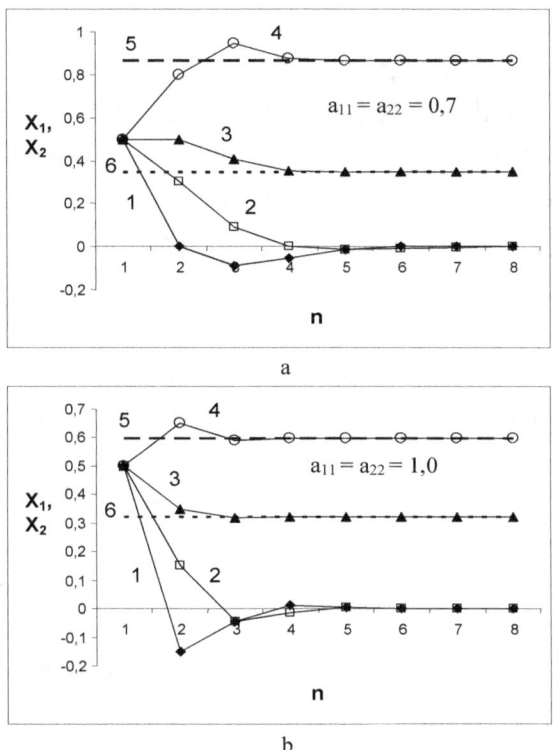

a

b

Fig. 2.5. Comparison of results of two interrelated processes' adaptation with absence of external influences, in the Method of system dynamics ($1 - x_1$, $2 - x_2$) and in the Method of Adaptive Balance of Causes ($3 - x_1$, $4 - x_2$, $5 - x_2^*$, $6 - x_1^*$): (a) $a_{11} = a_{22} = 0,7$; (b) $a_{11} = a_{22} = 1,0$

It is well-known that J. Forrester offered *SD*-method to visualize processes in complex systems in the form of streams of various substances flowing into the system from outside, and flowing out of the system (see paragraph 1.3 of Chapter I).

Adaptation of the processes to each other under these conditions means achieving total balance of flows in the system. If external influences – streams, altering levels, are terminated, then the flowing out streams direct the contents of all levels to zero. Therefore, the *SD*-method describes only the dynamic equilibrium of processes in complex systems, when it is forced by external influences.

ABC-method allows us to simulate both, static and dynamic balance of processes. Mutual influences of processes in this method should be interpreted as applying forces for mutual interaction between processes. In the steady state, we can see balance of forces which decline values of the processes from the state of total equilibrium and keep the processes in this state.

In *SD*-method, flows only occur in the system under existence of external influences, or external streams which come flowing into the system and out of it. Stationary flows always exist in *ABC*-method, as they reflect mutual influences between the processes. They disappear only under the condition when the system falls apart into a number of separate, independent modules.

In the example shown above in Fig. 2.5, the balance of forces defined by the equations of *ABC*-method (2.21), demonstrates deviations of stationary values of processes from neutral equilibrium ($x_1^* = x_2^* = 0,5$): $x_1^* = 0,32$ и $x_2^* = 0,59$. When external influences occur in the *ABC*-model, a kind of dynamic equilibrium is established. In this state, the system continuously restores balance of forces (influences), and adapts itself to any external influences.

2.4. Prediction of Development Scenarios in the *ABC*-Model of Environmental Management

The properties of *ABC*-Method described above, allow us to apply this method for regimes of static and dynamic balance modeling in complex natural-industrial systems of coastal zone areas. As an example of implementation of systems management methodology, let us set the task to achieve a balance between the amount of consumption of marine resources and the volumes of their reproduction in a coastal zone area. This balance, supported by means of environmental protection measures, can be presented in the conceptual model of ecological-economic system, as shown in Fig. 2.6.

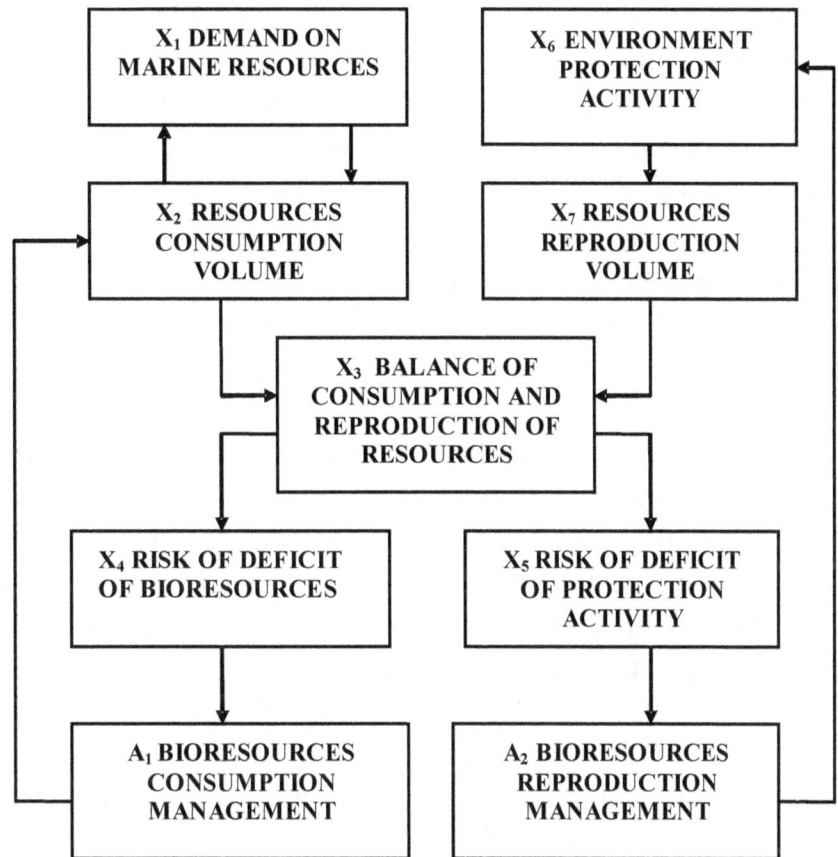

Fig. 2.6. Conceptual model of the ecological-economic system of coastal zone environmental management

In accordance with its goal, this model includes the following variables: x_1 – demand for marine resources, x_2 – consumption volume of bioresources, x_3 – balance index of consumption and reproduction of biological resources, characterized by the concentration of bio-resource in marine environment, x_4 – risk of deficit of bio-resource concentration, x_5 – risk of shortages of environmental protection activities, x_6 – volume of environmental protection activities, x_7 – volume of reproduction of bio-resource concentration.

Description of a Simple *ABC*-model of Environmental Management. The balance of consumption and reproduction of marine biological resources is characterized in this model by an integrated parameter – the concentration of bio-resource x_3. The process x_3 is under the influence of two oppositely directed tendencies. On the one hand, it is strongly affected by pollution of marine environment which is associated with consumption of biological resources x_2, in order to support interests of the economic system of production. On the other, – it experiences positive changes to the environmental conditions which occur due to implementation of environment protection activities, the amount of which is designated as x_6.

Grounding on fundamental concepts of systems thinking, in order to ensure balanced development of these tendencies, a number of closed chains of positive and negative feedback that would balance each other, must be introduced in our model of ecological-economic system. This condition is achieved, for example, by inclusion of processes that characterize the risk of bio-resource concentration shortage x_4, and the risk of shortage of environment protection activities. To control the balance of bioresources, management agents A_1 and A_2, controlling their consumption and reproduction, are included into the feedback chains.

In order to formalize the conceptual model shown in Fig. 2.12, we will use a set of equations (2.16) of the *ABC*-Method. First of all, it is necessary to ensure that all simulated processes are presented in a dimensionless form and are brought to a common interval of variability, such as (0, 10). To do this, it is convenient to use a linear transformation of the following form, which will also be applied in what follows

$$x_i = 5\frac{x_i^d}{\overline{x}_i^d}, \qquad x_i^d = 0,2x_i\overline{x}_i^d \tag{2.22}$$

In it, x_i^d denotes the value of the simulated dimensional process, presented in dimensionless form as x_i, and through $\overline{x}_i^d$ identifies the mean value of the dimensional process' variability interval.

Taking into account the scheme of reciprocal influences in the system of environmental management, we obtain the following dynamic equations for the dimensionless formal processes of the *ABC*-model

$$\frac{dx_1}{dt} = x_1[1 - 2(x_1 - a_{12}x_2 - F_1)],$$

$$\frac{dx_2}{dt} = x_2[1 - 2(x_2 - a_{21}x_1 - A_1(x_4)],$$

$$\frac{dx_3}{dt} = x_3[1 - 2(x_3 + a_{32}x_2 - a_{37}x_7)],$$

$$\frac{dx_4}{dt} = x_4[1 - 2(x_4 + a_{43}x_3)],$$

$$\frac{dx_5}{dt} = x_5[1 - 2(x_5 + a_{53}x_3)], \tag{2.23}$$

$$\frac{dx_6}{dt} = x_6[1 - 2(x_6 - A_2(x_5)],$$

$$\frac{dx_7}{dt} = x_7[1 - 2(x_7 - a_{76}x_6)].$$

External influences in this model, along with management agents A_1 and A_2, represent function F_1 which takes into account possible changes in demand for the marine resources.

Let us set the maximum permissible value of the risk of shortage in marine biological resources. This value reflects such a state of the marine ecosystem, which can lead to irreversible consequences in quantitative or qualitative composition of bioresources. A biodiversity index of marine environment, which decreases its value with the growth of potential risk, can serve as an integrated measure of the risk.

Let us denote the maximum amount of the admissible deficit value as x_4^*. A management agent A_1 should monitor the size of x_4 so that the risk level of biological resources deficit does not exceed the maximum permissible value x_4^*. When this state occurs, the consumption of biological resources should be limited. Therefore, taking into account the properties of the *ABC*-method equations, it is expedient to choose the following representation for the management agent A_1

$$A_1\left(x_4, x_4^*\right) = IF\left[x_4^* > x_4; a_{24}x_4 e^{-\alpha_1 \tau_1}; b_{24}\left(1 - e^{-\alpha_2 \tau_2}\right)\right]. \qquad (2.24)$$

Similar arguments can justify the choice of the second management agent A_2 in the model. Insufficient amount of environment protection activities inevitably results in growing pollution of the marine environment. It is contaminated with substances that affect concentrations of bioresources. Thus, pollution level can serve as an integrated feature of deficit in environment protection activities x_5. This value should not reach the critical level x_5^* – once this value is exceeded, no more measures to restore the bioresources concentration will be efficient anymore. Therefore, the formal representation of the agent A_2 should be as follows

$$A_2\left(x_5, x_5^*\right) = IF\left[x_5^* < x_5; a_{65}x_5 e^{-\alpha_3 \tau_4}; b_{65}\left(1 - e^{-\alpha_4 \tau_4}\right)\right]. \qquad (2.25)$$

The coefficients b_{24}, b_{65}, $\alpha_1 - \alpha_4$ allow to choose the intensities and rates with which the agents affect the development processes in the system, and thereby they ensure control over the balanced dynamics of the environment management.

Simulation Experiments with *ABC*-Model of Environment Management. For the purpose of computation experiments, the environmental management model (2.23) – (2.25) was presented in the Euler's finite-difference scheme, just as it was carried out in the general model of *ABC*-method (2.17). Then, using the function F_1, a time-variable demand for marine bioresources was simulated and influence coefficients of the processes influencing each other, were set up. Their values were chosen in such a way that the balance between consumption and reproduction of marine bioresources could significantly affect the magnitudes of the risks of bioresources shortage and the lack of environment protection activities. At the same time, their values did not exceed 0.5. As the practice of computing demonstrated, when the values exceeded this limit, solutions of the equations can demonstrate the regimes of chaos. Calculations were performed on 730 dimensionless time steps. The goal of these calculations was to limit consumption of biological resources when the risk of resources' deficit reaches the value $x_4^* = 3.3$.

The scenarios of processes within the system which were caused by simulated random fluctuations of the demand for bioresources are shown in Fig. 2.7, a and b. Simulated scenario of the risk of extremely high bioresources consumption caused by the high level of demand, exceeded the threshold of risk, equal to 3.3. Scenarios of resources consumption management, generated by the agent A_1, are shown in figure 2.7.

The control influences were supposed to "switch on" every time when the risk curve crossed the straight line $x_4^* = 3,3$ from below, and to "switch off" when the intersections were coming from above. According to the figure, the management of bioresources

115

consumption resulted in reduction of risks of bioresources shortage, as its average value fell down below the 3.3 mark.

The agent A_2 is supposed to ensure that the threat of insufficient environmental protection activities does not exceed the value $x_5^* = 3.5$. The risk scenarios of this deficit, and the control actions of the agent A_2 are shown in Fig. 2.7, b. The emphasis is placed on two periods of greatly enhanced environmental protection activities, predicted by the model for periods from 200 to 300 and from 390 to 460 time steps. A more active response to the lack of environment protection activities allowed us to reduce the risk of bioresources deficit in these periods
(see Fig. 2.7, a).

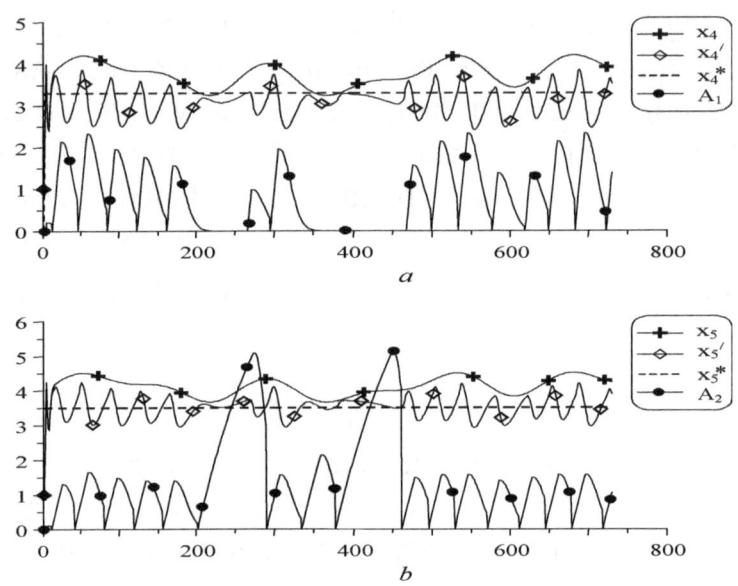

Fig. 2.7. Risks control over biological resources shortage and lack of environment protection activities in the model of environmental management: (a) x_4 – risk of biological resources shortage with no control (limit level of 3.3), x_4' – risk with control, x_4* – critical risk level of bioresources shortage, A_1 – function of agent that restricts consumption of bioresources; (b) x_5 – risk of deficit of environmental protection activities with no control (limit level of 3.5), x_5' – risk with control, x_5* – critical risk level of environmental protection activities, A_2 – function of agent that increase the level of environmental protection activities

Results of the management over the resource-oriented scenarios of processes in the system are shown in Fig. 2.8, a, b. The curve of risk of the bioresources' deficit, shown in Fig. 2.7 qualitatively repeats the curve of demand for bioresources prior to control actions. Comparing it with the curve of demand after the management, as shown in Fig. 2.8, a, b, we can see that management actions in the ecosystem affected the values of the demand, which became more volatile. Comparison of the scenarios x_2 (see graphs in Fig. 2.8, a) demonstrates how consumption of biological resources is reduced, compared to the situation that would have occurred if no management actions had been undertaken with the system.

Similar conclusions follow from the results of comparison of levels of the biological resources reproduction in calculations taken with and without control, as depicted in Fig. 2.8. The level of reproduction was significantly higher, although it acquired the character of a process oscillating about a mean value. These fluctuations are due to the deviations from balance between consumption and reproduction of biological resources x_3 in the management process of the marine ecosystem's state; they also takes into account the given risk values. Despite existing fluctuations of the balance curve in the graph, which characterize the state of the ecosystem, it goes substantially above the position which corresponds to the situation of the process which took place with no control actions (see Fig. 2.8, b).

The average level of environment protection was defined as $x_6 =$ 2.0 (see Fig. 2.8, b). The need to reduce the risk of the ecosystem's degradation led to the fact that, while limiting consumption of biological resources, it took an additional amount of measures to ensure their reproduction. A graph of these events, being predicted by the model, is reflected by curve 5 in Fig. 2.8, b. Thus, model calculations allowed us to provide approximate evaluation of total amount of environmental activities, and the periods of time when the environmental works should be improved.

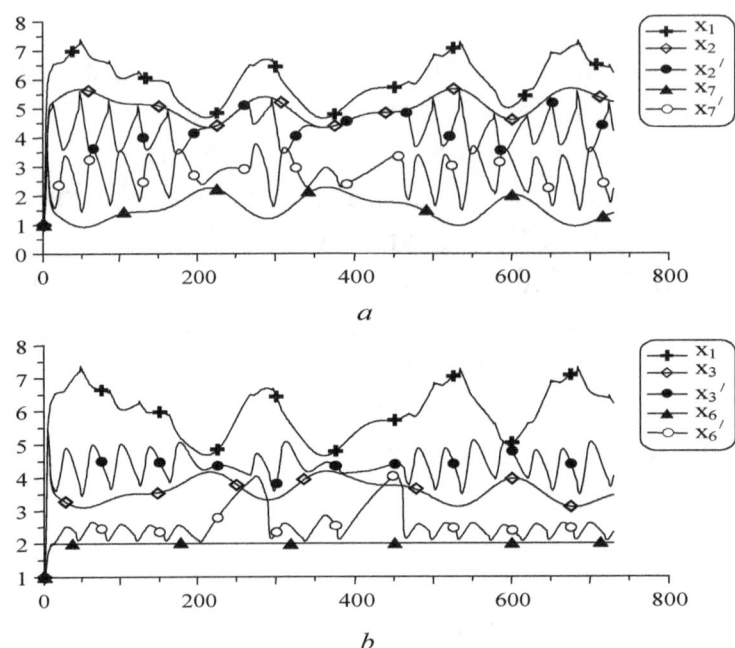

Fig. 2.8. Scenarios of processes in the system of management by the balance of consumption and reproduction of marine biological

118

resources: (a) x_1 – demand for bioresources, accounting management actions, x_2 – consumption volumes without control, x_2' – consumption volumes with control, x_7 – reproduction volumes with no control, x_7' –

reproduction volumes with control; (b) x_1 – demand for bioresources, accounting management actions, x_3 – concentration of bioresources without management, x_3' – concentration of bio resources with

management, x_6 – volumes of environmental protection activities with no control, x_6' – volumes of environ mental protection activities with

control

2.5. Management of Development Processes in a Model of a Regional Socio-Economic System

As another example of the use of systems management, we will consider an integrated model of socio-economic system of the coastal zone area. Let us introduce (as a state variable) ten major processes that characterize the socio-economic development processes in CZA: budget of the zone x_1, gross domestic product x_2, integrated demand x_5, bank's capital x_3, social guarantees to the population x_4, demand for manufactured goods and services x_5, inflation expectations x_6, standard of living x_7, production volume x_8, investment into economy x_9 and employment rate among population x_{10}.

Conceptual model of development processes. We assume that the main problem of the system's development management is the correct allocation of budget between investment into the economy x_9 made via banks capital x_3, and non-productive charges to implement social guarantees x_4. Cause-effect relationships between socio-economic development processes are depicted in Fig. 2.9, with negative effects are shown with dashed arrows.

Because of the limited budget, the increase in expenditure for social protection reduces total amount of investments into economy, which leads to decrease in employment rate x_{10} and lowers the standard of living. The index of population's living standard x_7 presents integrated demand for manufactured goods and services x_5, which directly affects production output x_8, and therefore, the gross domestic product, x_2, fill the budget of the area x_1. With increasing costs of production development, employment rates increase, but at the same time, social payments to the population (pensions, allowances, scholarships, etc.) experience reduction.

Thus, the living standard of population is formed in this model under the influence of two opposite trends, which are represented by a complex system of positive and negative feedback.

We assume that management of *CZA* socio-economic systems can be complicated by presence of inflation expectations x_6, associated with possibility of crisis phenomena in the economy. For such situations we utilized a simplified scheme of crisis development, where the economic growth of bank's capital for any reason loses its ability to invest funds into production.

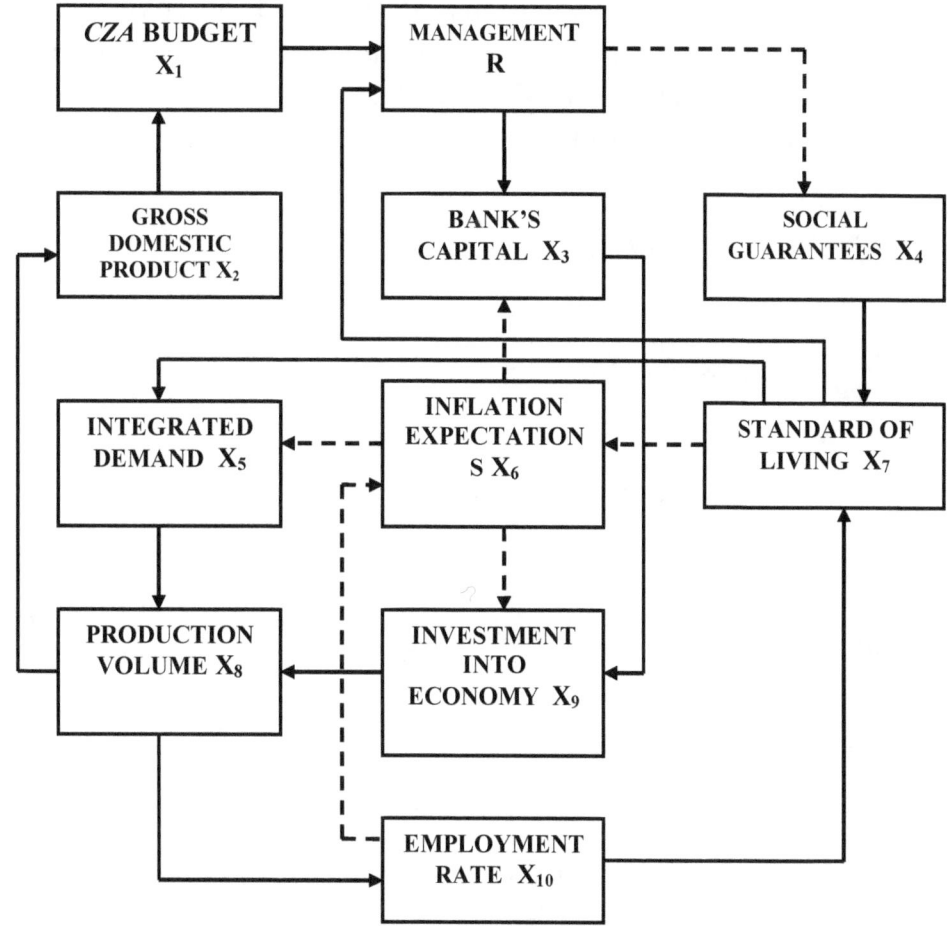

Fig. 2.9. Conceptual model of regional socio-economic system of a coastal zone area

Inflation expectations increase the development of crisis in the system through three circuits of positive feedback, which contain an even number of negative influences (2 dotted arrows). The first of these forms a closed cycle: $x_6 - x_5 - x_8 - x_{10} - x_7 - x_6$; the second is: $x_6 - x_9 - x_8 - x_{10} - x_7 - x_6$; the third is a cycle: $x_6 - x_3 - x_9 - x_8 - x_2 - x_1 - R - x_4 - x_7 - x_6$. Control function of R is included in the third cycle of positive feedback in order to restrain inflation expectations and the development of production crisis.

Prediction of Scenarios of Socio-Economic Development. Applying *ABC*-method to the conceptual model in Fig. 2.9, we obtain the following dynamic equations of modeled processes of socio-economic development of the coastal zone. This provides the function of redistribution of budgetary resources between social services and investment into the economy

$$\frac{dx_1}{dt} = x_1[1 - 2(x_1 - a_{12}x_2)],$$

$$\frac{dx_2}{dt} = x_2[1 - 2(x_2 - a_{28}x_8)],$$

$$\frac{dx_3}{dt} = x_3[1 - 2(x_3 - a_{34}x_4 + a_{36}x_6 - a_{3R}R)],$$

$$\frac{dx_4}{dt} = x_4[1 - 2(x_4 + a_{43}x_3 + a_{4R}R)], \qquad (2.26)$$

$$\frac{dx_5}{dt} = x_5[1 - 2(x_5 + a_{56}x_6 - a_{57}x_7)],$$

$$\frac{dx_6}{dt} = x_6[1 - 2(x_6 + a_{67}x_7 + a_{6/10}x_{10})],$$

$$\frac{dx_7}{dt} = x_7[1 - 2(x_7 - a_{74}x_4 - a_{7/10}x_{10})],$$

$$\frac{dx_8}{dt} = x_8[1 - 2(x_8 - a_{85}x_5 - a_{89}x_9)],$$

$$\frac{dx_9}{dt} = x_9[1 - 2(x_9 + a_{93}x_3 - a_{96}x_6)],$$

$$\frac{dx_{10}}{dt} = x_{10}[1 - 2(x_{10} - a_{10/8}x_8)].$$

Two computational experiments were conducted with this model on the 370 steps in time with the dimensionless scale of variability [0,10]. In the first experiment, it was thought that the management function R supports a constant level of funding for the social sphere.

At the initial stage of the experiment the population of the area had relatively high standard of living. Later, due to a number of conditions (e.g., due to creation of so-called "financial pyramids"), the bank's capital of the economic system was cut down, as shown in Fig. 2.10. However, due to a sufficiently high and constant level of social protection, high standards of living among the population, and positive integrated demand, continued to develop on the increase for some period of time. The buying capacity of the population supported the amount of production output and helped to fill local budget, despite reduction in investments by the bank sector.

Starting with step 70 of the computer simulation, the reduction of bank's capital investment into the economy has reached such a level (20%), when the decline in production, followed by reduction in employment rates and a slower growth in living standards, led to increase in inflation expectations, which altogether denoted the beginning of a crisis. In a view of the above three chains of feedback a sharp increase in inflation expectations had began accompanied with decreasing values of integrated demand. At the same time, collapse of production output and gross domestic product ultimately led to a sharp deterioration in living standards (see Fig. 2.10). Note that because of multiple positive feedback, the model reproduces a very rapid development of the crisis.

In the second computing experiment, some resources allocated to social safeguards were put into a proportional dependence on the size of the available budget. In this experiment, the results of which are shown in Fig. 2.11, the task was posed: to weaken the development of the crisis by means of the budget re-allocation. Management scenarios were in such a way that, due to temporary reduction in social expenditure, the bank's capital has been strengthened and investments into the economy eventually increased.

For this purpose, the following functions were given to the control operator R

$$R = IF\{x_9 > x_9^*; 0; b_R[1 - \exp(-\alpha_R t)]\}$$

(2.27)

In this formula x_9^* stands for the necessary amount of investment into the economy, in order to prevent the crisis; b_R and α_R are parameters needed to set the speed of the crisis development. As it can be seen in Fig. 2.11, with the collapse of investments in the economy (curve 3) and the reduction of production output (curve 2), failures began in social sector expenditure (curve 4), which significantly increased up to step 101 of the calculations, when a control function R switched on in the equations for bank's capital x_3 and social guarantees x_4.

The revised allocation of budget resources led to significant reduction in social expenditure in the period from step 101 to step 130 of calculations, and more smoothly- to reduction in the subsequent period of time (curve 4). Due to the ability to save on social sector expenditures during this period of time, banks' capital investments into the local economy were significantly increased (curve 3). This marked significant increase in the gross domestic product (curve 2) and in local budget (curve 1). Reaction processes in the social sphere of anti-crisis budget reallocation is shown in Fig. 2.11b. As expected, the standard of living of the population (curve 3) has provided a very strong influence on inflation expectations

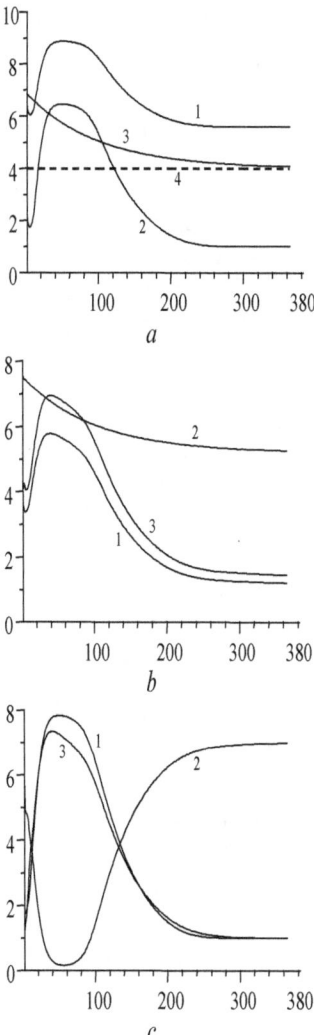

Fig. 2.10. Scenarios of development processes in a permanent social well-being conditions: (a) 1 – Budget x_1, 2 – gross domestic product x_2, 3 – bank's capital x_3, 4 – social guarantees x_4; (b) 1 – integrated demand x_5, 2 – inflation expectations $x6$, 3 – standard

of living x_7; (c) 1 – production output x_8, 2 – investments in the economy x_9, 3 – Employment x_{10}

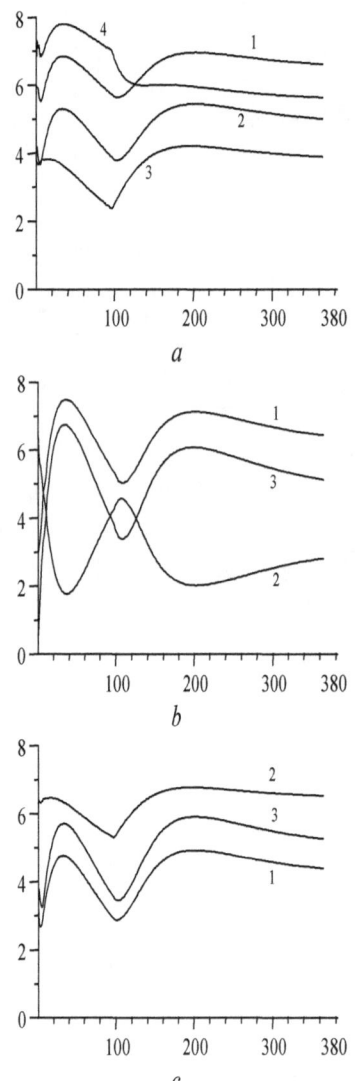

Fig. 2.11. Scenarios of development processes in terms of social security depends   on the *CZA* budget: (*a*) 1 – Budget x_1, 2 – gross

126

domestic product x_2, 3 – bank's capital x_3, 4 – social guarantees x_4; (b) 1 – integrated demand x_5,

   2 – inflation expectations $x6$, 3 – standard of living x_7; (c) 1 – production output x_8, 2 – investments into economy x_9, 3 – Employment x_{10}

(curve 2) and on integrated demand (curve 1). A significant decline in living standard was during the development of crisis (in the neighborhood of 100 steps of calculations), but then, from 101 to 180 step, the rise of it was observed, due to the growth of social security x_4, and thanks to general economic rehabilitation which happened by this time.

The status of economic processes in the crisis period are characterized by the scenarios, shown in Fig. 2.11, b. After a sharp fall in the period of crisis, the output of production (curve 1), investments in the economy (curve 2) and employment (curve 3) rose significantly from 101 to 180 step of the experiment, after which the growth was replaced by a smooth decrease in economic activity. It should be noted that, despite the reduction in social costs, the standard of living of the population in post-crisis period grew up compared to the before-crisis period, as we could notice the restoration of production volumes, increase in employment and reduction of inflation expectations.

The computational experiments which we conducted, showed that the systems methodology for modeling and forecasting processes of socio-economic development allows us to relatively easily obtain model development scenarios in complex socio-economic systems. Similar scenarios are needed to support administrative decisions, associated with management development. Application of local control functions (agents) in the equations of the *ABC*-models allows to optimize the use of development resources, since the agents monitor the dynamics of the projected processes and guide them in accordance with the conditions and tasks that were set.

2.6. *ABC*-Model "Daisy World" of the Global Temperature Adaptation

Today, when growing amounts of greenhouse gases enter the Earth's atmosphere due to development of transportation systems and energy consuming industries, the consequences of this phenomenon on the world's climate, economy, and environment are being studied intensely. Special interest is given to *GAIA*-theory [100, 105], which puts forward the hypothesis of adaptation of global dynamic processes in animate and inanimate nature. The roots of this theory can be found in the works of Vernadsky [193], who introduced the concept of "noosphere" and claimed that the noosphere is a complex system that includes terrestrial and oceanic flora and fauna, geological structure of the land's upper layer and the seabed of oceans, as well as geochemical structure of the atmosphere. This unified system develops in dynamic balance with external influences (forces) – specifically, with the sunlight.

According to the scientists supporting *GAIA*-theory, the global system "atmosphere – ocean – land surface", along with the living organisms inhabiting it, has an ability to adapt to changing external conditions through self-regulation, providing by living organisms. The natural property of adaptation is a reason for utilization of the method of adaptive balance of causes in the models which can give an integrated assessment of the global self-regulation processes. In this formulation the *ABC*-model of global development processes was studied in [180].

The aim of the work was to apply the method to the global natural social ecological-economic system. This approach can be seen as a development of *GAIA*-theory. In this paper, the assumption that human society, being the highest form of living on the Earth empowered with intelligence, has more developed self-organizing properties than any other living beings. The global social ecological-economic system and its natural environment is regarded to be in a state of adaptive balance that ensures its sustainable development.

***GAIA*-Theory and the Effects of Self-Regulation.** A British scientist James Lovelock [100] suggested that there is a global self-regulation of natural processes, and due to this fact, average surface temperature of the Earth fluctuates within a narrow range of values: $10 - 22\ ^oC$, despite the growth coming from solar energy. For about 4 billion years, the average temperature on the Earth has remained within these limits, while solar radiation for the period increased by one third. To explain the large-scale self-regulation temperature, we would like to offer the two statements that follow.

First of all, microorganisms, lichens, plant roots and other pro-starts, cause destruction and grinding and, as a sequence – weathering of rocks. These processes increase as the average temperatures grow, and it results in appearance of additional nutrients that many of the organisms can use for their growth and dissemination on the surface of the land. The increase in green mass of plants, in its turn, reduces the content of CO_2 in the atmosphere and reduces the average surface temperature of the Earth.

The second mechanism of temperature self-regulation is related to the oceans. With increasing temperatures, the volume of marine algae also grows. They absorb atmospheric CO_2, thereby lowering the average temperature. Note that the ocean has an impact on local sea-surface atmospheric layer temperatures. As the result of marine algae extinction a gas (dimethyl sulfide) is formed, which is released into the atmosphere to form sulfate aerosols in it. The latter serves as condensation nuclei for clouds formation, which reduces solar radiation and lowers the temperature of the ocean surface, as well as the sea-surface layer of the atmosphere.

The influence of living organisms inhabiting the Earth is seen in the fact that chemical composition of the atmosphere is permanently constant, although it is kept in a stable, non-equilibrium state. The Earth atmosphere contains very little carbon dioxide compared with the neighboring planets Venus and Mars, where in condition of chemical equilibrium its volume exceeds 95%. The Earth atmosphere contains 79% of nitrogen, 20.7% of oxygen and only 0.03% of carbon dioxide. Such a mixture can remain stable only due to existence of living organisms, which compensate the shortage or withdraw extra concentrations of these gases in the atmosphere. The greatest contribution to the concentration of CO_2 in the atmosphere is made by volcanic eruptions.

Complex bacterial processes maintain a constant ratio of concentrations of nitrogen and phosphorus in sea water, phytoplankton and zooplankton. Despite the intake of huge amounts of minerals into the sea from the land surface, the salt composition of the oceans remains constant. This is important for the cells of living organisms, as they begin to die with increasing salinity. Salinity is regulated by the processes of evaporation and desalination of the ocean, which largely depend on local temperature gradients and precipitation, associated with formation of clouds.

There is no doubt that the Earth's biosphere behaves as a self-regulating system. This emergent property of the biosphere is the result of countless acts of living organisms seeking to survive in the constantly changing external conditions of their existence. The GAIA-theory supporters regard the Earth as a single organism, which controls and maintains itself in its tendency to provide necessary life conditions for all existing living organisms populating it.

The effects of self-regulation confirm Darwin's theory of evolution. In the process of natural selection, only those of the organisms that are capable of self-regulation, obtain a selective advantage over the others. They survive and preserve the ability of self-regulation, and then they tend to pass it on to their heritage. Similar processes take place at the ecosystem level. Those of the ecosystems which had a total negative feedback connections providing self-control, acquired the ability to adapt to external influences. Thus, in course of evolution, the world processes of self-regulation acquired a global character [193].

L. Margulis suggested that atmosphere, hydrosphere and lithosphere have the ability for self-correction, so that their states tend to keep certain scenarios developing over the time [105]. The same idea of existence of some dynamic balance between processes developing in complex systems is used in the method of adaptive balance of causes, proposed for modeling of complex systems developing in time.

Modeling Self-Regulation Effects in the Model "Daisy World". One of the first models for simulating global processes of temperature self-regulation was built in response to criticism of the *GAIA*-theory. A model called "Daisy World" [199], introduced a solar system planet inhabited only by two species of living organisms: black and white daisies. In it, both species had identical conditions of growth and breeding (distribution) on the planet, because they had the same dependence between the rate of reproduction *GF* and the temperature *T*

$$GF = 1 - 0.003265(22.5 - T)^2$$
(2.28)

The white daisies reflect solar radiation to a greater extent than the black ones. This is why the black daisies are warmer than the whites. Their dominance on the planet would be expressed in elevated temperature of the planet. The model was built to study the process of adaptation of plants on the planet to gradually increasing external impact of the growth of solar radiation.

At the beginning of the experiment the temperature of the planet was so low that only a few of the daisies could survive. It follows from (2.28) that the best condition for this is the temperature $T = 22.5$ ^{o}C . As the temperature of the planet grew up, the advancing conditions of homeostasis became favorable for both, the white and the black daisies. The growth factor (2.28) determined the range of temperatures, at which the black and the white daisies were able to disseminate all over the planet. Each type of daisies acted under its own rules, but due to being combined into a common system, they acquired a new quality – temperature stability, which was supported by a mechanism of adaptation.

Let us consider the processes of global temperature self-regulation in the *ABC*-model Daisy World [180], as shown in Fig.

2.12. The average temperature of the planet T_m depends on the amount of heat absorbed by each of the three parts of the planet's surface: part x_2, taken with white daisies and having a temperature TW, part x_3, occupied by black daisies and having a temperature TB, and part G of the whole surface which remained unoccupied and has a temperature TG. We assume that the part x_2 has albedo 0.75, the part x_3 has albedo 0.25, and the part G has albedo 0.5. Total area of the planet is denoted as G_0, and the intensity of solar radiation is $SR(t)$, where t is time.

In contrast to the experiment described in [199], instead of a parabolic curve identifying growth factor (2.28), we will use a Gaussian curve W. This approach seems to be more natural to describe limits of the planet's temperature changes, at which the existence of daisies is possible.

$W = \exp[-0,1(22,5-T)^2]$, where T is equal TW or TB
(2.29)

Under these conditions, the equation of the ABC-method (2.16) lead to the following dynamic model Daisy World

$$\frac{dx_2}{dt} = x_2\{1 - 2[x_2 - a_{23}(G_0 - x_3)\exp[-0,1(22,5 - TW)^2]]\},$$

$$\frac{dx_3}{dt} = x_3\{1 - 2[x_3 - a_{32}(G_0 - x_2)\exp[-0,1(22,5 - TB)^2]]\},\qquad(2.30)$$

$$G = G_0 - x_2 - x_3,\quad TW = 0,25(SR)G_0^{-1},\quad TW = 0,75(SR)G_0^{-1},$$

$$T_m = SR(0,5G_0 + 0,25x_3 + 0,75x_2)G_0^{-1}$$

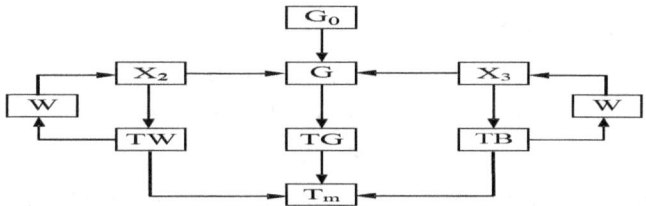

Fig.2.12. Conceptual *ABC*-model Daisy World

Computational experiments conducted with this model were to show reaction of the model to growth of solar radiation that affects the planet. For this purpose, the linear increase in solar radiation *SR* was simulated, accompanied with increase in temperatures of surfaces x_2 and x_3. In order to simplify the situation, the average temperature of the planet was taken in proportion to the amount of heat absorbed by its area G_0. Graphs of the temperature growth on the planet occupied by black and white daisies, are shown in Fig. 2.13a.

Scenarios of resettlement of the black and the white daisies over the world were constructed. Equations (2.30) allowed us to calculate the dynamics of the area occupied with each of the species in conditions of growing solar radiation. Scenarios of the species' resettlement, obtained with accounting of limiting growth factor (2.29) are shown in Fig. 2.13, b. For better clarity, the curves x_2 and x_3 are placed in one figure together with the proper time pace of the planet's average temperature scenario T_m.

133

It becomes clear from the figure that coexistence of the two forms of life on the planet in the time period of the experiment (from 200 to 400 steps of calculations) led to stabilization of the average temperature dynamics on the planet. With significant increase of solar radiation, the average temperature of the planet virtually did not change. Despite the simplicity of the constructed model, experimenting with it suggested a possibility of using the *ABC*-method for study of global adaptation processes in animate and inanimate nature.

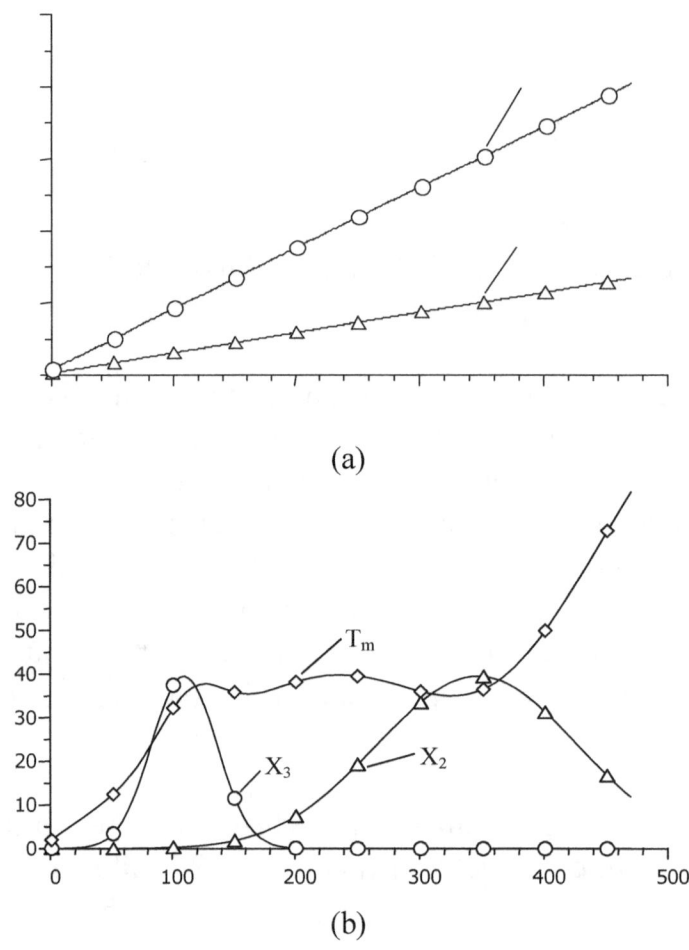

Fig. 2.13. Results of calculation experiments with the *ABC*-model Daisy World.

2.7. Modeling of Global Self-Organization Processes in the Nature and the Society

Development of processes in the global social ecological-economic system is taking place simultaneously with the natural adaptive processes, which are studied by the *GAIA*-theory. In accordance with it the dominating desire of the biota to maintain favorable conditions for life on the Earth leads to changes in global climatic processes. In particular, self-regulation of the Earth surface's average temperature causes dramatic change in atmospheric processes, which cause hurricanes and floods. Natural disasters inevitably affect social, environmental and economic processes of global development. So, it seems expedient to integrate natural and socio-economic processes into a single global system.

Following the systems principles of *ABC*-models construction, it is necessary to determine the most important processes of the world development as the first step of the modeling. The principal objective of sustainable development is achieving certain standards of living of the Earth's population and ensuring security and safety of life. The way towards achievement of these goals depends on many factors. If we try to select the most meaningful of them, we need to assume that the material standard of living of the population depends primarily on the development of industrial production, as well as on the population numbers and the rate of employment among the population. Development of production is determined by education, science, technology and use of the resource potential of the Earth. By the later we will mean natural, economic, informational and other opportunities to improve people's lives by means of the development of industrial production.

These considerations were used in [180] to construct conceptual model of global development processes, as shown in Fig. 2.14. Let us denote processes included in the conceptual model, as follows: x_1 – concentration of CO_2 in the atmosphere, x_2 – mean surface temperature, x_3 – level of environment pollution (contamination), x_4 – intensity of resources' consumption, x_5 – frequency and intensity of hurricanes, x_6 – resource potential, x_7 – gross production output, x_8 –

level of political confrontation in the world due to competition for resources, x_9 – number of population, x_{10} – state of health, x_{11} – development of science and technology, x_{12} – standard of living , x_{13} – level of self-organization of the world community, x_{14} – level of education (public education), x_{15} – level of unemployment, x_{16} – level of social unrest, x_{17} – level of environmental risk, x_{18} – frequency and intensity of floods.

Production growth reduces unemployment, but it also affects the quality of environment through its pollution with industrial, agricultural, energetic and transport industry waste. Along with the need to resolve the problem of global warming, special focus must be made on emission of carbon dioxide CO_2 and other greenhouse gases into the atmosphere. We will regard these problems as proportional to the level of gross production output.

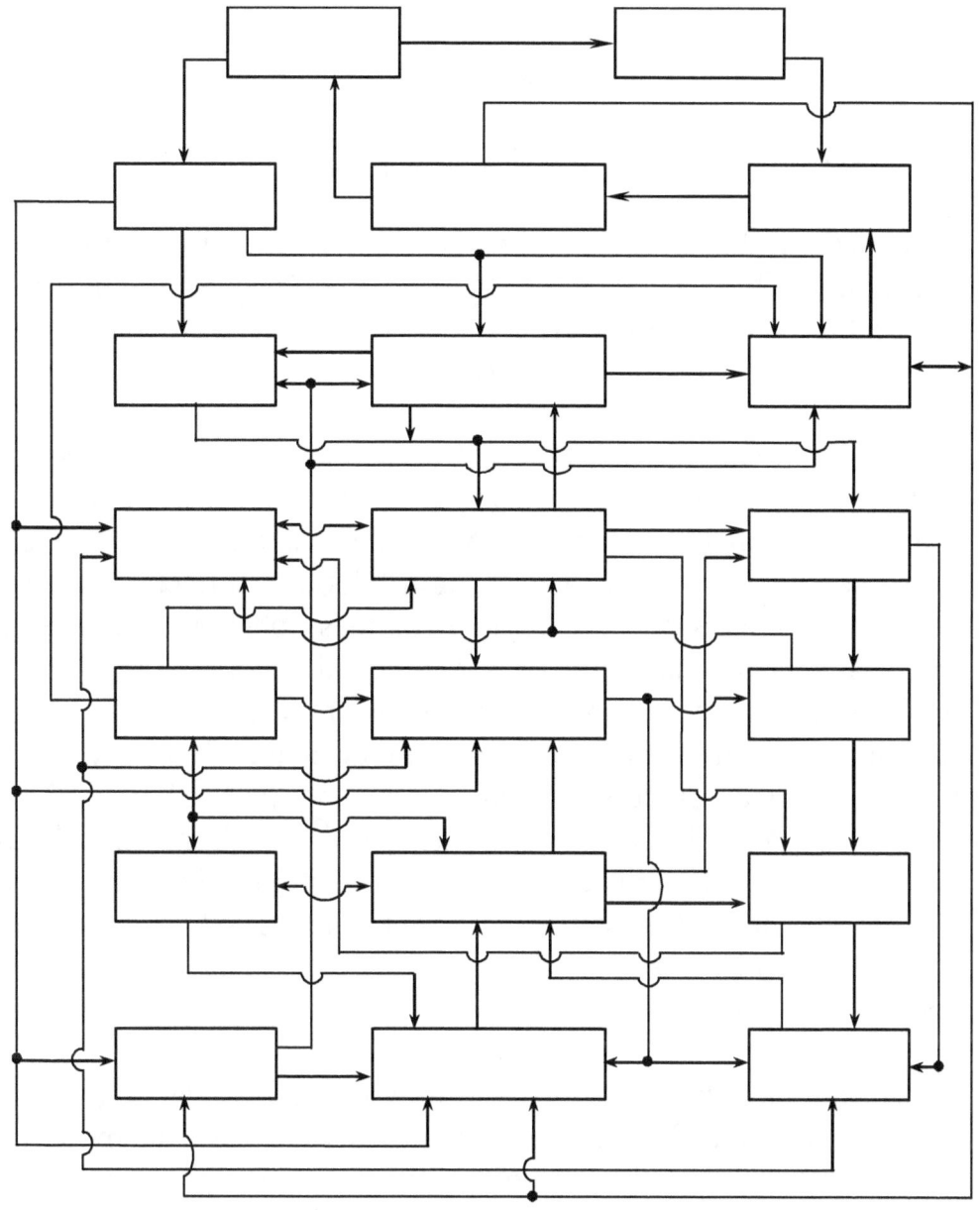

Fig. 2.14. An integrated model of global natural and socio - economic processes

Greenhouse gases' content in the atmosphere increases globally averaged surface temperature. However, the above-mentioned mechanisms of self-regulation come into force and compensate for this increase. To account for the effects of temperature self-

137

regulation, we will introduce into the model a feedback mechanism that provides such compensation.

Oceans are considered to be among the main factors of natural self-regulation of greenhouse gases. We will take into account the influence of the world's oceans by introducing in the model the negative feedback between fluctuations of CO_2 concentration in the atmosphere and the difference between average temperature of the atmosphere and of the Earth's surface.

An important side effect of the temperature self-regulation is the disastrous weather events, such as hurricanes and floods. Their impact on global development will be presented in the model by processes marked as "hurricanes" and "floods", reflecting integrally both, the numbers and the strength of these phenomena. The most important for human security are ecological processes in the natural environment and socio-political processes in the society. With a view to simplifying the system, we assume that mankind is threatened by two integrated processes: "environmental hazard" and "social danger". Among the factors that increase the environmental hazard, we mention natural disasters and environment pollution. They oppose the development of education and health, reduction of diseases and human mortality, and as a sequence, they reduce the level of ecological danger.

Among the main factors forming the social danger, we will consider struggle for resources, standard of living, unemployment and the state of health. It is obvious that we have selected a very limited list of reasons that could cause social unrest. The struggle for resources means wars, socio-political revolts, violation of democratic rights and civil liberties, terrorism, distribution of drugs. Among the factors increasing social danger, we'll identify low living standards, high unemployment, poor health care.

Similarly to the *GAIA-* theory which puts forward the hypothesis that self-regulation of the average surface temperature helps save life forms on the Earth, we assume that there are mechanisms of self-organization of society, able to prevent increase of environmental and social dangers. The driving forces of these mechanisms are science and education. Due to the global movement towards informational society, the international community has new opportunities for sustainable development management through worldwide dissemination of knowledge and advanced technologies of production, health improvement and environment protection.

An important role is played by such universally accepted model of socio-political organization of people as "open (civil) society" [161]. By including a process called "self-organization of society" into our model, we ensure resistance against growing environmental and social risk due to the effect of achievements in education, science and systems thinking [166, 183].

The notations introduced above allow us to write the following equations of the dynamic *ABC*-model of global social ecological-economic system

$$\frac{dx_1}{dt} = x_1\{1 - 2(x_1 - a_{13}x_3 + b_1 IF[x_2 < T_m; 0; 1 - \exp(-\alpha_1 t)])\},$$

$$\frac{dx_2}{dt} = x_2[1 - 2(x_2 - a_{21}x_1)],$$

$$\frac{dx_3}{dt} = x_3[1 - 2(x_3 - a_{35}x_5 + a_{3/11}x_{11} + a_{3/18}x_{18} - a_{34}x_4)],$$

$$\frac{dx_4}{dt} = x_4[1 - 2(x_4 - a_{45}x_5 - a_{4/18}x_{18} - a_{47}x_7)],$$

$$\frac{dx_5}{dt} = x_5\{1 - 2[x_5 - a_{52}(x_2 - T_m)]\},$$

$$\frac{dx_6}{dt} = x_6[1 - 2(x_6 + a_{64}x_4 + a_{65}x_5 + a_{6/18}x_{18} - a_{6/11}x_{11})],$$

$$\frac{dx_7}{dt} = x_7[1 - 2(x_7 - a_{76}x_6 - a_{7/12}x_{12} - a_{7/11}x_{11} - a_{19}x_9)],$$

$$\frac{dx_8}{dt} = x_8[1 - 2(x_8 + a_{86}x_6 - a_{87}x_7 + a_{8/13}x_{13} - a_{84}x_4)],$$

$$\frac{dx_9}{dt} = x_9[1 - 2(x_9 - a_{9/10}x_{10} + a_{98}x_8)],$$

$$\frac{dx_{10}}{dt} = x_{10}[1 - 2(x_{10} - a_{10/7}x_7 - a_{10/11}x_{11} - a_{10/14}x_{14} + a_{10/17}x_{17})],$$

$$\frac{dx_{11}}{dt} = x_{11}[1 - 2(x_{11} - a_{11/7}x_7 - a_{11/14}x_{14})], \qquad (2.31)$$

$$\frac{dx_{12}}{dt} = x_{12}[1 - 2(x_{12} - a_{12/9}x_9 + a_{12/15}x_{15} - a_{12/7}x_7 + a_{12/16}x_{16} + a_{12/5}x_5)],$$

$$\frac{dx_{13}}{dt} = x_{13}[1 - 2(x_{13} - a_{13/14}x_{14} - a_{13/11}x_{11} + a_{13/16}x_{16} - a_{13/17}x_{17})],$$

$$\frac{dx_{14}}{dt} = x_{14}[1 - 2(x_{14} - a_{14/13}x_{13} - a_{14/11}x_{11})],$$

$$\frac{dx_{15}}{dt} = x_{15}[1 - 2(x_{15} + a_{15/9}x_9 + a_{15/7}x_7 + a_{15/13}x_{13})],$$

$$\frac{dx_{16}}{dt} = x_{16}[1 - 2(x_{16} - a_{16/15}x_{15} + a_{16/10}x_{10} - a_{16/17}x_{17} - a_{16/8}x_8)],$$

$$\frac{dx_{17}}{dt} = x_{17}[1 - 2(x_{17} + a_{17/14}x_{14} - a_{17/18}x_{18} - a_{17/5}x_5 - a_{17/8}x_8)],$$

$$\frac{dx_{18}}{dt} = x_{18}[1 - 2(x_{18} - a_{18/2}x_2 - a_{18/5}x_5)].$$

Adaptation of Development Scenarios in the Global Social Ecological-Economic System. As it was noted above, it is convenient to use dimensionless processes $x_1 - x_{18}$ reduced to a common scale of variability, e.g., $(0,10)$. Perform a linear transformation of the initial dimensional processes by formulas

(2.22) and choose such influence coefficients in equations (2.31), which ensure mutual adaptation of all processes and brings the system into a steady-state balance of influences. If we apply some external influence, the system enters a regime of adaptive adjustment to these influences. Global development process will be adapted to each other and to the external influence due to interaction between positive and negative feedbacks in the structure of the model (2.31).

The *GAIA*-adjustment mechanism of CO_2 concentration in the atmosphere is presented in the right part of the model's first equation (2.31) as a logical operator which ensures that current average surface temperature x_2 does not exceed its long-term norm T_m. When concentration of greenhouse gases in the atmosphere increases to a dangerous level, the average surface temperature x_2 exceeds the threshold value T_m and a control agent "switches on" exponential function in the right-hand side of the equation for x_2, which is increasing with saturation. This leads to decrease in atmospheric CO_2 concentration x_2, which is simulates the *GAIA*-effect.

Calculations with this model were performed for 500 time steps of the dimensionless time scale. In the first computational experiment, solar radiation was set to grow; this growth was followed by increase in concentration of CO_2 in the atmosphere. The *GAIA*-adjustment mechanism of CO_2 concentration in the atmosphere was started only after 101 steps of calculations, in order to assess its impact on the development scenarios, which are shown in Fig. 2.15.

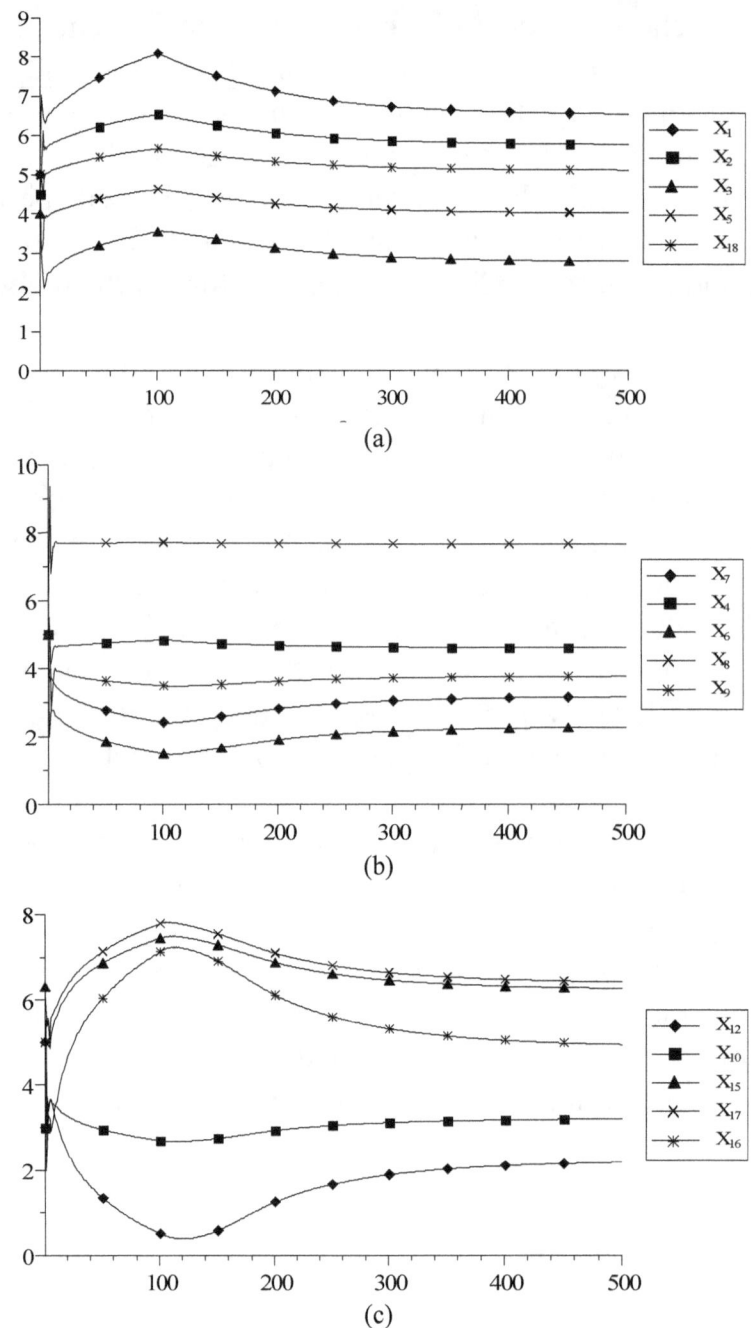

Fig. 2.15. Processes that accompany increase of CO_2 concentration in the atmosphere.

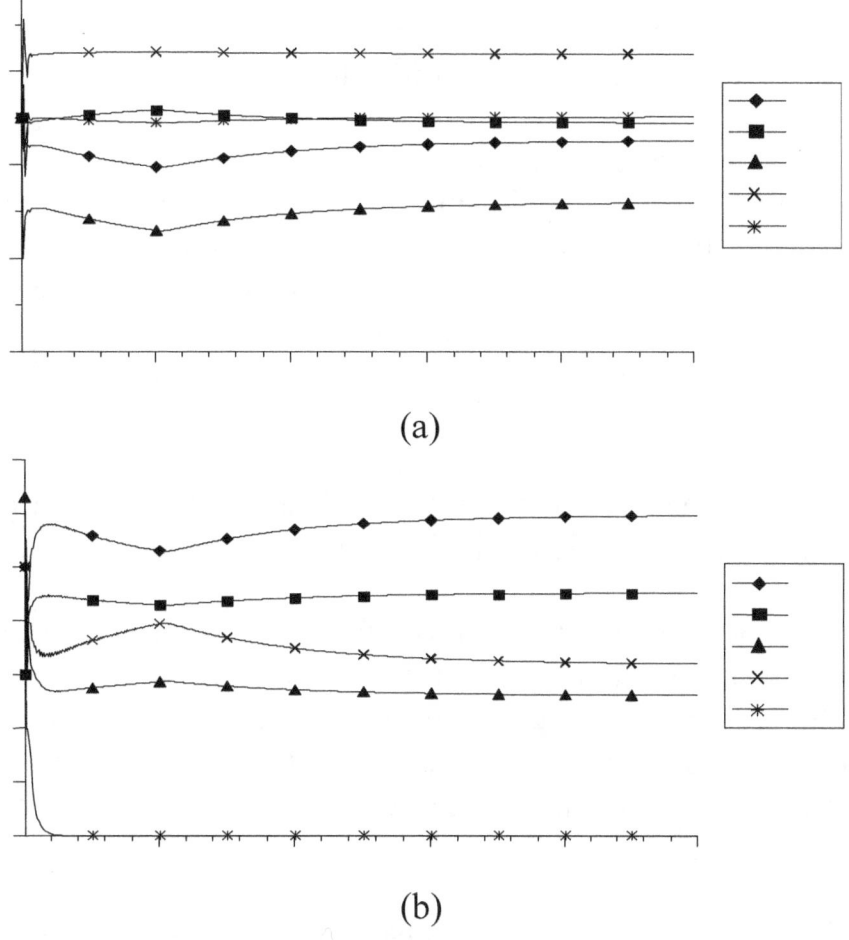

(a)

(b)

Fig. 2.16. Influence of society's self-organization processes on sustainable development: starting from step 200, termination of struggle for resources was imitated

As follows from the figure, at the initial time period from step 1 to step 100, along with increasing concentrations of greenhouse gases x_1 and the average temperature x_2, the intensity of hurricanes x_5 and floods x_{18} increased and environmental pollution x_3 grew up, too. The resource potential x_6 reduced and gross production output x_7 started to fall down. Unemployment x_{15} gave a sharp rise, and risks of environmental x_{17} and x_{16} social danger increased significantly.

143

After inclusion (at step 101) of self-regulation mechanisms for average temperature, these negative trends were offset along with normalization of the average temperature.

In the experiment described above, the socio-political mechanisms of self-regulation of development processes have not been used: the process x_{13} – "self-organization of society" has not been switched on. In the above discussion we assumed that development and dissemination of knowledge inevitably leads to appearance of response reaction to the growing human ecological and social danger.

Self-organization of society should reduce the level of danger to human life and should raise their level of well-being.

To obtain relevant scenarios, the calculations described above were repeated with inclusion of the self-organization process x_{13}. The results are shown in fig. 2.16a, b. Comparing the scenarios of the processes shown in Fig. 2.15c and 2.16b, one can notice a sharp decrease in the environmental, and almost complete disappearance in the social danger.

In the following computational experiment random fluctuations in concentration of greenhouse gases in the atmosphere were imitated, being caused by massive emissions of CO_2, which originated from volcanic eruptions and large forest fires. In this experiment, the increase in competition for resources x_8 was simulated in the first place, and then, after step 200, it was terminated. Relevant scenarios are shown in Fig. 2.17a – c.

At the first 100 steps of calculations, in the absence of GAIA-adjustment mechanism for average temperature, the levels of environmental danger x_{17} and social danger x_{16} sharply increased, competition for resources x_8 became sharper, rapid decline in the resource potential x_6 was noticed, and standard of living x_{12} went down. After switching on the effect of *GAIA*-adjustment, the tendency to reduce the average temperature x_2 had appeared. However, the increase of pollution x_2 was saved. Decline of the resource potential x_6 and standard of living x_{12} were continuing, albeit slow. Ecological and social risks remained high.

After 200 time steps, gradual weakening of struggle for resources as a sequence of self-organization of the society was imitated. This clearly demonstrates the graph of the curve x_8 in Fig. 2.17b. It is clearly related to the behavior of all scenarios after 200 steps of calculations. Despite the fact that the level of environment pollution x_3 remained high, the tendency to reduce ecological danger x_{17} appeared. The social danger x_{16} begun to reduce noticeably, a rise in production output x_7 was registered, and the standard of living x_{12} started to grow. Thus, the natural mechanisms of the *GAIA*-control, reinforced by mechanisms of self-organization in the society, are able to control the global development scenarios, even in situation of significant environmental pollution.

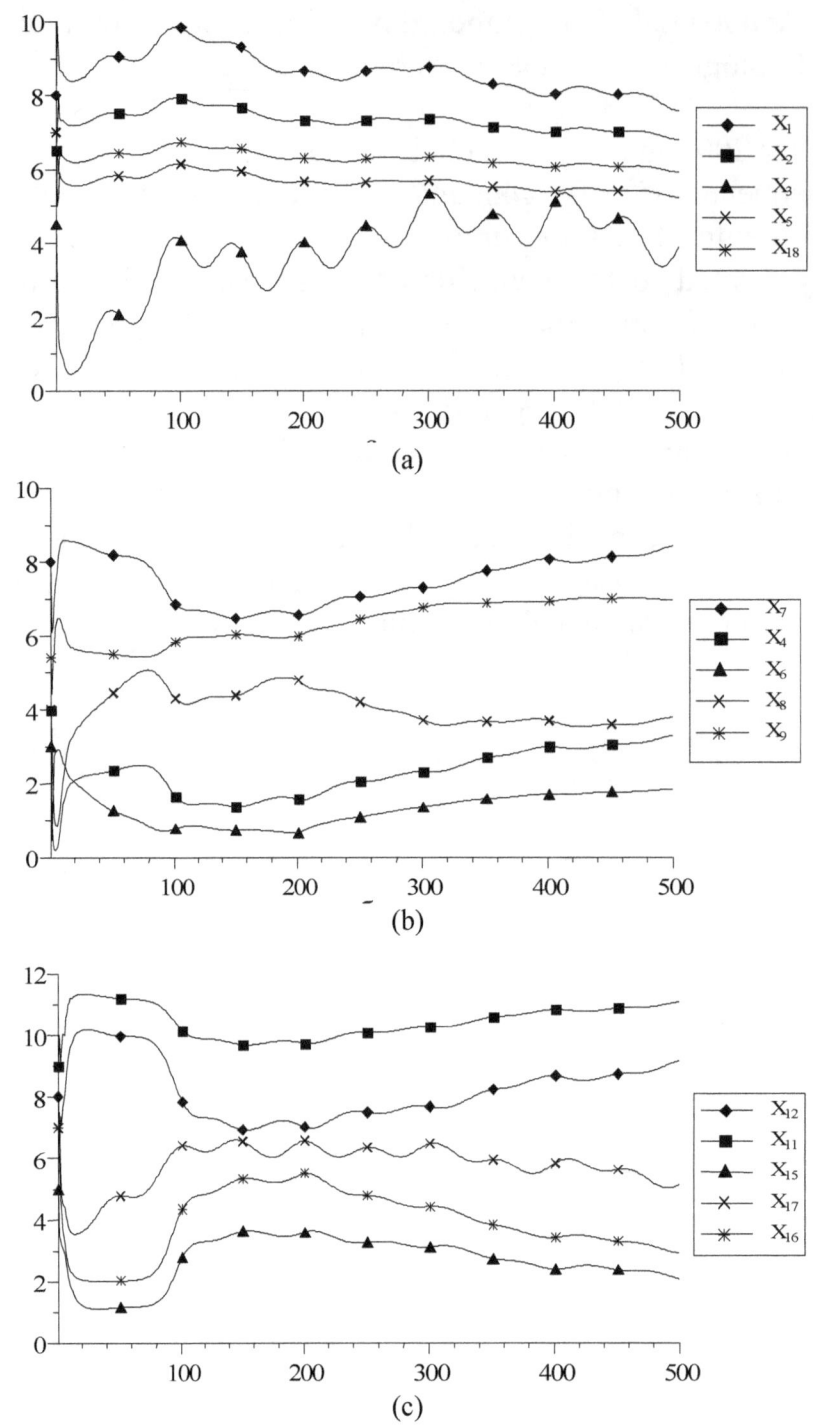

Fig. 2.17. Scenarios for random fluctuations in concentration of greenhouse gases in the atmosphere. The *GAIA*-adjustment of temperature was switched on after 100 steps of calculations;

simulation of the weakening of the struggle for resources was started after 200 steps

It was of interest to conduct an experiment in which natural *GAIA*-adjustment mechanisms failed to completely exclude growth of average surface temperature

of the Earth. Is it possible in these conditions that the effects of the society's self-organization would weaken environmental and social risks and raise the material standard of the population of the Earth? This situation illustrates the scenarios shown in Fig. 2.18a, b, c. As it follows from graph in Fig. 2.18b, fluctuating trend of environment pollution x_3 was accompanied by a slight increase in average temperature x_2 and related high-intensity hurricanes x_5 and floods x_{18}.

Such trends have been observed during the time corresponding to 200 steps of calculations, after we started one of the compensatory mechanisms of society's self-organization- the growing influence of science and new technologies x_{11} (see Fig. 2.18c) on production x_7. This led to a sharp rise in production x_7 (see Fig. 2.18b), the standard of living x_{12} and some reduction of social danger x_{16} (see Fig. 2.18c). Nevertheless, the level of ecological danger x_{16} remained high, as well as the level of competition for resources x_8 (see Fig. 2.18b). In order to reduce environmental risk, it is necessary to realize that along with facilitation of production growth, science must create advanced technologies which would be environmentally friendly and could ensure reduction of all kinds of pollution (contamination).

The results of computer simulation showing the role of clean production technologies, are shown in Fig. 2.19a, b, c. In this experiment, under the same conditions as the previous one, new research technologies were aimed at reducing environmental pollution and, in particular, at reduction of greenhouse gases' emissions into the atmosphere. Environmentally friendly technologies were put in place, starting with 150 steps of calculations, which led to a sharp drop in the level of pollution x_3 and concentration of CO_2 in the atmosphere x_1 (see Fig. 2.19a).

Despite some growth of competition for resources x_8 due to the increase in production output x_7 (see Fig. 2.19b), the level of social danger x_{16} significantly decreased, and the standard of living x_{12} significantly grew (see Fig. 2.19c). The relatively simple model of global social ecological-economic system considered above, was built to show the possibility of forecasting development scenarios. The obtained results underline the importance of such dynamic models for management of sustainable development. The right decisions, concerning the use of natural, environmental and economic resources of the Earth, must be based on predicted model scenarios of global processes in the light of adaptive balance of possible causes and effects.

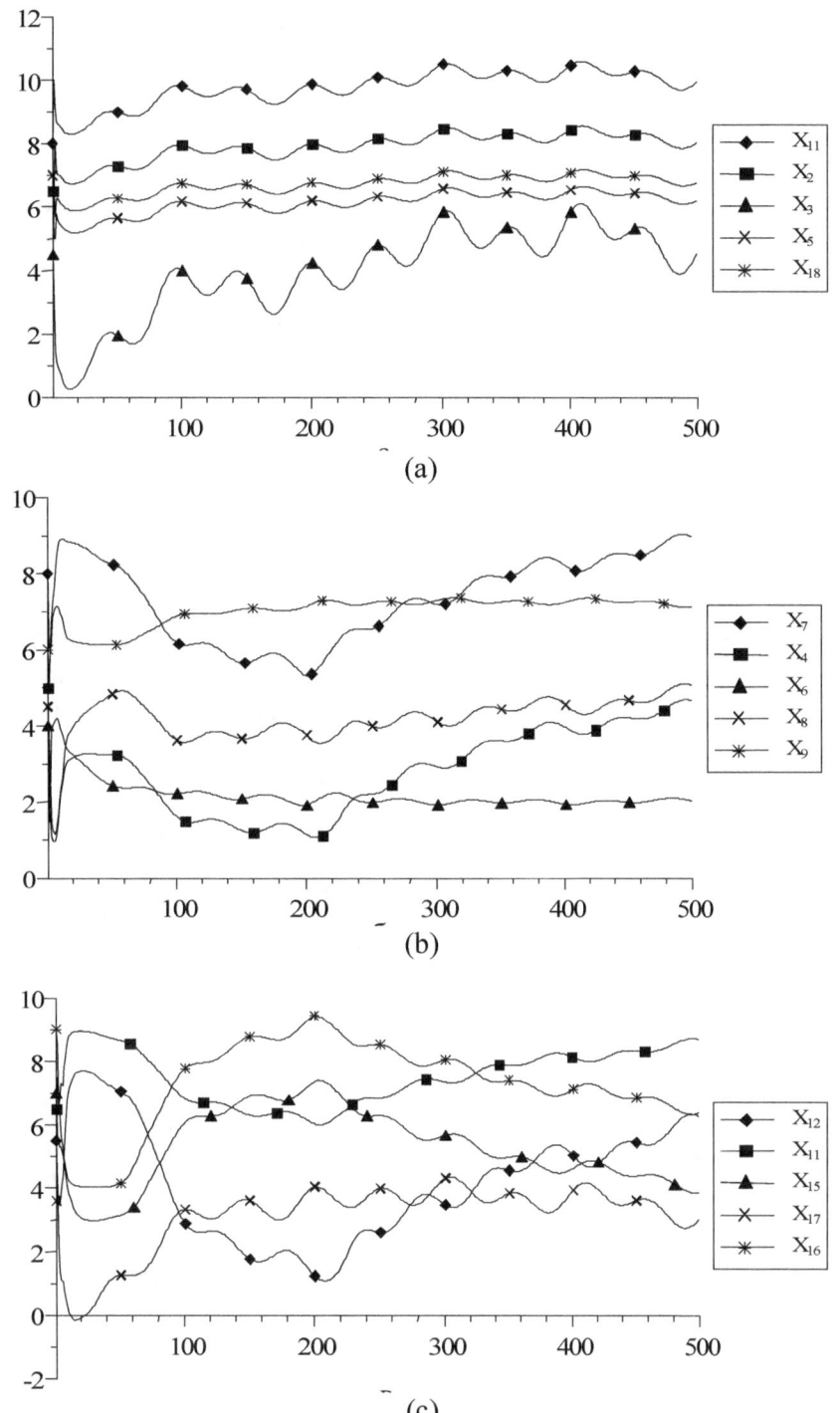

Fig. 2.18. Development Scenarios for the case when the *GAIA*-adjustment is not in a position to compensate the growth of greenhouse gases in the atmosphere. After step 200, an increase in the influence of science and new technologies on the production output was simulated

149

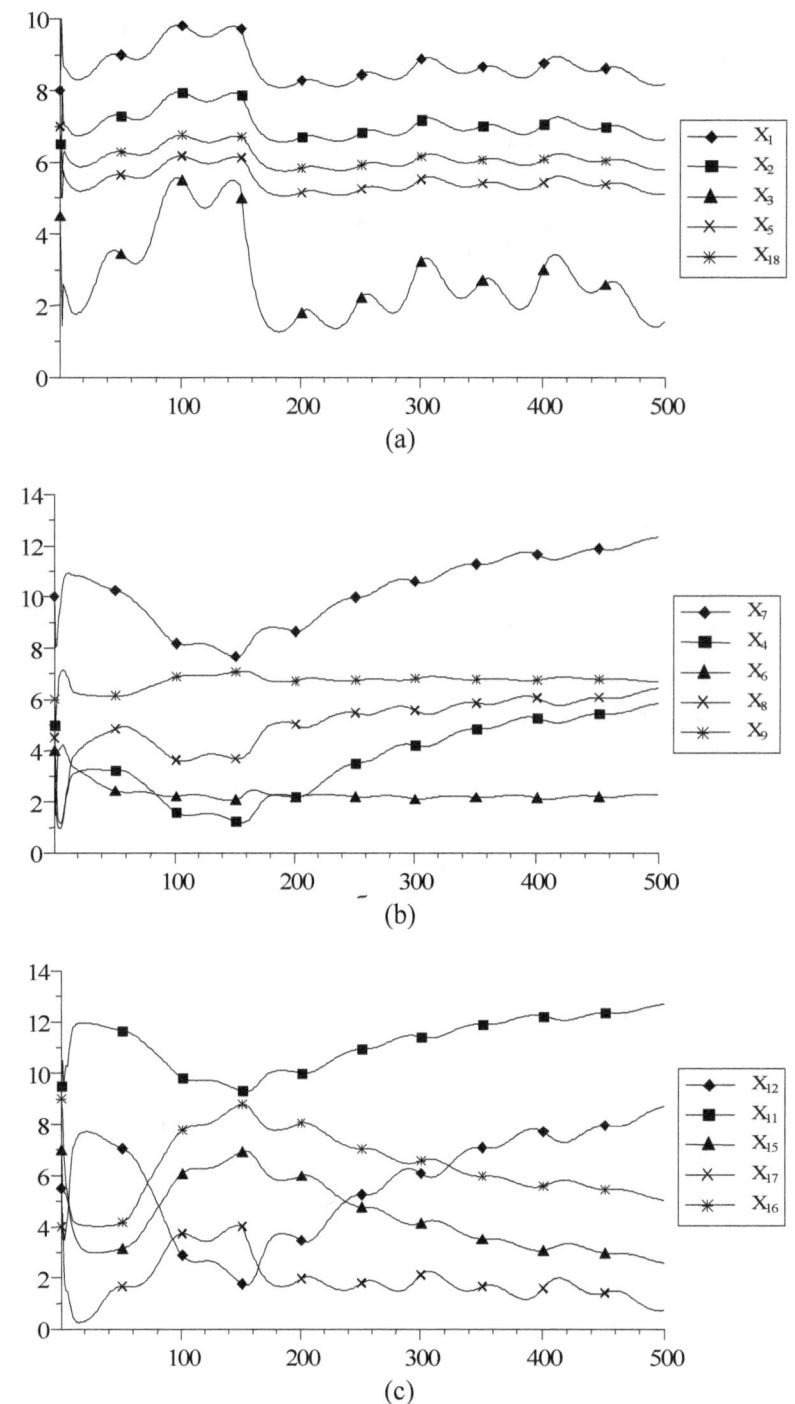

Fig. 2.19. Effect of clean production technologies on concentrations of greenhouse gases in the atmosphere (switched on after 150 steps of calculations) of production development x_7 and

life standard level x_{12} fell (see Fig. 2.18b). Curves of environmental x_{17} and social x_{16} danger showed a significant increase (see Fig. 2.18c).

2.8. Stochastic Identification of Equations Coefficients in *ABC*-Models

One of important aspects of the method of adaptive balance of causes (*ABC*-Method) for prediction of scenarios, is its ability to provide influence coefficients in the equations of the *ABC*-models and to process them. In general, when the simulated system contains n interrelated processes, the dynamical equations' system takes the form

$$\frac{dx_i}{dt} = x_i\left[1 - 2\left(x_i - \sum_{j=1}^{n-1} a_{ij}x_j - f_i\right)\right], \ (i,j = 1,2,...,n), \ (i \neq j).$$

(2.32)

Influence coefficients a_{ij} in these equations can be identified in two ways: according to assessments by experts, and based on statistical analysis of observational data of the simulated processes. The first way suggests to set any numbers of coefficients provided by an expert analysis of the observed scenarios, and to include information about environmental conditions predicted by experts. This method is used when the available information about behavior of a system is very limited. The values of influence coefficients are selected on the basis of subjective judgments of experts who compare the degree of influence of some processes on each other.

The second method can be used in the presence of time-series observations of the development processes, carried out during some time and stored in archives. Focusing on the construction of a computer technology for management decisions, it is recommended to prefer the second method and to carry out backups of the observed and projected scenarios, to acquire objective evaluation of the influence coefficients.

151

Statistic Evaluation of Influence Coefficients in *ABC*-Models.
For statistic evaluation of coefficients in *ABC*-models (2.32), we assume that the functions of external influences f_i are random fluctuations f' around zero mean values. They cause deviations x_i' of development scenarios from a steady state $\overline{x}_i$, which occurs when external influences are equal to zero. In this steady state, the variables of the system (2.32) satisfy the conditions

$$\overline{x}_i = 0,5 + \sum_{j=1}^{n-1} a_{ij}\overline{x}_j$$

Then, for deviations of processes from average state, we have

$$x_i' = \sum_{j=1}^{n-1} a_{ij}x_j' - f_i' \, . \tag{2.33}$$

Let us introduce coefficients of mutual correlation

$$R_{kl} = E\left\{x_k' x_l'\right\} \text{ and } G_{mn} = E\left\{x_m' f_n'\right\}.$$

(2.34)

By multiplying (2.33) in turns on x_k', then by performing averaging of the obtained expressions and taking into account (2.34), we'll obtain

$$\sum_{j=2}^{n} a_{1j}R_{jk} = G_{1k} - R_{1k}, \left(j \neq 1\right);$$

$$\dots\dots\dots\dots\dots\dots\dots\dots\dots\dots \tag{2.35}$$

$$\sum_{j=1}^{n-1} a_{nj}R_{jk} = G_{nk} - R_{nk}, \left(j \neq n\right).$$

The expression (2.35) represents n systems of linear algebraic equations for finding the $n(n-1)$ unknown coefficients of the *ABC*-model equations (2.32), containing n dynamic equations. Thus, utilizing sets of archived observational data about the development processes x_i and external influences f_i, with the help of equations (2.35) it is possible to obtain an objective assessment of influence coefficients a_{ij}.

Let us note that the equations for influence coefficients evaluation through correlation coefficients are identical in form, with well-known equations of optimal interpolation of stationary random functions, obtained by Kolmogorov [90]. In the *ABC*-method, weight coefficients of optimal interpolation act as factors of influence between the processes. For example, in order to express coefficient of influence a_{1j} in explicit form, it is enough to put $k = j$ in the first equation (2.35)

$$a_{1j} = \frac{G_{1j}}{R_{jj}} - \frac{R_{1j}}{R_{jj}} - \sum_{p=2}^{n} a_{1p} \frac{R_{pj}}{R_{jj}} \quad . \tag{2.36}$$

In [181] it was suggested that the influence coefficients, linked with each other by equations (2.35), form a complex system themselves. In this system, they perform the role of variables connected to each other by correlation links. Variability of influence coefficients is a sequence of non-stationary character of the development processes and of the external factors applied to the system. This assumption allows us to regard the influence coefficient a_{1j} as a steady state of the *ABC*-model of the influence coefficients' system, which in this case can be represented by the following equations

$$\frac{da_{ij}}{dt} = a_{ij} \left[1 - 2 \left(a_{ij} - \frac{G_{ij}}{R_{jj}} + \frac{R_{ij}}{R_{jj}} + \sum_{p=1}^{n-1} a_{ip} \frac{R_{pj}}{R_{jj}} \right) \right], \tag{2.37}$$

$$(i, j = 1, 2, \ldots, n), \ (i \neq j)$$

Our *ABC*-model (2.37) complements the model (2.32), because, when it is being used, there is no more need in expert evaluations of values and signs of influences that exist in the modeled system of development processes. Equations (2.37) allow us to evaluate the influence coefficients by the correlation matrices of observed processes and external influences and, logically, to assimilate objective data about the nature of mutual influences between processes into a model.

Example of Statistic Evaluation of Constant Influence Coefficients. As an example of statistic evaluation of constant coefficients, let us consider the problem of forecasting of random components for natural processes. Following this purpose, in [178] a simulated time series of observations was utilized. Its correlation function is shown in Fig. 2.20, *a*. In a computational experiment which we performed, we used deviations from deterministic time series component, which was selected by "sliding averaging" of the time series [163, 204].

Let us agree to call values of a process' deviations from its mean value "samplings". Then, "sampling points" are runtimes of observations, and "steps" in time are time intervals between observations. Let us set a task to forecast time series of deviations for 20 steps forward on two, three and four known deviations, measured with the interval of 20 steps from each other. The validity of using a similar interval between the measurements will be considered below.

To assess influence coefficients, a segment of the simulated time series of observations was selected. It contained 1000 samples, and its correlation matrix $R_{kl} = E\{x'_k x'_l\}$ was computed on its values in 200 initial points of the series. With this matrix, assessments of constant coefficients linking the starting points with the forecast point of the series were calculated from the system of equations (2.36) (see Fig. 2.20b).

Prediction was carried out by moving a number of reference points along the time series row, starting with 220 reference points, as shown in Fig. 2.20b. The accuracy of forecasting characterized by graphs of prediction error's variance, calculated by comparing the estimates obtained by the model, with true values of the

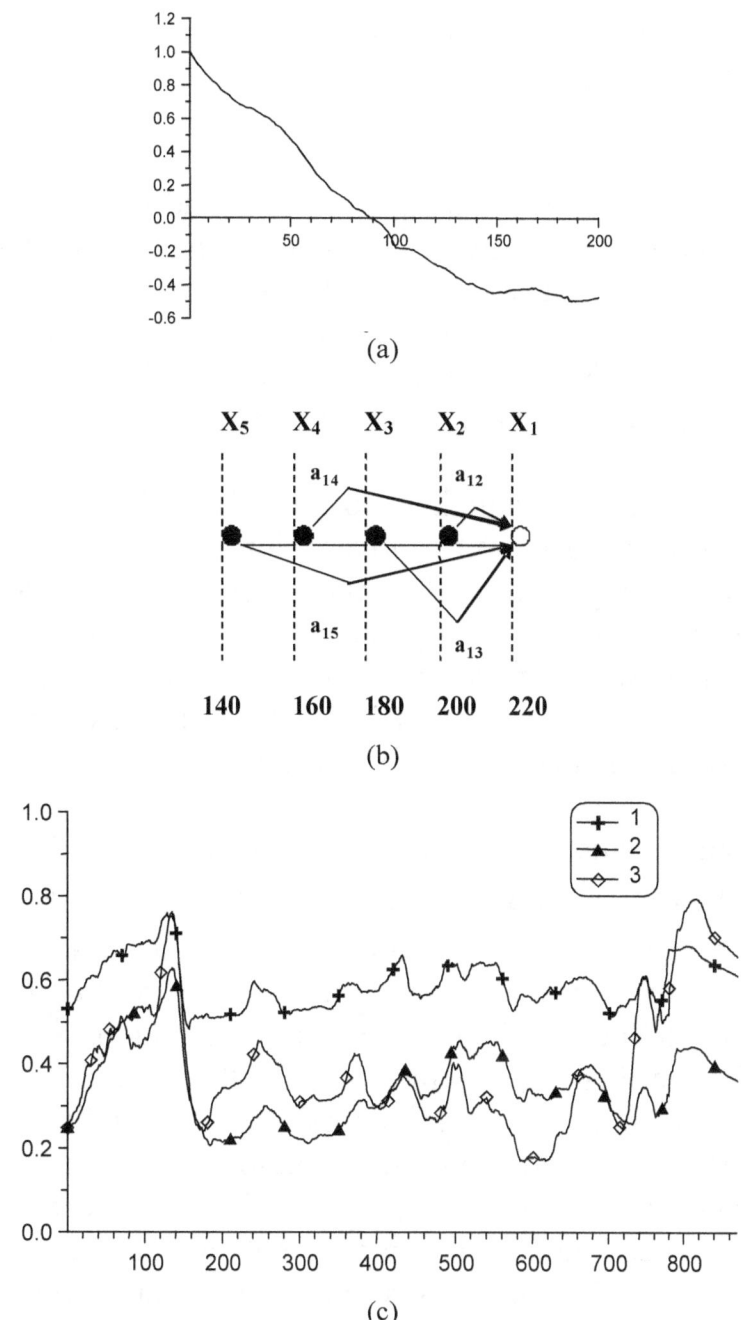

Fig. 2.20. (a) Correlation function of the time series of observations; (b) position of points of measurement (black) relative to the point of prediction (white); (c) Variances of forecasting errors, normalized to the variance of the observations (averaging at step 200), with the use of:

155

1 – measurements x_1 and x_2, 2 – measurements x_1, x_2 and x_3, 3 – measurements x_1, x_2, x_3, and x_4

series. In Fig. 2.20c these graphs show the variability of sample error variance of forecasts, as it moves along a row of forecasting operations. The variances were calculated by averaging over 200 points of the current series and normalized to the sample variance of true values of this series. Fig. 2.21a and b allow us to judge the quality of predictions obtained in this experiment.

The accuracy of prediction proved to be different, as the groups of four measurements were moved along the simulated series of observations. Judging from the middle series of observations in the row, the values of relative error variance, the quality of forecasts for three samples was considerably higher than for two. However, the effect of including an additional (the fourth) measurement did not prove to be as efficient as the previous ones.

Constant influence coefficients which were identified through a system of equations (2.36), have the following values: $a_{12} = 0.15$, $a_{13} = 0.21$, $a_{14} = 0,5$ $a_{15} = 0,5$. Judging by their values, a major role in shaping the future values of the process was played by the third and fourth measurements, whereas the highest correlation with the points of predictions was fixed at the first and the second measurements. The explanation of this effect is possibly due to the fact that in this experiment, we used constant sample correlation coefficients, constructed from the first 200 points of the series. Therefore, it was interesting to continue further experiments with a variable matrix of sample correlation coefficients, along the entire time series of observations.

Building an Adaptive Prediction Model with Variable Influence Coefficients. It follows from the general algorithm of influence coefficients' evaluation (2.36) that the coefficients do not depend on the observations used in the forecast, as they are determined solely by the values of correlation coefficients between the point of the forecast and the measurement points [11]. If we change the location of measurements, correlation coefficients will also change, and consequently, the weighting coefficients of extrapolation of observations will change, as well. These cause-effect relationships are reflected in the *ABC*-model of variable influence coefficients (2.37) in cases when variability of influence coefficients results from non stationary behavior of sample correlation coefficients $R_{kl} = E\{x'_k x'_l\}$.

In current example of prediction of time series observations, the influence coefficients in the equations of the *ABC*-model (2.37) are tuned to a pattern of correlations, created by choice of measurement points' location. Therefore, we can use the dynamic *ABC*-model of influence coefficients in the following form [181, 183]

$$\frac{da_{ij}}{dt} = a_{ij}\left\{1 - 2\left[a_{ij} - R_{jj}^{-1}\left(R_{ij} - \sum_{k=1}^{n} a_{i,k} R_{k,j}\right)\right]\right\},$$
(2.38)

$$(i, j = 1, 2, ..., n), \ (k \neq i)$$

Equations (2.38) introduce variable influence coefficients in the *ABC*-model (2.32) and hence, they open an opportunity to supply the model with information about statistic variability between the point of the forecast and the measurement points. In other words, the prognostic model (2.32), (2.38) becomes adaptive because sample correlation coefficients are used in it. They take into account non-stationary random components of natural processes, which take place in practice quite often. In order to evaluate the role of variable influence coefficients in adaptive prediction of stochastic processes, a number of computational experiments was carried out. The results of these actions are shown in Fig. 2.22 and 2.23.

Random time series of deviations, shown in Fig. 2.21 was used for the prediction of its values, first at 40steps, and then at each 60 steps forward. Prediction was carried out at each step of calculations, starting with sample point 200. For prognostic estimates were taken four consecutive reference samplings, located in 20 steps from each other. On each time step the correlation matrix was calculated on 200 points of the series, which preceded the group of 4 points of measurement used for predictions.

A pattern of 200 samples was moved along the row with 4 samples, which were used to carry out the forecast. The initial 200 samples simulated the available archive data. The following values simulated arrival of new observational data, which were added to the archive for recalculating matrix of sample correlations at each step.

Therefore, in the first experiment, prognostic estimates were made at 240, 241, ..., 900 points of the series, while in the second experiment they were made at 260, 261, ..., 900 points. At each step, the following system of equations with variable influence coefficients was solved, which is a special case of the general model (2.32) (2.38). The system of equations of the ABC-model (2.39) ensures adaptation of the influence coefficients a_{ij} to each other and to the variable elements of the correlation matrix R_{kl} at each step of calculations. In turn, the variable coefficients influence formation of prognostic estimates obtained from the equation for x_1, and thereby provides an adaptive prediction of the modeled process.

The system of equations of the ABC-model mentioned above has a form

$$\frac{dx_1}{dt} = x_1 \left[1 - 2 \left(x_1 - \sum_{j=1}^{4} a_{1\,j+1} x_{j+1} \right) \right];$$

$$\frac{da_{12}}{dt} = a_{12} \left\{ 1 - 2 \left[a_{12} - R_{22}^{-1} \left(R_{12} - \sum_{k=2}^{4} a_{1,k+1} R_{k+1,2} \right) \right] \right\};$$

$$\frac{da_{13}}{dt} = a_{13}\left\{1 - 2\left[a_{13} - R_{33}^{-1}\left(R_{13} - \sum_{\substack{k=1 \\ k\neq2}}^{4} a_{1,k+1}R_{k+1,3}\right)\right]\right\};$$

$$\text{(2.39)}$$

$$\frac{da_{14}}{dt} = a_{14}\left\{1 - 2\left[a_{14} - R_{44}^{-1}\left(R_{14} - \sum_{\substack{k=1 \\ k\neq3}}^{4} a_{1,k+1}R_{k+1,4}\right)\right]\right\};$$

$$\frac{da_{15}}{dt} = a_{15}\left\{1 - 2\left[a_{15} - R_{55}^{-1}\left(R_{15} - \sum_{k=1}^{3} a_{1,k+1}R_{k+1,5}\right)\right]\right\}.$$

The dynamics of the influence coefficients' adaptation is shown in Fig. 2.22b and 2.23b. It follows from these figures that the coefficients have experienced significant changes due to non-stationarity of sample correlations. At each step of forecasting we performed redistribution of roles of each of the four measurements which had been used to develop the prognostic evaluation process. The advanced prediction of the second experiment (made 20 steps earlier than in the first experiment) did not lead to significant changes in the scenarios of each of the influence coefficients.

In both experiments, the greatest influence on formation of predicted values was provided by the sample x_2, since its coefficient of influence reaches its maximum value. These results allow us to suggest that the main role in shaping of the prognostic evaluation is not given to the proximity of the measuring points to the prediction point, but to the joint (emergent [87]) effect of the system of statistic relations, presented in a matrix of sample correlations.

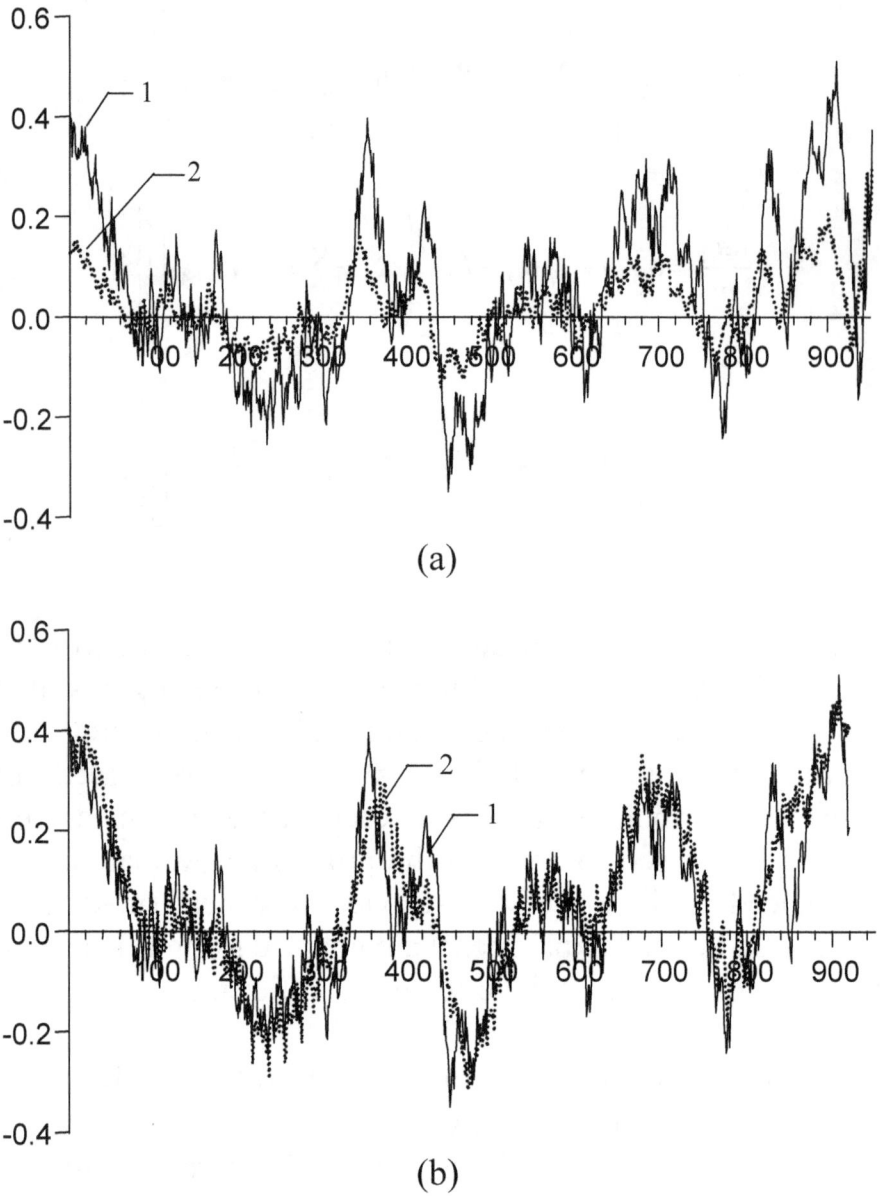

Fig. 2.21. Prediction results for 20 steps forward taken at 20-step interval between   measurements (1 - a true process, 2 – a predicted process): (a) along two dimensions; (b) along four dimensions (see Fig. 2.20b)

The qualitative comparison of forecasts for 40 and 60 steps with the true values of the process can be performed on their time graphs shown in Fig. 2.22a and 2.23a.

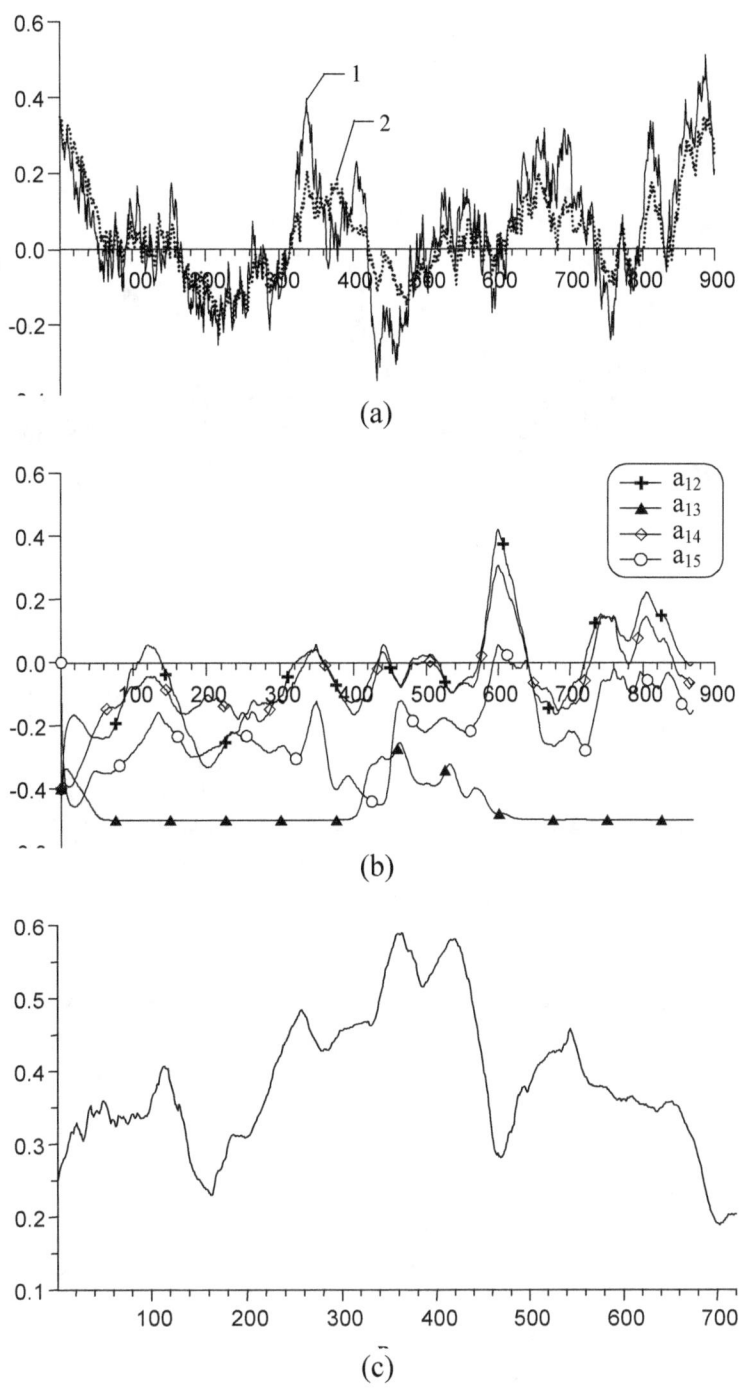

(a)

(b)

(c)

Fig. 2.22. Adaptive forecast made for 40 steps ahead: (a) 1 – original time series, 2 – the forecast; (b) influence coefficients

161

dynamics in the process of   adaptation: a_{12}, a_{13}, a_{14}, a_{15}; (c) dispersion of prediction error at

forecasting for 40 steps forward, normalized to the variance of the   predicted series (averaging on 200 points)

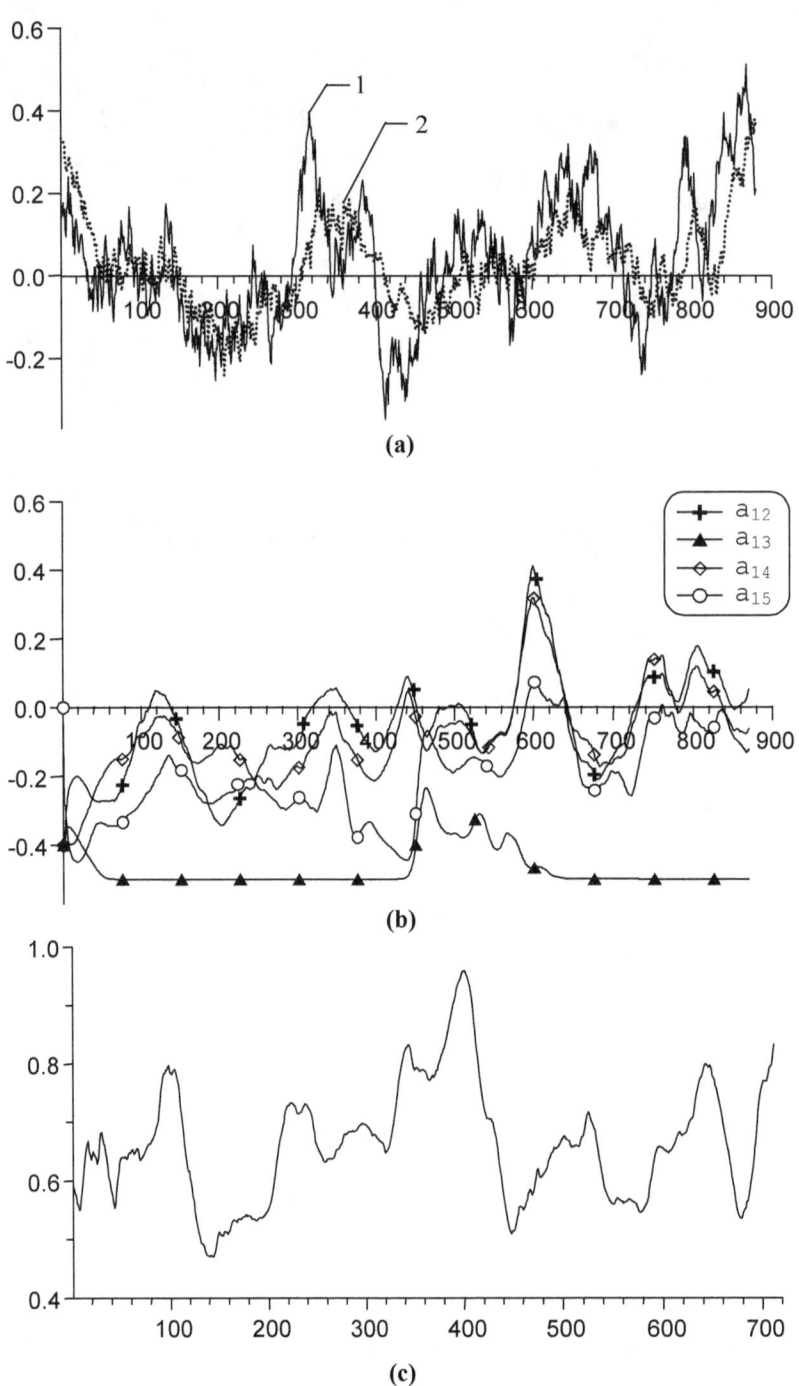

(a)

(b)

(c)

Fig. 2.23. Adaptive forecast for 60 steps ahead: (a) 1 – original time series, 2 – forecast; (b) dynamics of influence coefficients in the process of adaptation: a_{12}, a_{13}, a_{14}, a_{15}; (c) dispersion of prediction error at

forecasting for 40 steps forward, normalized to the variance of the predicted series (averaging at 200 points)

The quantitative assessment – sampling variances of forecast errors, normalized on proper variances of the initial time series, show a noticeable difference in the accuracy of prediction (see Fig. 2.22c and 2.23c). If at the 40-step forecast these estimates vary at the interval of values 0,2 – 0,6, then the advanced forecast (up to 60 steps) drifts them into the interval 0,5 – 0,9.

Studying features of statistical methods of the influence coefficients' evaluation which we have performed has proven the perspective of creating so-called dynamic-stochastic models of the *ABC*-complex systems [170]. Rapid adaptation of the modeled processes to each other and to external influences, applied to the system, which is characteristic of the *ABC*-method, is supplemented in *ABC* dynamic-stochastic models with the objective evaluation of the influence coefficients, which is of practical importance.

2.9. Prediction Improvement by Data Assimilation in *ABC*-Models

The example of consequent prediction of a random component of a natural process at the points of future observations, which we studied in the above paragraph, demonstrated to us that it is possible to accumulate long-term time series of forecast errors or discrepancies, (which become evident when compared to the data of observations). A thorough study of such discrepancies allows us to use the information contained in them to adjust the prognostic results for each process. Considering the discrepancies of prediction in points of observation as the "measurements of the forecast errors component", one can predict their values and add them to the prognostic evaluations of the process itself. This operation of observations data assimilation should serve to increase the accuracy of prognostic evaluations of the process, due to the new information contained in the "measurements" of the discrepancies.

The optimal prediction of discrepancies should be supplied with a correlation matrix of discrepancies. If the available observational data are sufficient to evaluate the correlation matrix of the forecasts' discrepancies, the *ABC*-method can be used for statistic evaluation of influence coefficients (2.38), and the basis for it is laid by Kolmogorov's method of optimal interpolation [90]. It is natural to call this procedure "assimilation of observational data according to the Kolmogorov's optimal filter scheme".

Observations Data Assimilation According to Kolmogorov's Optimal Filter Scheme. In the above case of the time series' prediction at 40 steps ahead, using the data of four measurements, the correlation matrix of the discrepancies' forecasts could be calculated directly on a number of observations of the forecast' errors, which were accumulated over a period of time, for example, during the time covered by 200 steps. Therefore, the use of Kolmogorov's scheme, based on the *ABC*-model (2.38), did not present any difficulty.

For an example, let us return to the problem of optimal forecasting of simulated time series of observations which we discussed in the previous section. Let us set a goal to adjust the results of forecasting by assimilation of four discrepancies of forecasts, which appear in the prediction time points (see Fig. 2.22b) each time when prediction time comes. Correlation function of the time series of discrepancies, calculated at the first 200 points of the series, is shown on Fig. 2.24a.

As we can see from this figure, the interval of significant correlation (radius of correlation) of discrepancies was significantly shorter than that of the original series, because the first crossing of the horizontal axis by the correlation function has shifted from 90 to 22 time-step. By setting the prediction interval at 40 steps prior to the actual event, it was necessary to increase the correlation interval, so that its value exceeded the interval of prediction. With this purpose, a sliding averaging at 250 points was applied to the time series of discrepancies. This operation increased the correlation interval for discrepancies, so that the first zero crossing of the correlation function's graph shifted up to the time-step 57 (see Fig. 2.24a).

It is well-known that the accuracy of prediction of random processes depends not only on how the correlation interval exceeds the interval of forecasting, but also on the accurate choice of the intervals between the measurements which are used for the prediction. In accordance with the Kotelnikov-Shannon theorem [92], for each random process, there is some upper frequency limit ω_c corresponding to its function of spectral density $S(\omega)$, which is connected with the maximum acceptable interval between samples of a process Δt_{opt}. If the interval between measurements is increased so that it exceeds this value, this will cause appearance of a mean-square error of forecasts $\varepsilon(\omega_c)$. Its average value can be estimated by the formula

$$\overline{\varepsilon}(\omega_c) = \int_{\omega_c}^{\infty} S(\omega)d\omega \qquad (2.40)$$

In this case, in order to correct forecasts through assimilation of observations, we need to perform an operation of prefiltering [120, 138] of discrepancies, the meaning of which is to reduce some of the discrepancies, and to bring the upper boundary frequency of spectral density of this series into accordance with the selected interval between samples.

In order to verify the correctness of the above-chosen interval of 20 time steps between samples in the original series, as well as of the interval in the series of discrepancies, it is sufficient to construct curves of spectral densities for the corresponding correlation functions. Since the correlation functions depicted in Fig. 2.24a belong to the type of exponential-cosine functions, they can be approximated by the dependencies of the form

$$R(\tau) = e^{-\alpha\tau}\cos\beta\tau \qquad (2.41)$$

Estimation of parameters in the formula (2.41) by the least squares method gave the following results: for the correlation function of the original series $\alpha_1 = 0,01$;

$\beta_1 = 0,03$; for the correlation function of discrepancies of predictions $\alpha_2 = 0.15$;

$\beta_2 = 0.08$ for the correlation function of smoothed series of discrepancies

$\alpha_3 = 0,05$; $\beta_3 = 0,018$.

By carrying out the Fourier's transformation of correlation function (2.41), we can obtain the following expression for spectral densities of temporal series

$$S(\omega) = \frac{1}{2\pi}\left\{\frac{\alpha}{\alpha^2 + (\omega + \beta)^2} + \frac{\alpha}{\alpha^2 + (\omega - \beta)^2}\right\} \qquad (2.42)$$

The spectral density plots built with this formula are shown in Fig. 2.24 (b). From this figure we can quickly find the following dimensionless estimates $\omega_{ci} \cdot 5 \cdot 10^{-3}$ – upper boundary frequency of their spectra: for original time series $\omega_{c1} = 15$, for discrepancies of forecasts $\omega_{c2} = 40$, and for smoothed discrepancies of forecasts $\omega_{c3} = 20$.

According to the theorem of Kotelnikov-Shannon, the intervals between discrete measurements of the process, used to restore the values of this process at any point Δt_{opt}, should not exceed the amount that can be determined by the formula

$$\Delta t_{opt} 2\omega_c = 2\pi \qquad (2.43)$$

Therefore, for the original series $\Delta t_{opt1} = 42$ time steps; for discrepancies of forecasts $\Delta t_{opt2} = 16$ time steps; and for smoothed discrepancies of forecasts $\Delta t_{opt3} = 31$ time steps. The interval between samples of the original series, selected for the prediction, was 20 steps, and therefore, it was less than the acceptable maximum.

However, it exceeded the maximum permissible value of 16 steps for time series of the discrepancies. So, prefiltering of the discrepancies time series was performed, which increased the maximum admissible interval between discrepancies up to a value of 31 steps.

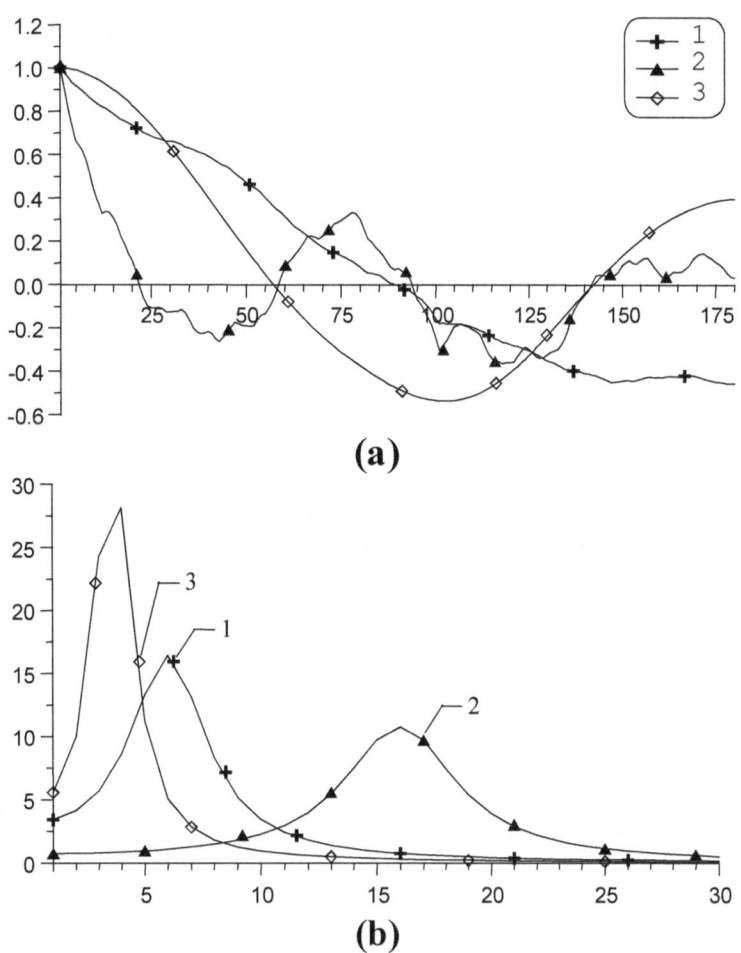

Fig. 2.24. Statistic characteristics of time series of observations. (a) correlation functions: 1 – the original series, 2 – forecasts' Discrepancies to pre-filtration, 3 – forecasts' discrepancies after preliminary filtration;

(b) spectral density (scale on the horizontal axis $5 \cdot 10^{-3}$ dimensionless units): 1 – forecasts' discrepancies after preliminary filtration,

2 – forecasts' discrepancies to pre-filtering, 3 – of the initial series

To implement the Kolmogorov's filter algorithm at each step of the prediction with formulas (2.34), a correlation matrix of forecasts' discrepancies was calculated. Its elements changes significantly over the time, as it can be seen on Fig. 2.25a, which shows the time graphs of weight coefficients of the observations data assimilation, found from the system of equations (2.38) and identified as g_{ij}.

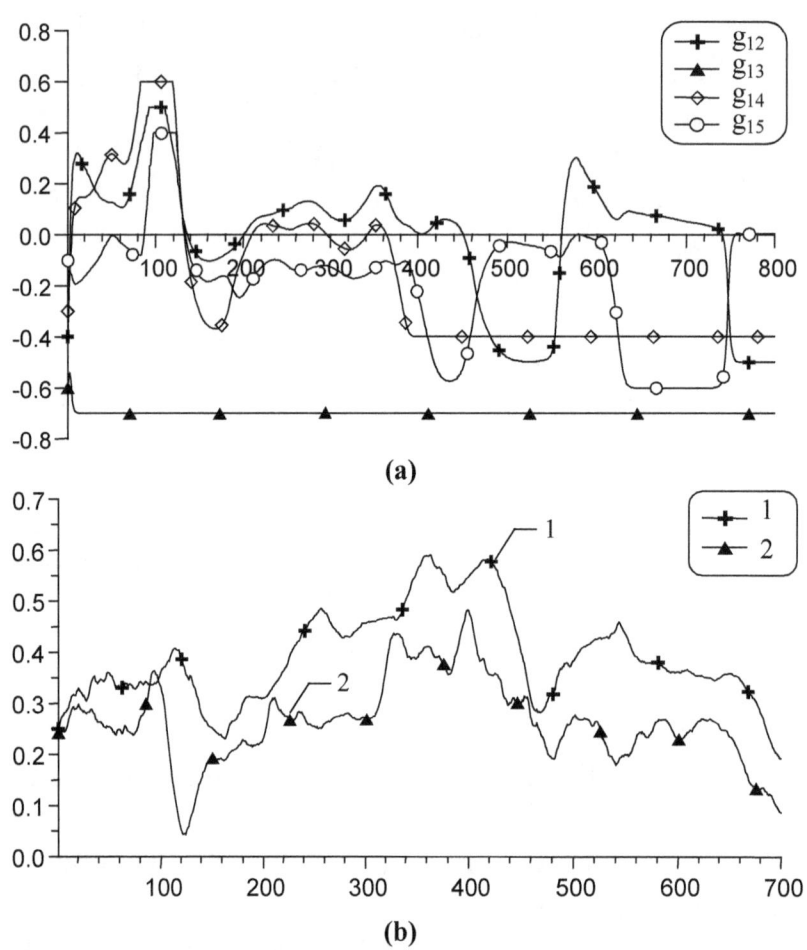

(a)

(b)

Fig. 2.25. (a) Dynamics of the weighting coefficients of observations data assimilation; (b) Dynamics of mean-square errors of forecasts before assimilation (1) and after assimilation (2)

All weight coefficients of prediction discrepancies' assimilation, except the coefficient g_{13}, show significant variability, which is explained by the sample scenarios of correlation coefficients. Weighing of four assimilated discrepancies of forecasts with these coefficients allowed us to conduct an optimal prediction for 40 steps forward. In fig. 2.25b, the mean-square prediction errors with data assimilation on the Kolmogorov's filter scheme, are compared with the errors of forecasts made without data assimilation.

It follows from this figure that the weight coefficients of assimilation, despite their seemingly disordered temporal course, in every single moment of time have been fairly well agreed between themselves, which led to the refinement of predictions, as we had expected.

Observations Data Assimilation in Accordance with the Optimal Kalman filter Scheme. At this step, let us consider a general case, when information about the actual errors of forecasts is not sufficient to directly evaluate the elements of the discrepancies' correlation matrix on a time series of observations. The solution to this situation was found by R. Kalman [88], who suggested to construct prediction models to evaluate elements of the predictions discrepancies' correlation matrix, by means of using a dynamic model of the original process itself. This is the meaning of the well-known Kalman filter, which is widely used in applied sciences [89, 91, 120, 145, 170, 171].

Now, let us denote predictions' discrepancies as z_j at the time points t_j and weighting coefficients, with which they must be added to the forecasted process, at the time t_i – as g_{ij}. For the coefficient of correlation of the discrepancies in the points t_i and t_j we will use the notation P_{ij} again. Let us suppose that n_i samples of the process are used for the prediction at the point t_i, and n_j samples are used for the prediction at the point t_j. To correct the forecast at the point t_i the assimilation of m observational data, prior to this point, is executed.

Then, based on the prediction model for the process (2.32) (2.38), we can construct the following dynamic *ABC*-model for assimilation of observational data at m points; we can implement Kalman filter through forecasting of elements of correlation matrix of forecasts' discrepancies at these points

$$\frac{dP_{ij}}{dt} = P_{ij}\left[1 - 2\left(P_{ij} - R_{ij} + \sum_{k=1}^{n_i} a_{ik}R_{jk} + \sum_{l=1}^{n_j} a_{jl}R_{il} - \sum_{k=1}^{n_i}\sum_{l=1}^{n_j} a_{ik}a_{jl}R_{kl}\right)\right],$$

$$(i = 1,2,...,n),\ (i \neq k),(j \neq l) \tag{2.44}$$

$$\frac{dg_{ij}}{dt} = g_{ij}\left\{1 - 2\left[g_{ij} - P_{jj}^{-1}\left(P_{ij} - \sum_{k=1}^{m} g_{ik}P_{kj}\right)\right]\right\},\ (j = 1,2,...,m),$$

$$(i \neq j),(i \neq k), \tag{2.45}$$

$$\frac{dx_i^{opt}}{dt} = x_i^{opt}\left\{1 - 2\left[x_i^{opt} - (x_i + \sum_{l=1}^{m} g_{il}z_l)\right]\right\},\ (i \neq l) \tag{2.46}$$

$$\frac{dP_{ik}'}{dt} = P_{ik}'\left[1 - 2\left(P_{ik}' - P_{ik} + \sum_{j=1}^{m} g_{jk}P_{jk}\right)\right],\ (j \neq k) \tag{2.47}$$

Let us explain the sequence of operations for observations data assimilation with Kalman filter scheme in the *ABC*-model (2.32) (2.44) – (2.47).

1. According to correlation matrix $\{R_{ij}\}$ of the process x, which is familiar to us from the analysis of archival data, and which connects

170

the point of the forecast t_i with the observational data points t_j, the influence coefficients a_{ij} are determined from the system of equations (2.38), and the adaptive prediction estimate of x_i in the future time instant t_i is calculated from the equation (2.32).

2. The *ABC*-model of influence coefficients (2.38) and the correlation matrix $\{R_{ij}\}$ are used for prediction of elements of correlation matrix for prediction estimates' deviations x_i from observational data, or discrepancies of the process forecasts. This task is resolved by the adaptive equations (2.44), which provide adjustment of future values of the matrix $\{P_{ij}\}$ elements to the elements of matrix $\{R_{ij}\}$ values.

3. Using this correlation matrix $\{P_{ij}\}$ for assimilation of targeted groups of observational data (forecasts' discrepancies), the coefficients of assimilation g_{ij} are calculated with the system of equations (2.45).

4. From equation (2.46), we can obtain an optimal prognostic evaluation of the process x_i^{opt} in the future time instant t_i, adapted to the n observations of the process in past and improved by assimilation of m observations of errors (discrepancies) of forecasts made in the previous times.

5. With the help of equations (2.47), prognostic evaluation of correlation matrix $\{P_{ij}\}$, is corrected, which can then be used at the next stage of data assimilation. When $i = k$ these equations allow us to evaluate errors of the data assimilation related to the choice of assimilation coefficients g_{ij}.

Let us consider the example of observations data assimilation with Kalman filter scheme (discussed above), when prediction of a process is executed at the time period of 40 steps in advance, using four observations separated by 20 steps from each other. In order to simplify this case study, we will assume that only two of the discrepancies of current forecasts – at points 2 and 3 (see Fig. 2.20b), can be used for predictions. Adaptive forecast with assimilation of the two previous forecasting discrepancies was provided by equation (2.46), which now takes the form

$$\frac{dx_1^{opt}}{dt} = x_1^{opt} \left\{ -2\left[x_1^{opt} - (x_1 + g_{12}z_2 + g_{13}z_3) \right] \right\} \tag{2.48}$$

The weight coefficients of assimilation g_{12} and g_{13} must satisfy the system of equations (2.45), which in this case is reduced to

$$\frac{dg_{12}}{dt} = g_{12} \left\{ -2\left[g_{12} - P_{22}^{-1}(P_{12} - g_{13}P_{32}) \right] \right\}, \tag{2.49}$$

$$\frac{dg_{13}}{dt} = g_{13} \left\{ -2\left[g_{13} - P_{33}^{-1}(P_{13} - g_{12}P_{23}) \right] \right\}.$$

The coefficients of mutual correlation of prediction discrepancies, which are needed to solve this system of equations, could be found from the adaptive *ABC*-equations (2.44) of the form

$$\frac{dP_{12}}{dt} = P_{12}\left[1 - 2\left(P_{12} - R_{12} + a_{23}R_{13} + a_{12}(R_{22} - a_{23}R_{23}) + a_{13}(R_{32} - a_{23}R_{33}) \right) \right],$$

$$\frac{dP_{13}}{dt} = P_{13}\left[1 - 2\left(P_{13} - R_{13} + a_{32}R_{12} + a_{12}(R_{23} - a_{32}R_{22}) + a_{13}(R_{33} - a_{32}R_{32}) \right) \right],$$

$$\frac{dP_{32}}{dt} = P_{32}\left[1 - 2\left(P_{32} - R_{32} + a_{23}R_{33} + a_{32}(R_{22} - a_{23}R_{23}) \right) \right], \tag{2.50}$$

$$\frac{dP_{22}}{dt} = P_{22}\left[1 - 2\left(P_{22} - R_{22} + a_{23}(2R_{23} - a_{23}R_{33}) \right) \right],$$

$$\frac{dP_{33}}{dt} = P_{33}\left[1 - 2\left(P_{33} - R_{33} + a_{32}(2R_{32} - a_{32}R_{22}) \right) \right].$$

To determine prognostic evaluation of correlation coefficients from the system of equations (2.50), we need to build another system of equations allowing adapt coefficients a_{ij} to the sample correlation coefficients of the original time series. Taking into account current scheme of assimilation of the two observations that preceded the prediction, the *ABC*-system of equations for the influence coefficients takes the form

$$\frac{da_{12}}{dt} = a_{12} \left\{ -2\left[(a_{12} - R_{22}^{-1}(R_{12} - a_{13}R_{32})) \right] \right\},$$

$$\frac{da_{13}}{dt} = a_{13} \left\{ -2\left[(a_{13} - R_{33}^{-1}(R_{13} - a_{12}R_{23})) \right] \right\},$$

$$\frac{da_{23}}{dt} = a_{23} \left\{ -2\left[(a_{23} - R_{33}^{-1}(R_{23} - a_{21}R_{13})) \right] \right\},$$

$$\frac{da_{21}}{dt} = a_{21} \left\{ -2\left[(a_{21} - R_{11}^{-1}(R_{21} - a_{23}R_{31})) \right] \right\},$$ (2.51)

$$\frac{da_{32}}{dt} = a_{32} \left\{ -2\left[(a_{32} - R_{22}^{-1}(R_{32} - a_{31}R_{21})) \right] \right\}.$$

Thus, the systems of equations (2.48) – (2.51) form an adaptive predictive model of the Kalman filter scheme, built by the *ABC*-method.

The results of this algorithm of assimilation of observational data are shown in Fig. 2.26 and 2.27. Figures 2.26a and b illustrate the noticeable variability of influence coefficients: $a_{12}, a_{13}, a_{23}, a_{21}, a_{31}, a_{32}$ in the process of moving along the series of observations of the group of samples used in predicting the values of this series. In Fig. 2.26c the results of predictions of elements of correlation matrix of predictions' discrepancies $P_{12}, P_{13}, P_{32}, P_{22}, P_{33}$ are presented. It follows from this figure, that the correlation coefficients of predictions' discrepancies had a general tendency to decreasing in magnitude as the increase in the number of observations assimilated.

The results of assimilation illustrate graphs of the processes shown in Fig. 2.27. The weighting coefficients of observations' assimilation g_{12} and g_{13} shown in Fig. 2.27a have significant variability along the time series of forecasts' discrepancies.

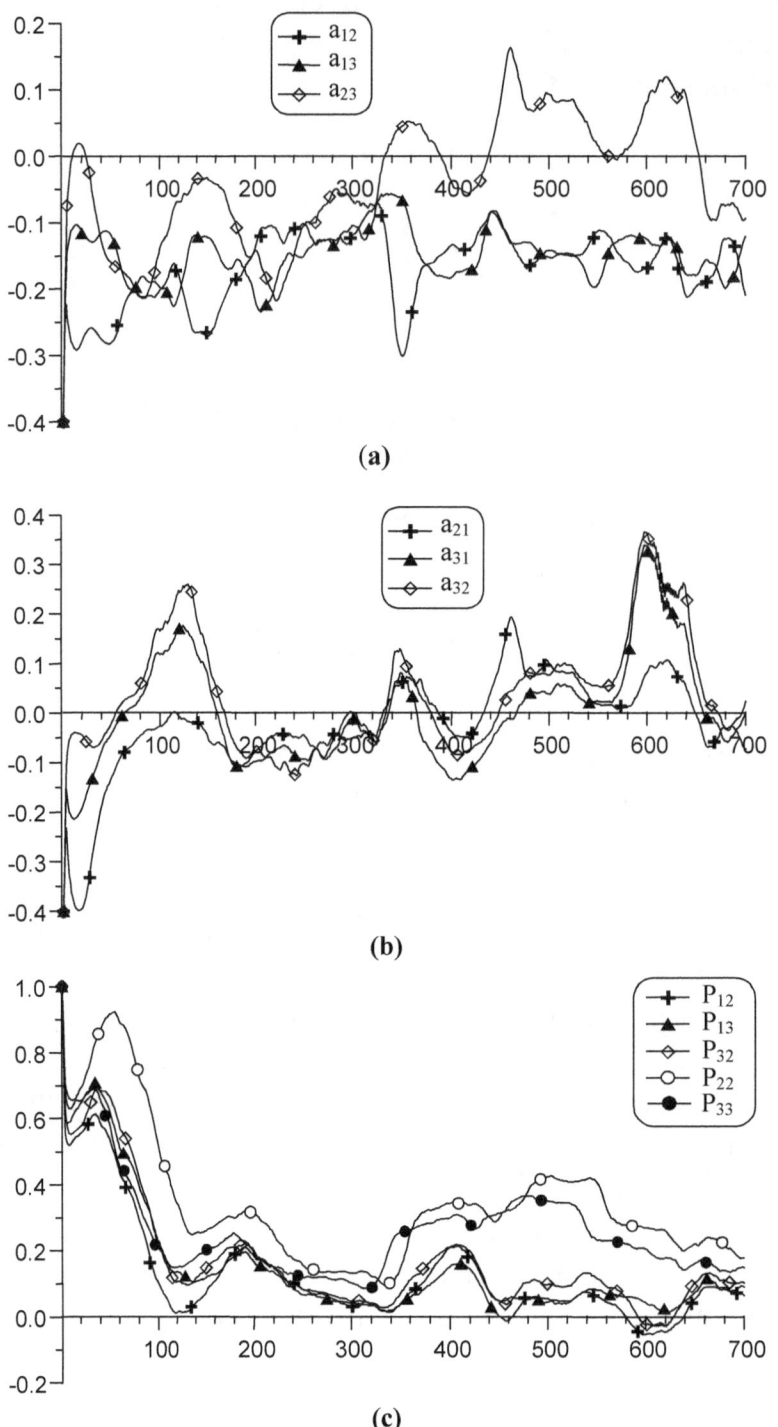

(a)

(b)

(c)

Fig. 2.26. Variability of influence coefficients: (a) a_{12}, a_{13}, a_{23} ; (b) a_{21}, a_{31}, a_{32};   (c) The results of forecasting elements of covariance matrix of

discrepancies' forecasts: P_{12}, P_{13}, P_{32}, P_{22}, P_{33}

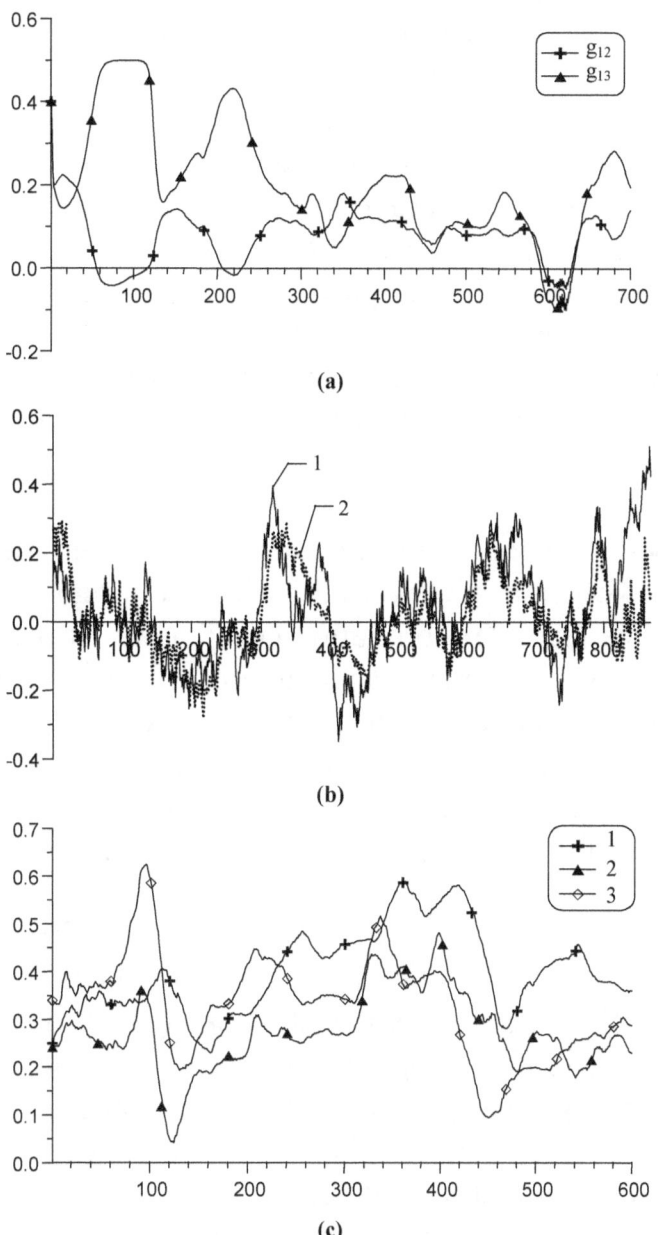

Fig. 2.27. Results of observations data assimilation: (a) variability of the

175

weighting coefficients of assimilation g_{12} and g_{13}; (b) 1 – original series   of observations, 2 – forecast for 40 steps ahead using four observations with assimilation of two preceding samples, (c) – dynamics of the mean-   square forecast errors: 1 – initial series prediction for 40 steps ahead, using four observations from the time series without assimilation of   observations, 2 – the forecast with assimilation of four observations with   Kolmogorov filter's scheme, 3 – the forecast with the assimilation of two observations with Kalman filter scheme

After adding two discrepancies with these weights to the forecast, some improvement in prognostic assessment was noticed. The resulting predictions' accuracy with data assimilation is characterized by Fig. 2.27b and c. The first of them gives a qualitative overview of the forecasting process results for 40 steps ahead; the second one shows how much the sample mean-square error of prediction reduced as the result of assimilation of two observations with Kalman filter scheme.

The best prediction of all was the forecast performed with assimilation of four observations with Kolmogorov's scheme by using variable (sample) correlation coefficients of the time series of observations. The accuracy of predictions of two measurements with Kalman filter assimilation scheme were turned to be something less. Note, however, that in the latter case the correlation function of discrepancies' prediction was calculated by the model and only two observations of the process were assimilated instead of four.

The experimental results suggest the following general conclusions. Systems modeling of natural processes, with random variability, allow create practical schemes of observations data assimilation, based on optimal methods of Kolmogorov's and Kalman filtering. Theoretically, these methods provide the best accuracy of prediction of random processes under the condition of known correlation matrices of these processes from observations or from model calculations. *ABC*-model (2.32) (2.38) can be successfully applied to the prediction of processes with the assimilation of observational data on the Kolmogorov's optimal filter scheme, when there is a possibility to calculate the sample correlation coefficients of the processes and sample correlation coefficients of forecasts' discrepancies directly from observations.

The *ABC*-model (2.32) (2.44) – (2.47) allows to forecast development of processes with assimilation of single measurements, when there is no possibility to directly evaluate correlation coefficients of forecasts' discrepancies, and when the statistics of discrepancies is calculated by a dynamic model (2.44), (2.45). The results are applicable to the systems of interrelated processes, developing in integrated models of environment.

CHAPTER III. MODELING OF DEVELOPMENT PROCESSES IN COASTAL ZONE MARINE ECOSYSTEMS

3.1. Integrated Model of Plankton Concentration at the North – West Shelf of the Black Sea

Let us consider a system of chemical and biological processes developing in the upper layer of the north-west shelf of the Black Sea, which are conventionally assigned to the center of this region. Our goal is to build a simple model of plankton community development for this region, in approximation of integrated assessment. This assessment of processes means spatial averaging over the volume of marine environment. We will choose a characteristic scale of the temporal variability of simulated processes, which will allow us to follow the annual course of these processes most efficiently.

To evaluate the external climatic conditions that form variability of the processes which take place in the ecosystem, we selected a number of criteria: annual variation of temperature and illumination of the sea, intensity and duration of wind exposure, and inflow of nutrients from the river runoff. For better efficiency of modeling, we took into account such factors as the features of cold-loving and heat-loving zooplankton's behavior, and characteristics of fish larvae and fish. These data have been collected through many years of observations of the north-west Black Sea shelf region. Some of the data used for the modeling was obtained during a number of marine expeditions, the other data was taken from published scientific literature.

Grounding on the research objectives stated above, we chose integrated assessments of phytoplankton concentrations, zooplankton concentrations and bioresources as basic characteristics of the ecosystem, with understanding that the bioresources, the fish larvae and the fish itself, which inhabit the surrounding area, have considerable commercial value. These biological resources are a sensitive indicator of all natural consequences of human economic activity in the north-western Black Sea shelf, as the biomass of the fishes' food is determined by productivity of phytoplankton and zooplankton formation.

Observational data about the known patterns of plankton communities' formation in this area, give some knowledge about the state of the ecosystem in the area. For example, we have information about dependence between annual variation of phytoplankton biomass' scenarios and the wind regime. If the number of storm days in January and February (when the wind speed exceeds 10 m/sec) exceeds 26, the scenario of phytoplankton's biomass has a sole maximum, which comes subsequently in July – August. In a relatively quiet winter period, when the number of stormy days is less than 26, two maxima are observed in the annual biomass of phytoplankton [159]. The first (large) maximum is in April, the second (a smaller one) happens in June and July. These experimental data are shown in Fig. 3.1, a. There are also intra-annual dependences characterizing the biomass of different species of zoo-plankton, phytoplankton and larvae of the main fish resources type – anchovy on temperature of sea water. An example of such dependence is shown in Fig. 3.7, b.

When the water temperature grows up to 19 ^{0}C, the heat-loving fish begins to spawn, with maximum values of the process observed in July-August. From January to March, the primary production increases, giving rise to the abundance and biomass of phytoplankton at the expense of diatom species of algae. In the intra-annual cycle of zooplankton biomass, the cold-loving zooplankton takes the leading role in the winter-spring period. Its development is influenced by total biomass of phytoplankton, sea water temperature and concentration of fish larvae.

The heat-loving species of zooplankton reach their maximum abundance in the water of the north-western part of the sea in summer.

Fluctuations of phytoplankton's productivity from year to year produce significant inclinations in stocks of biological resources. The main anchovy spawning ground is the north-west part of the Black Sea. In coastal areas, the life time of anchovy larvae is 4 – 5 hours, whereas, for example, the life time of horse mackerel is 1 hour. The optimal thermal background for development of anchovy is within the temperature range of 14 ^0C – 23 ^0C. The nutrients' concentration in the marine environment is formed by decomposition of detritus and due to penetration of nutrients and detritus coming in with the river runoff, which plays significant role in the processes of the northwest Black sea shelf zone, due to continental runoff of the Danube and the Dnieper.

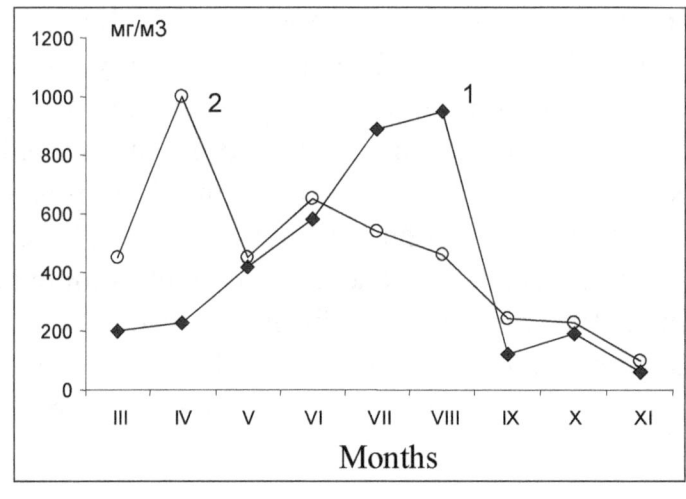

(a)

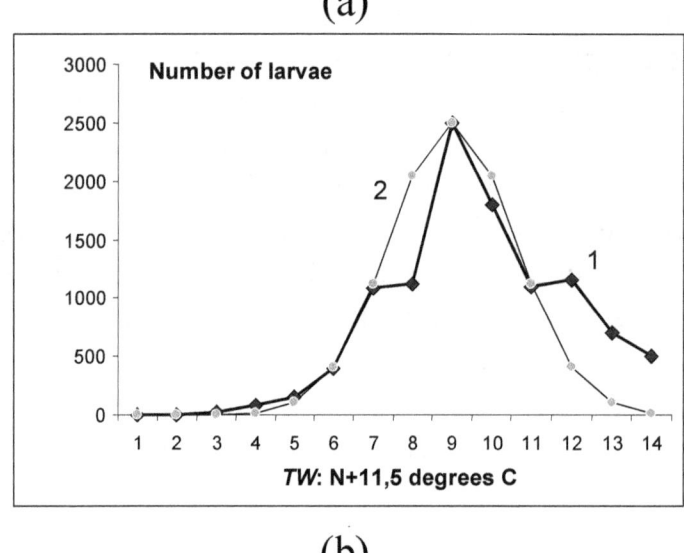

(b)

180

Fig. 3.1. (a) Scenarios of inter-annual variability of plankton's biomass when the number of stormy days in January – February are: 1 – more, then 26, 2 – less, then 26; (b) Number of anchovy larvae in one 10-minute fishing, depending on the sea water temperature: 1 – according to data of observations [159], 2 – approximation by the formula (3.12)

The analysis shows that the simplified model of plankton community of the Black Sea northwest shelf must include processes that represent concentration of these substances: phytoplankton – *PP*, cold-loving zooplankton – *ZPC*, heat-loving zooplankton – *ZPH*, larvae of anchovy – *LF*, anchovy – *BR*, biogenic elements – *BG* and detritus – *DT*. The relationships between these processes, as we can see from observations, make it possible to build a conceptual model of the ecosystem, which is shown in Fig. 3.5.

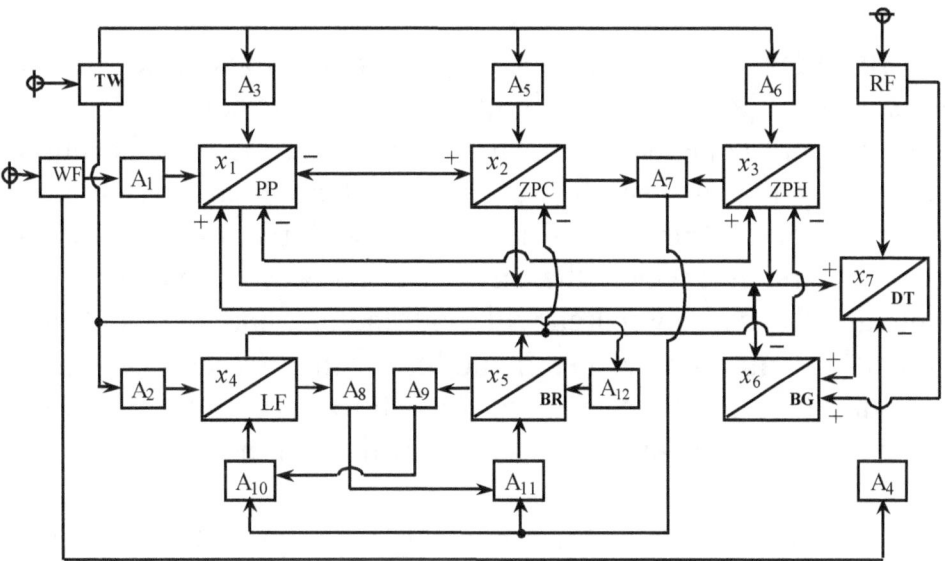

Fig. 3.5. Conceptual model of marine ecosystem of the Northwest shelf of the   Black Sea.

In the proposed model, the anchovy biomass *BR* is determined by cold-loving zooplankton, heat-loving zooplankton, fish larvae, and

181

by availability of nutrients. It also depends on factors of external impact on the marine environment, which are introduced in our study by using some management agents, designated as A_i, ($i = 1, 2 ,..., 12$). In particular, with the given composition of anchovy's food, positive correlation between number of their generations, amount of forage and water temperature in the spawning period, were introduced by operators: A_{10}, A_{11}, A_{12}.

The operator A_1 models annual variation of phytoplankton concentration, which is known from observations. The phenomenon is caused by action of the wind in winter time, in the beginning of each year (in January and February). As we noted above, if during this period of time the wind has a speed exceeding 10 m/sec during 26 days and more, the annual course of phytoplankton concentration has one maximum, which falls on August (the 225-th day of the year). Otherwise, the curve of annual variability of phytoplankton has two maxima: in April (the 165-th day of the year) and July (the 205-th day of the year).

In order to take into account the conditions necessary for the existence of larvae, we set limitations on their concentration (taking into account dependence on the amount of food and the number of larvae producing fish. It was agreed that at any given time, the concentration of larvae was determined with the influence of those of two factors, for which the smallest values were taken. Then, the operator A_{10} was compared with both factors, and we determined that it had the minimal value. In addition, the operator A_9 took into account the finite time required for fish to develop form larvae. Such dependency was also introduced for limitation the number of fish.

To construct the *ABC*- model of the ecosystem, we will use notations of concentrations, which are shown in Fig. 3.5: x_1 – phytoplankton *PP*, x_2 – cold-loving zooplankton *ZPC*, x_3 – heat-loving zooplankton *ZPH*, x_4 – fish larvae *LF*, x_5 – bioresources (fish) *BR*, x_6 – biogens *BG*, x_7 – detritus *DT*. In order to simplify circuit connections in Fig. 3.5, some of them (for example, the influence of fish larvae and fish on the concentration of detritus) were omitted. Applying the standard equation of the *ABC*-method and using the assumptions made above, we obtained the following system of dynamic equations of the ecosystem's model

$$\frac{dx_1}{dt} = x_1\{1 - 2[x_1 + a_{12}x_2 + a_{13}x_3 - a_{16}x_6 - A_1(WF) - A_3(TW)]\},$$

$$\frac{dx_2}{dt} = x_2\{1 - 2[x_2 - a_{21}x_1 + a_{24}x_4 + a_{25}x_5 - a_{26}x_6 - A_5(TW)]\},$$

$$\frac{dx_3}{dt} = x_3\{1 - 2[x_3 - a_{31}x_1 + a_{34}x_4 + a_{35}x_5 - a_{36}x_6 - A_6(TW)]\},$$

$$\frac{dx_4}{dt} = x_4\{1 - 2[x_4 - A_{10}\{A_7(x_2,x_3); A_9(x_5)\} - A_2(TW) - a_{46}x_6]\}, \qquad (3.10)$$

$$\frac{dx_5}{dt} = x_5\{1 - 2[x_5 + A_{11}\{A_7(x_2,x_3); A_8(x_4)\} - a_{56}x_6 - A_{12}(TW)]\},$$

$$\frac{dx_6}{dt} = x_6\left[1 - 2\left(x_6 + a_{61}x_1 + a_{62}x_2 + a_{63}x_3 + a_{64}x_4 + a_{65}x_5 - a_{67}x_7 - a_{6/RF}RF\right)\right],$$

$$\frac{dx_7}{dt} = x_7\{1 - 2[x_7 - a_{71}x_1 - a_{72}x_2 - a_{73}x_3 - a_{74}x_4 - a_{75}x_5 + a_{76}x_6 -$$
$$- a_{7/RF}RF + A_4(x_7,WF)]\}$$

In the right side of the equation for phytoplankton x_1 a management agent A_1 is included, which takes into account the influence of winter winds on the intra-annual scenario of phytoplankton concentrations. According to the above conditions, this scenario should take the form of curve 1 or curve 2, which are depicted in Fig. 3.4, a. Therefore, we selected the following representation for it

$$A_1(WF) = IF(H < 26; J; K), \qquad (3.11)$$

$$J = J_1 \exp[-\alpha_{j1}(t - 205)^2] + J_2 \exp[-\alpha_{j2}(t - 165)^2],$$

$$K = K_k \exp[-\alpha_k(t - 225)^2],$$

where N is the number of stormy days in winter time.

Except for the wind, the concentration of phytoplankton is influenced by the temperature of the sea water. The best conditions for phytoplankton's growth are the temperature of 16 degrees C. To account for these conditions, the management agent $A_3(TW)$ was utilized in the form

$$A_3(TW) = A_{1/TW} \exp[-\alpha_{1/TW}(TW - 16)^2].$$ (3.12)

Similarly, the influence of annual variations in the sea temperature was taken into account in the equations for the cold-loving zooplankton x_2 by the agent $A_5(TW)$, and for the heat-loving zooplankton x_3 by the agent $A_6(TW)$

$$A_5(TW) = A_{5/TW} \exp[-\alpha_{5/TW}(TW - 12)^2],$$ (3.13)

$$A_6(TW) = A_{6/TW} \exp[-\alpha_{6/TW}(TW - 23)^2].$$

Parameters $A_{n/TW}$ and, $\alpha_{n/TW}$ (as well as similar parameters in formulas (3.11) - (3.12)), were selected experimentally. For example, in the Fig. 3.4, b, approximation of the dependence between temperature values and fish larvae numbers was presented, which was introduced in the equation for x_4 by the management agent of the form

$$A_2(TW) = 2500 \exp[-0,2(TW - 20,5)^2]$$ (3.14)

The conditions for formation of fish larvae concentration, depending on the amount of food and the number of caviar-producing fish were determined by management agents $A_7(T)$, A_9

and A_{10}, with accounting of the finite time needed for growth of fish larvae

$$A_{10} = \min[A_7(x_2, x_3), A_9(x_5)], \quad A_7 = x_2 + x_3,$$

$$A_9 = IF\{x_5(t) > x_5(t - \Delta t); A_9(x_5); x_5(t) + [x_5(t - \Delta t) - x_5(t)]\exp(-\alpha_{45}\tau)\},$$
(3.15)

Similarly, the conditions affecting the formation of fish concentration were simulated

$$A_{11} = \min[A_7(x_2, x_3), A_8(x_4)],$$

$$A_8 = IF\{x_4(t) > x_4(t - \Delta t); A_8(x_4); x_4(t) + [x_4(t - \Delta t) - x_4(t)]\exp(-\alpha_{54}\tau)\}.$$

In order to simplify the model of ecosystem, the processes of respiration of living organisms were not represented in it. However, the driving influence of wind on the concentration of detritus was used in the model, in order to indirectly take into account the increase in concentration of dissolved oxygen and increase of the wind mixing over the upper sea layer waters. It was believed that, with increasing wind speed, the concentration of detritus decreases due to enhanced oxidation, while the concentration of nutrients increases at the same time. Therefore, the management agent $A_4(WF)$ was taken in the following form

$$A_4(x_7, WF) = a_{7/WF} x_7 \exp(-\alpha_{7/WF} WF)$$

To carry out calculations with the model, it was needed to determine the coefficients of influences included in the right parts of equations (3.10). In cases where there is archived data on the modeling processes, the *ABC*-method offers a possibility to objectively evaluate the magnitudes of these coefficients by statistic analysis of the archival data, as it was shown in chapter II. In our study, we are using a different approach to resolve this problem. It is based on expert judgments about the degree of the ecosystem's processes influence on each other.

3.2. Applying the Method of Analytical Hierarchy Process to Assessment of Influence Coefficients and Execution of Simulation Experiments with the Ecosystem Model

To estimate the values of coefficients in a model built by the *ABC*-method, the method of Analytical Hierarchy Process (*AHP*-method) can be applied. Then, the further development of this method – the Analytical Network Process (*ANP*-method), can also be utilized. Both methods were proposed by T. Saaty, and have been widely known as the "Method of Analysis of Hierarchies" [150]. The Method of Analysis of Hierarchies is oriented at expert decision-support, and assists to resolve the task of difficult choice between existing alternatives. The essence of the method can be described as follows.

Assessment of Influence Coefficients by *AHP*-Method. Suppose that there is a linear dependence between a process z and other interrelated processes y_i ($i = 1, 2, \ldots, m$), developing in a complex system

$$z = b_1 y_1 + b_2 y_2 + \ldots + b_m y_m.$$
(3.16)

Let us further suppose that the processes y_i are themselves linear functions of another system of interrelated processes x_j ($j = 1, 2, \ldots, n$)

$$y_i = a_{i1} x_1 + a_{i1} x_2 + \ldots + a_{i1} x_{in} \qquad (3.17)$$

Then a hierarchical system of influences arises: on the first (upper) level of the hierarchy the process z is set; on the second (average) level, processes y_i are set; on the third (lower) level, the processes x_j come up. The practical problem is to choose the values of the coefficients b_i and a_{ij}, which identify the degree of influence (priorities) of the relevant factors. Such choice in the analysis of complex socio-economic systems is to be done by experts. However, their estimates often contain errors and may not be consistent. In the hierarchy analysis procedure, objectification of the expert's assessments is proposed, by means of their global harmonization within the hierarchy.

To assess the effect of the lower level of the hierarchy influence on elements of upper level Saaty had suggested a pair-wise comparison between the expert opinions on the influence processes x_j for each of the processes of y_i, and then conduct a similar assessment of the impact of processes y_i on the process z on the next upper level. Let us suppose, for example, that A_j – factors (processes) x_j affect some process y_i, which is located at a higher level of the hierarchy. Let us introduce w_j as those weights, which several experts estimate as the power of these influences. Let us assume that the degrees of influences can be selected within the scale [0,10]. Then, for each process y_i a paired-comparison matrix A can be composed, and its elements will be the expert-set relationship ratings w_j/w_k.

Since the elements of the matrix A satisfy the condition $a_{ij}a_{ji} = 1$ by definition, this matrix is inversely symmetric. Let us imagine a matrix of paired comparisons in the form of (3.17). To determine the relative value, desirability, or probability of each object in a lower level of the hierarchy towards the object located on a higher level of the hierarchy, one needs to calculate a set of eigenvectors of the matrix of paired comparisons (3.17), i.e. to solve the following problem [158]

$$Aw = \lambda w,$$

where $a_{ij}a_{ji} = 1$ and $a_{ik} = w_j/w_k$.

Let us assume that the experts correctly identified the ratio w_j/w_k for all influence opinions. Because of the inverse symmetry properties of the matrix it has rank 1, and its eigenvalue $\lambda_{\max} = n$. The corresponding eigenvector provides consistent estimates of the lower-level factors' effects on the upper level.

$$
\begin{pmatrix}
a_{11} & a_{12} & \cdots & a_{1n} \\
a_{21} & a_{22} & \cdots & a_{2n} \\
\vdots & \vdots & \ddots & \vdots \\
a_{n1} & a_{n2} & \cdots & a_{nn}
\end{pmatrix}
\text{, where } \quad a_{ji} = \frac{1}{a_{ij}}
\tag{3.17}
$$

In the *AHP*-method, to assess the degree of influence of a pair of processes x_i, and x_k on the process y_j can be difficult, and the result can turn out to be inaccurate. In this case, the matrix A will no longer be strictly consistent, and therefore, it may have more than one eigenvector. However, if the expert evaluation errors are insignificant, among the eigenvalues of the matrix there will be one (maximum) number λ_{max}, which is close in magnitude to the value of n. In comparison with it, the other eigenvalues will demonstrate smaller quantities [151]. Thus, the inclination $\lambda_{max} - n$ may be a measure of consistency of the matrix A. It indicates the level of error in the choice of the eigenvector, corresponding to the number λ_{max}, as an assessment of the impact of processes x_j on process y_j.

To assess the degree of consistency the index of consistency ε and the consistency ratio η are used. Later, we can choose the ratio of index ε to the value μ of the index, which it takes when the expert evaluations are absolutely inconsistent with each other (random)

$$
\varepsilon = \frac{(\lambda_{max} - n)}{n - 1} , \qquad\qquad \eta = \frac{\varepsilon}{\mu}
$$

The random consistency coefficient μ depends on the dimension of the matrix A. For $n = 3$ it is 0.58, for $n = 5$ it is 1.12, for $n = 7$ it is 1.32, for $n = 10$ it is 1.49 [151]. The value of η (the ratio of consistency) should be the order of 10% or less.

Application of the *AHP*-method has made it possible to assess the coefficients of influences in the equations of ecosystem model (3.10) with consistency ratio η about 5%. In each of the model's equations, small groups of positive and negative coefficients were compared with each other. The matrixes of paired comparisons had low dimensionality. The calculated values of the model coefficients are shown in Table 3.4. To determine influence coefficients which were included into the equations of the model (3.10) with the help of operators A_i, the parameters of the latter (the variables in formulas (3.11) - (3.15)) were selected in such a way that their effects corresponded to the conceptual model of the ecosystem.

Table 3.4. The values of some coefficients of influences

a_{ij}	1	2	3	4	5	6	7
1	1	0,33	0,33			0,15	
2	0,45	1		0,1	0,2	0,2	
3	0,4		1	0,15	0,2	0,2	
4				1		0,1	
5					1	0,1	
6	0,1	0,1	0,1	0,1	0,1	1	0,5
7	0,1	0,1	0,1	0,1	0,1	0,5	1

Conducting Simulations with the Model. With the constructed formal model (3.10) - (3.15) computational experiments were carried out. During the computation, our task was to compare the scenarios of integrated processes in the marine ecosystem between each other. To perform this, all the simulated processes were represented in dimensionless form and reduced to a single scale of variability [0,10] dimensionless units. To return the results of calculation to actual dimensional units, as shown in the figures below, they must be multiplied by the following factors: wind $WF \sim 5$ m / sec, temperature $TW \sim 30\ ^{0}C$, concentration of phytoplankton $PP \sim 0,5$

g/m³, concentration cold-loving zooplankton *ZPC* ~ 0,1 g/m³, concentration of heat-loving zooplankton *ZPH* ~ 0,1 g/m³, concentration of fish larvae *LF* ~ 0,005 g/m³, concentration of nutrients *BG* ~ 2.5 g/m³, concentration of detritus *DT* ~ 6.10-5 g/m³. The time unit equal to one day was chosen for the experiment. Calculations were performed for 400 dimensionless time units (steps of calculations).

In the first experiment, scenarios of integrated processes for weak winds (not more than 15 m/sec) during the year period were constructed. Simulated annual variations of the sea-surface wind speed modulus *WF* and temperature of the upper layer of the sea *TW* are shown in Fig. 3.6a.

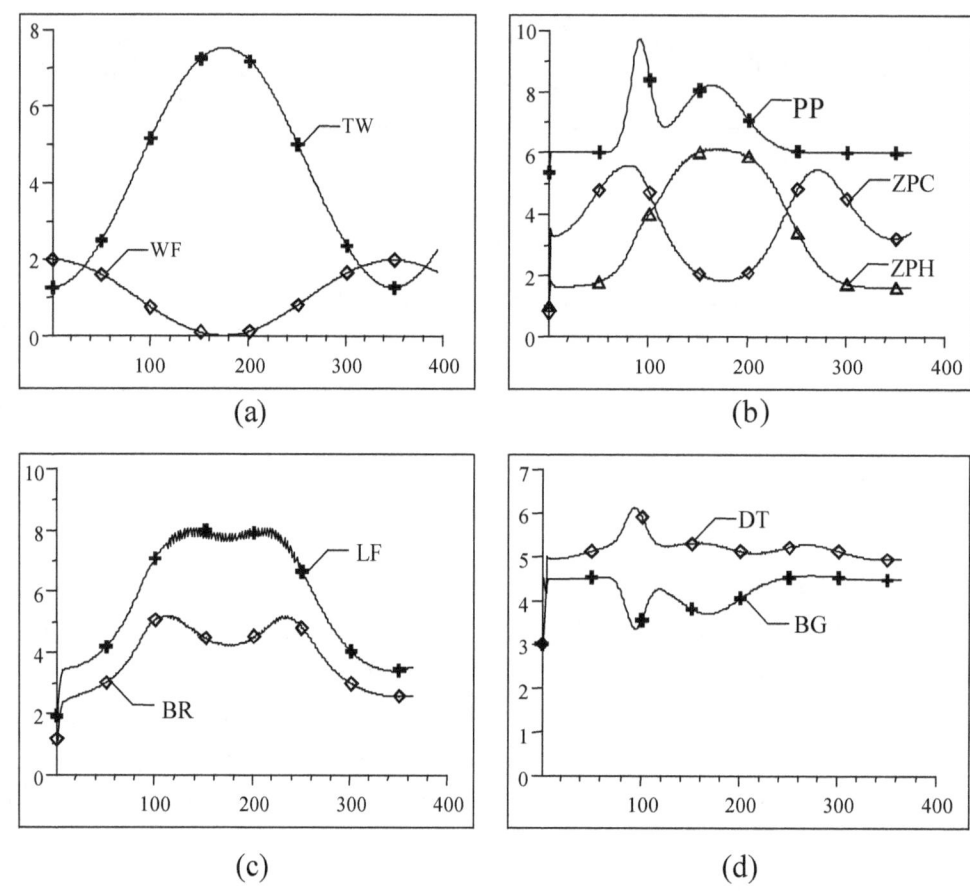

Fig. 3.6. Scenarios for integrated processes in the marine ecosystem of the north- west shelf of the Black sea in conditions of weak winter winds (January- February)

It is known that the maximum value of the thermal effect on the ecosystem is typical for July – August, and the minimum value is typical to happen in January or/and February. As it was mentioned above, if a particular year is characterized by relatively weak winds, the phytoplankton has two peaks in its development: the more intense one happens in the March, and the other happens in summer. In winter and spring period, active development of diatom complex species takes place. The low values of productivity are very characteristic for the first half of the winter period. It follows from the calculation results shown in Fig. 3.6b, that in period from late January to March, the primary production increases.

The first maximum depends on meteorological conditions. If a rapid warming of water happens in early spring, then maximum production occurs in March. At slow heating of the sea water, the peak of production shifts to the end of spring. After the spring maximum (which lasts for about a month) the rate of phytoplankton production is reduced, because concentration of nutrients lowers, and heterotrophic organisms appear (see curve *PP* in Fig. 3.6b).

The behavior of calculated curve of phytoplankton concentration in the second half of the year can be interpreted as follows. In the end of July or in the first half of August, the second maximum in the development of phytoplankton is reached due to the development of pirofit algae. In these circumstances, production of the summer maximum of phytoplankton can exceed the spring production. After the end of summer period, in September and October, the late-autumn development comes with an increase in primary production. In December, the primary production is reduced to minimum and remains at this level throughout the whole winter.

In the winter time, there is minimum of photosynthesis. At this period of time the synthesis is dominated by algae respiration, which happens due to insufficient light and the impact of winter vertical circulation, which captures the surface layer and the power of this process exceeds critical depth level where photosynthesis is possible. At the calculated curve of phytoplankton concentration, we can see two peaks (Fig. 3.6b). However, the third autumn maximum was not reproduced by the model. This revealed some of the limitations of the ecosystem's integrated processes parameters, which were used in the model.

The results of dynamic modeling of cold-loving and heat-loving zooplankton are shown in Fig. 3.6b (graphs *ZPC* and *ZPH*). They can be considered as quite adequate. They reflect plausible growth of zooplankton biomass, depending on the season; the data of observations were used. The ratio between organisms *ZPC* and *ZPH* in total biomass of zooplankton in the north-west shelf zone area of the Black sea also looks plausible.

The largest growth in anchovy population results from development of new generation in July. It basically defines the amount of one-year (yearlings) anchovy in September. The comparison of percents of survival amounts of caviar and larvae in July shows that for the caviar they differ from year to year in 1,1 times, and for the larvae the difference is more than 6 times. The amplitude of inter-annual fluctuations in the number of larvae is close to the amplitude of fluctuations in the number of yearlings (modified by an order of magnitude). Survival in the larvae period is a major determinant of these fluctuations. The model development scenarios are shown in Fig. 3.6c to demonstrate correspondence between the numbers of anchovy larvae and yearlings.

As we can see from Fig. 3.6d, variation of nutrients concentrations in seawater is in anti-phase with the progress of intra-annual phytoplankton concentrations, which is shown in Fig. 3.6b. This dependence demonstrates the significant effect which the growing concentration of phytoplankton makes on the consumption of nutrients. The positive correlation between concentration of phytoplankton and concentration of detritus is expressed in the model scenarios less clearly (see Fig. 3.6d).

In the second experiment, the effect of wind forcing on the marine environment has been increased four times. The calculated scenarios corresponding to it are shown in Fig. 3.7. In this case, the maximum value of the phytoplankton concentration curve is shifted to the summer time (see Fig. 3.4a). It is interesting to note, that a significant change of scenario of the phytoplankton concentration had practically no effect on intra-annual variability of each species of zooplankton, fish larvae, and fish. This conclusion follows from comparison of corresponding graphs in Fig. 3.6b, and 3.7a, b.

This similarity can be explained with the fact that concentration of phytoplankton was not the limiting factor in the development processes of the basic food chain. For the chosen parameters of external influences on the ecosystem, the dynamics of these processes was mainly determined by the intra-annual course of temperature of the marine environment.

Significant changes have occurred in the scenarios of concentrations of nutrients and detritus. Low concentrations of nutrients correspond to the time of maximum development of phytoplankton (see Fig. 3.7c), which (under conditions of strong winds in January and February) comes subsequent to the month of July. As for the concentration of detritus, its annual variability was largely determined by the dynamics of cold-loving zooplankton and wind effects. Therefore, the concentration of detritus has a minimum, which falls on the month of May.

In the third computational experiment, which was also carried out in conditions of strong winds, the flow of nutrients into the sea with river runoff was simulated. Particularly, we simulated an increase which was timed in spring and autumn seasons. Scenarios for the processes in the marine ecosystem are shown in Fig. 3.8. In the graph of the annual course of phytoplankton concentration, another maximum appeared, which fell on the period of the spring river runoff (see Fig. 3.8b).

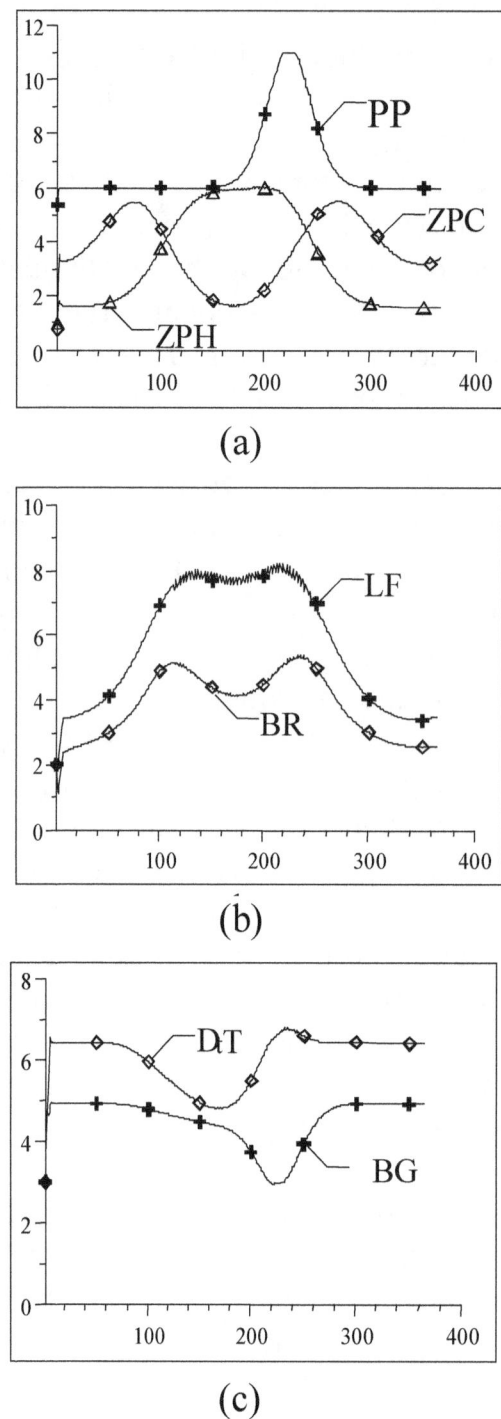

Fig. 3.7. Scenarios for integrated processes in the marine ecosystem of the north- west shelf of the Black sea in conditions of strong winter winds (January- February)

194

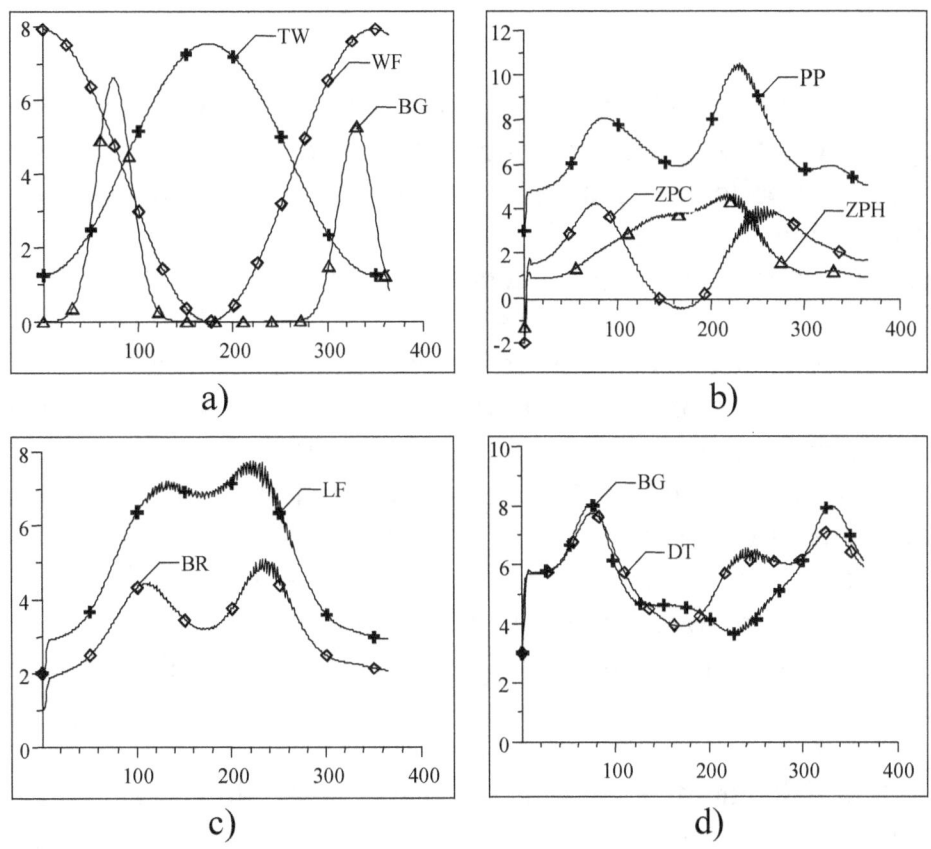

Fig. 3.8. Influence of the nutrients coming into the sea with river runoff on the   processes in the marine ecosystem

The scenario of cold-loving zooplankton remained almost unchanged (curve of this scenario in Fig. 3.8b is shifted down into 2 marks of vertical scale). But the high concentrations of the heat-loving zooplankton shifted in time for August. The summer increase in the concentration of this kind of zooplankton was influenced by the minimum concentration of phytoplankton in June.

Graphs of the fish larvae and the fish concentrations have the same form as in the absence of the nutrients inflow (see Fig. 3.7b and 3.8c), that indicates no direct influence applied on these processes by the coming nutrients. The scenario for concentration of nutrients is shown in Fig. 3.8d. It shows two main peaks, formed due to the imitated inflow of nutrients from the river runoff, and a minimum in the month of August, falling to the peak of phytoplankton concentration (see Fig. 3.8b).

The graph of the detritus content' annual variations is shown in Fig. 3.8d. It demonstrates increase in concentrations of detritus during the periods of nutrients inflow. These maxima resulted from introduction into the model of marine ecosystem (3.10), of the detritus' concentration dependence on the river flow.

The conducted computational experiments confirmed that it is possible to develop formal models of marine ecosystems, based on the expert knowledge about the behavior of ecosystems which was accumulated during long-term observations of the marine environment in a particular area. The verbal description of the marine ecosystem can give an opportunity to describe integrated cause-effect relationships between hydrochemical and hydrobiological processes in the marine environment in a form of conceptual model of the ecosystem.

The *ABC*-method used in this case-study, allowed to construct a formal dynamic model of the ecosystem of the north-west Black sea shelf zone, and to take into account long-term observations of plankton community formation specifics in this area. The utilization of the *AHP*-method of analysis made it possible to justify the choice of influence coefficients in the equations of the ecosystem's dynamic model, by bringing additional knowledge about the degrees of influences of some processes on each other. Thus, we can make a general conclusion which follows from this study: the dynamic models of integrated processes developing in marine ecosystems, allow us to predict possible scenarios of inter-annual variability of these processes in various non-typical external influences of the climatic or anthropogenic origin.

3.3. Statistical Analysis of Chlorophyll and Temperature Fields by the Data of Sea Surface Remote Sensing

To assess the resource properties of natural environment, overall monitoring of land and water areas, carried out by means of remote and contact measurements, is needed. The *ABC*-method of modeling makes it possible to use all available information in one model, and to indicate presence of biological, mineral and other types of resources, including minerals [177], in this sector of the environment. Such information can be obtained, for example, from results of remote sensing of the *CZA* land and marine areas, along the flight trajectory of an airplane or a satellite.

Profiles of spatial realizations of space-time fields observed along the trajectory, represent the ensemble of scenarios of related functions that could be referred to a single point in time, if the observations are carried out with a fairly high rate, compared with the temporal variability of the fields. This situation occurs in the remote sensing of sea surface fields of temperature and chlorophyll. Really massive bulks of observational data become available today, with modern development of operational oceanography [91]. We will consider the technique of satellite data processing on the example of analysis of statistical structure of the sea surface fields of temperature and chlorophyll concentrations in the upper layer of the Black Sea, using satellite observations obtained with a scanning optical spectrometer MODIS (Aqua) [143].

Analysis of Fields Statistical Structure Based on Observational Data. The problem of observational data interpolation from the points where they were obtained, to all the nodes of the grid covering the mapping area, arises in meteorology and oceanography, when the maps of spatial fields are being constructed. The basic method to solve this problem is a probability method of optimal interpolation, proposed by Kolmogorov [90] and developed in numerous studies on objective analysis of atmospheric fields and ocean [89, 170]. Many meteorological and oceanographic studies have shown that the optimal interpolation method provides high accuracy in constructing maps, provided that careful analysis of observations and accounting of probabilistic properties of these fields had been performed. These include the verification of homogeneity and isotropy of interpolated fields, with respect to the basic elements of their statistical structure: the mean values, variances and correlation coefficients, which are used in the method of Kolmogorov.

These issues are of relevance in connection with the development of technology of operational diagnosis and prognosis of the marine environment fields. Probability methods for data assimilation of the Earth remote sensing (*RS*), used in practice of operational diagnosis and prognosis of the Black Sea fields, also rely on the method of optimal interpolation of forecast errors in relation to their observations [12,120].

The assimilation of data is directly related to the problem of the data pre-filtering. To reduce interpolation errors, the following is necessary. If the fields' components spectra can not be resolved by the grid of measurements, the information about those components of fields should be excluded from the observational data. With this purpose, a set of preliminary smoothing (pre-filtering) of observations [120,138] is needed.

In the mapping of fields using contact measurements, the acquired information, as a rule, is not sufficient to perform prefiltering. Therefore, the errors that occur are not analyzed. A different situation takes place during processing of remote sensing data. Satellite observations provide a considerable amount of observations of various sea surface parameters, which opens possibility to correctly determine the upper limit of their spectra and to assess the degree of the real fields averaging, which is a process that accompanies mapping.

In [143], second-level processing data, obtained from *GSFC* archive (http://oceancolor.gsfc.nasa.gov) were used. They were further on processed with MHI NASU software. Among the tasks of the study were: selection of random components of surface temperature fields and chlorophyll concentrations from remote sensing data sets; analysis of their statistical properties; construction of maps by the method of optimal interpolation; using reduced data sets, and interpolation accuracy assessment, taking into account differences of random components from real fields of homogeneous and isotropic models.

Initial measurements were placed uniformly on the grid with steps of 1 km.

Data series along parallels and meridians on the north-west Black Sea shelf (*NWBSS*) had up to 370 measurements in one profile, which allowed to make an analysis of statistical structure of the fields. Preliminary check of the data content was made to find accidental peak and absent values.

The basis of statistical structure analysis of the investigated spatial field lies in evaluation of its homogeneity with respect to the first two moments of probabilities distribution of its values: the averages and spatial correlation functions over some areas of the field. Performing spatial averaging of the field by moving integration with a certain scale L

$$\hat{f}(x) = \frac{1}{L} \int_{x}^{x+L} f(x)dx, \tag{3.18}$$

where $\quad f(x) = \hat{f} + f',$ (3.19)

we can divide it into two parts. One of them may be conventionally taken as the deterministic component $\hat{f}$ and used in relation to its deterministic interpolation methods (linear, polynomial, etc.). Another component, f' can be obtained by subtracting $\hat{f}$ from the field f, and can be regarded as random. The validity of optimal interpolation method should be analyzed in relation to this very component. Varying the scale L of the moving averaging operation of the field, we can acquire the best match between the random field component and the hypothesis of its homogeneity and isotropy.

Let us denote the best in terms of the homogeneity of the random components, scales of observations of temperature L_{OT} and chlorophyll concentration L_{OCH} averaging. To find them, we will use the following requirements: mean values M_f, variances D_f, as well as correlation coefficients $K(x_1, x_2)$

$$K(x_1, x_2) = E\left\{ f'(x_1) f'(x_2) \right\},$$

calculated at fixed distances between points x_1 and x_2 in various parts of the field, and at different orientations of the interval $[x_1, x_2]$ in the plane $\{ x_1, x_2\}$ must have a minimal variation with respect to some middle values, which correspond to optimal scales of averaging.

The above analysis of satellite measurements of surface concentration values of chlorophyll and temperature of the Black Sea was carried out on selected sections of chlorophyll concentration and temperature along the parallels from 44°00' to 46°30' N and the meridian from 28°40' to 33°50' E, in the water of the north-west Black Sea shelf. Examples of the dependencies which we obtained are shown in Fig. 3.12 for the section along 44°30' N

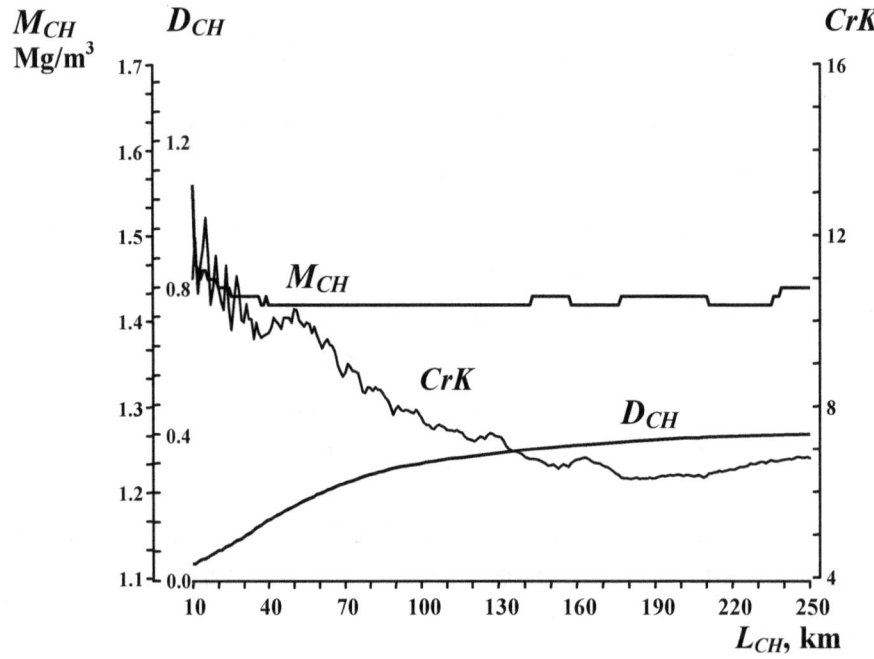

Fig. 3.9. Dependences of mean M_{CH}, variance D_{CH} and parameter $CrK = D_{f'_{CH}} \left| \overline{f}'_{CH} \right|^{-2}$ on the moving averaging scale L_{CH} at the section along 44°30' N in the field of chlorophyll concentration

It follows from the dependencies in this figure, that the best separation of chlorophyll concentrations field on deterministic and random components is achieved with a moving scale of averaging not exceeding 100 km. Fig. 3.10 allows us to estimate the dependence of spatial correlation functions of the random field components of chlorophyll and surface temperature on the moving scale of averaging.

The figure 3.10 shows that the optimal (in the above indicated sense) averaging scales L_{OT} and L_{OCH} for chlorophyll concentration fields and temperature differ from each other on the same profile. This also follows from Table 3.5, which shows radiuses of spatial correlations r_T and r_{CH} , variances D_T and D_{CH} , and number of measuring points N_T and N_{CH} along parallels and meridians in the north-west Black Sea shelf region at the optimal averaging scales of fields.

With increasing latitude and longitude L_{OT} and L_{OCH} decrease and the differences between them changes unevenly. This is due to the fact that with increasing latitude and decreasing longitude in *NWBSS* region, the heterogeneity of chlorophyll and temperature distributions seem to grow in different patterns. In the deep sea areas, where distribution of chlorophyll and temperature is more uniform, their optimal scales of averaging L_{OT} and L_{OCH} are almost the same, closing to nearly 200 km. Then, moving to the north and to the west, as they approach the shoreline, the optimal averaging scales are reduced, and at the section 46°00' N they are: for the temperature L_{OT} = 125 – 150 km, and for chlorophyll L_{OCH} = 75 –110 km.

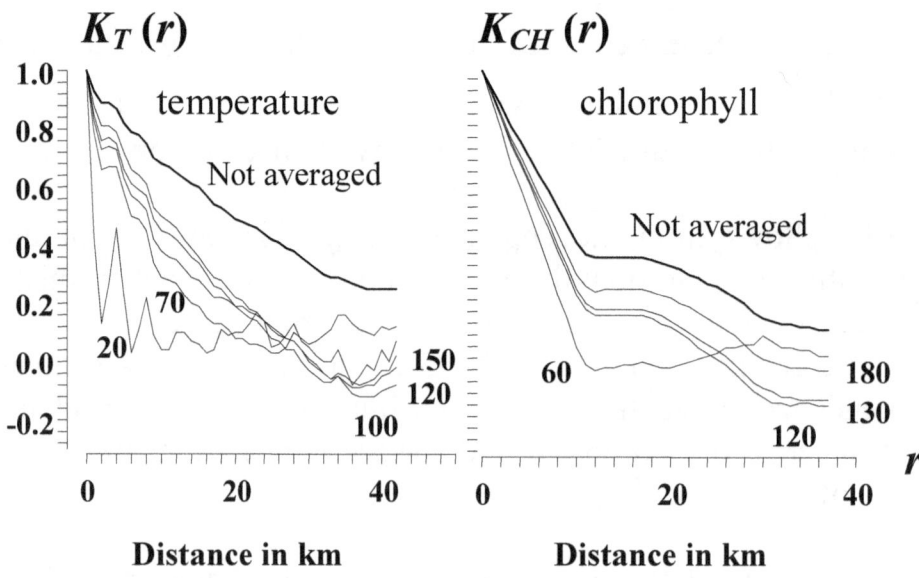

Fig. 3.10. Correlation functions of temperature and chlorophyll concentration on the section 45°30' in *NWBSS* region at different scales of smoothing, the numbers on the curves correspond to the scales of averaging L_{OT} and L_{OCH} in km.

Table 3.5. Statistical characteristics of chlorophyll concentration and temperature fields along different sections in *NWBSS* region at the optimal averaging scales of fields 6.07.2006.

Section along	r_T KM	r_{CH} KM	D_T	D_{CH}	N_T	N_{CH}
44° N	12	11,5	0,25	0,39	372	361
44,5° N	13	26	0,7	0,91	323	290
45° N	18	18	0,86	1,43	277	254
45,5° N	13,5	9,8	0,5	0,24	216	187
46° N	10,3	24	0,18	4,8	181	153
46,5° N	10	10	2,8	1,8	71	60
30° E	4,5	5,3	0,12	3,6	174	160
30,5° E	6,5	12	0,08	2,7	197	194
31° E	9	9,8	0,08	2,3	208	208
31,4° E	12	8,3	0,08	1,7	238	237
32° E	24	4,6	0,17	0,04	212	211
32,4° E	22	4,6	0,28	0,76	238	237

From the analysis of variability of the correlation functions shown in Fig. 3.10, it follows that they have the least variability when their averaging scales are approaching the optimal values of $L_{OCH} = 120 - 180$ km for the chlorophyll and $L_{OT} = 120 - 150$ km for the temperature. Radiuses of spatial correlations considered for $K(r) = 0,3$ had values from 4,5 to 26 km, while for $K(r) = 0,1$, they rose by about a half. The variability of these correlation distances along parallels and meridians was observed. However, there was a certain similarity in behavior of spatial correlation functions of each of the random fields, which allowed us to use these correlation functions to perform mapping of chlorophyll concentrations' fields and of surface temperatures by the method of optimal interpolation.

One of the purposes of field structure statistical analysis is to determine the best step (the distance between adjacent nodes) of interpolation grid for calculation of the field map. It is well-known that the main idea for choice of the optimal grid's step is to link this parameter with the width of the field spectrum base. We assume that the interpolation grid is square and has a step a. By analogy with the correlation interval, we will take the upper boundary wave number k_c

for the width of the field spectrum base. Then, the curve of the spectral density falls down to 0.1 of its maximum value.

Petersen and Middleton [138] showed that for homogeneous and isotropic random fields (in an infinite space of the grid area), there exists the Fourier transform, which introduces a quadratic "spectral" grid in the corresponding space of wave numbers. Let us agree that the spectral grid's step will be b. Then, for two-dimensional random fields, represented in the nodes of a square grid, we have the following generalized form of Kotelnikov-Shannon theorem [92]

$$a^2 b^2 = (2\pi)^2 \qquad (3.21)$$

The formula (3.21) is valid for those cases when circumferences of homogeneous and isotropic field's spectrum bases (the centers of which are located in neighboring nodes of the spectral grid) contact, but not intersect each other. Let us denote the circle radius of the spectrum base as k_c. Since $b = 2\,k_c$, from the formula (3.21) we will find $a = \pi k_c^{-1}$. The upper boundary of the spatial wave-number spectrum of the random field component essentially depends on the scale of the field averaging by the formula (3.18). For each of the scales Lo used above, we can found the best interval of spatial discretization of the field $a_0 = \pi k_{c0}^{-1}$.

Let us approximate correlation functions of random field components of temperature and chlorophyll concentration, calculated along the measured profiles of fields (see Fig. 3.13), by the exponential dependence

$$K(r) = De^{-\alpha|r|}. \qquad (3.22)$$

Then, the spectral density of random components can be approximately represented in the following form

$$S(k) = \frac{D\alpha}{\pi(\alpha^2 + k^2)} \qquad (3.23)$$

Using formulas (3.22) and (3.23), it is easy to estimate the values of the upper wave number of k_c for each of the used scales Lo. Fig. 3.11 shows the dependence of parameter k_c for fields of surface temperature and chlorophyll concentration on the scale of the averaging for the section 43° N.

It follows from this figure that at the scales of surface temperature and chlorophyll concentration averaging exceeding 100 km, parameter k_c takes the values 0.0123 km^{-1} and 0.0264 km^{-1}, respectively. The best intervals of the fields' sampling (steps of measurement grid) for this case take the values: $a_{0T} = 41,6$ km for the temperature field, and $a_{0CH} = 19,2$ km for the field of chlorophyll concentration. If we increase the intervals of sampling in comparison with these values (at the optimal interpolation), then interpolation errors will arise. Its average value can be estimated by the formula [138]

$$\overline{\varepsilon} = \int_{k_0}^{\infty} S(k)dk,$$

(3.24)

Let us assume that the value of the average interpolation error for chlorophyll concentration fields and temperature (according to the *RS* data for the sections in the *NWBSS* region) should not exceed 20% of the variance of the random field components. Now, let us estimate distance between the samples, i.e. a step of the measurement grid, required for this purpose. Examples of spectral functions of chlorophyll concentration and sea surface temperatures at the section along the latitude 45°30□ N at different scales of smoothing are shown in Fig. 3.12a and b. As it can be seen from these figures, the spectral functions are almost independent of the choice of smoothing scale, when their values lie within the range of $L_{OCH} = 100 - 150$ km for chlorophyll and $L_{OT} = 120 - 180$ km for the surface temperature.

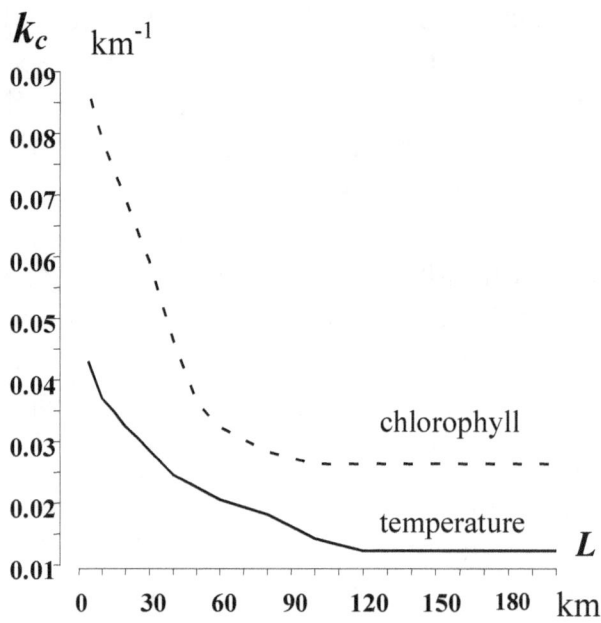

Fig. 3.11. Dependence between upper wave-numbers k_c of surface temperature spectrums (solid line) and chlorophyll concentration (dashed line) on the scale of averaging for the section 43^0 N

Under these conditions, to ensure a 20%-accuracy of reconstruction of these fields a choice of the measurement's grid cell with the size about 30 km seems rational. However, in a view of the actual inclination of the fields' statistical structure from homogeneous and isotropic model, there exists the dependence of spectral functions, and hence the optimal steps of measurement grids a_{0T} and a_{0CH} on the latitude. The results are shown in Fig. 3.13.

Maximum values of optimal steps and of maximum differences between a_{0CH} and a_{0T} were found on the south of the *NWBSS* region (55 km and 65 km), however, these estimates came close to each other at moving to the north, reaching 32 km at the latitude of 46^o30 N.

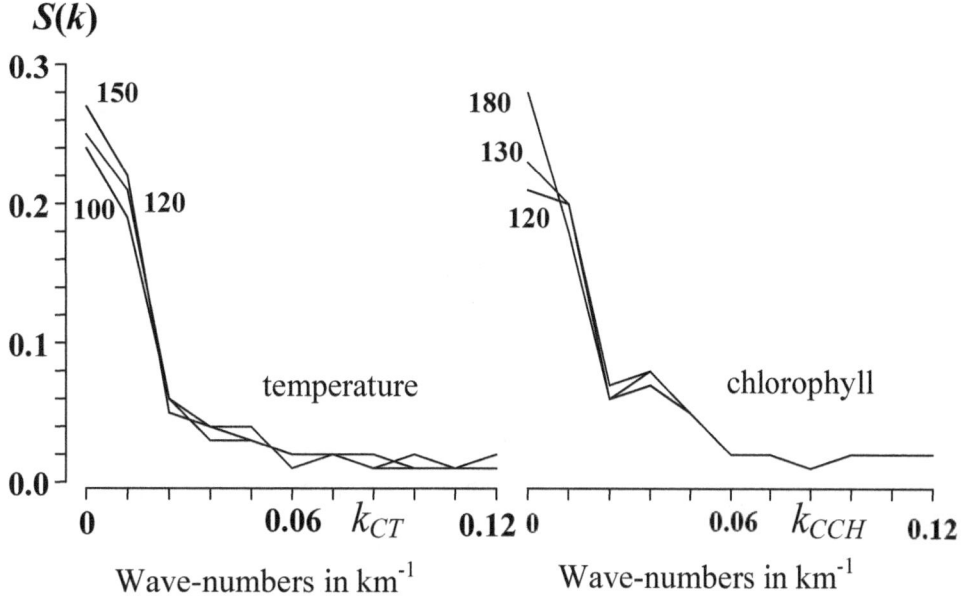

Fig. 3.12. Spectral function of temperature and chlorophyll concentration in 6.07.2006 for section 45° 30' N of the *NWBSS* region at the optimal aver   aging scales of fields 100 –180 km.

Thus, to ensure a 20%-accuracy of the fields mapping, it is necessary to use the measurement grid with steps of about 30 km. In general, the measurement data at the grid nodes should be furtherly smoothed by means of moving averaging – prefiltering, which should reduce the upper limit of wave number spectral density of the field to a value determined by the formula (3.21), i.e. $k_c' = \pi(30)^{-1}$km^{-1}. In our case the upper wave-numbers of spectral densities were significantly lower than this value. In accordance with the graphs shown in Fig. 3.12, they had the order of 0.05 - 0.07 km^{-1}. Therefore, operation of data pre-filtering was not required.

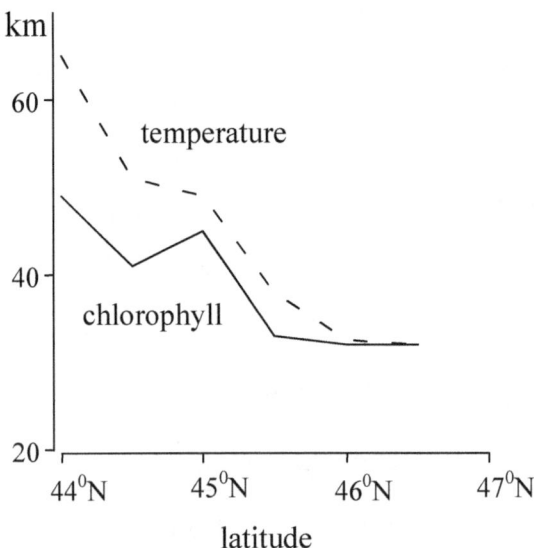

Fig. 3.13. Dependence between optimal step of measuring chlorophyll

       concentration and temperature of latitude in the *NWBSS* region.

Algorithm for Optimal Interpolation. Let us study a random component for the field of environment $f(x)$. We'll assume that x – vector of spatial coordinates defining a precise point of the field. We will also assume that measurements are made without any field averaging, but in general, they contain errors $v(x)$. This allows us to represent a formula for making measurements in the following form

$$z(x) = f(x) + v(x) \qquad (3.25)$$

Usually, the measurement procedure is performed so that the error does not depend on magnitude of the field itself. In addition, the measurement errors at various points are not linked, since the measurements are performed independently of each other. Therefore, using the coefficients of correlation, it is possible to introduce the following probabilistic model for measurement errors

$$M\{v(x_n)v(x_m)\} = R(x_n, x_m)\delta(x_n - x_m) \qquad (3.26)$$

$$M\{v(x_n)f(x_m)\} = M\{v(x_n)\} = 0 \, ,$$

(3.27)

where $R(x_n, x_m)$ denotes the variance of the measurement error, $\delta(x_n - x_m)$ – Dirac delta function, which equals the unity when $n = m$, and vanishes when $n \square m$. Thus, it is assumed that the measurement errors are not correlated with each other for individual points of the field. However, dispersions of measurement errors are the functions of coordinates at various points in the field. The principle for calculation of field maps is the removal of a specific measurement from the point of interpolation, as measurements that are significantly at a distance, will have weak statistic relationships with this point and their use in the interpolation is impractical.

An estimate of the field $\hat{f}(x)$ can be found as a linear combination of measurements of the field

$$\hat{f}(x) = \sum_{p=1}^{N} g(x, x_p) z(x_p)$$

(3.28)

in which the weighting coefficients $g(x, x_p)$ must take into account the contribution made by each observation into reconstruction of the field (where N is the number of points enclosed in the correlation circle). The problem lies in the choice of these coefficients, where the field reconstruction error will become minimal. It is convenient to use mean square error of interpolation as a measure for the reconstruction error

$$\varepsilon(x) = M\{[f(x) - \hat{f}(x)]^2\}$$

(3.29)

To find optimal values of weighting coefficients in the formula (3.28), we will substitute this formula into the formula (3.29), and elevate to the square the expression in square brackets. The minimum values of interpolation error can be achieved in a case when the derivatives on $g(x, x_p)$ ($p = 1, 2, \ldots, N$) from (3.29) become zero. If we follow these steps, we will arrive to the following system of algebraic equations for finding optimal interpolation weights [171]

$$K(x, x_n) = \sum_{p=1}^{N} g(x, x_p)[K(x_p, x_n) + R(x_p, x_p)], \quad (n = 1, 2, \dots, N). \quad (3.30)$$

We have acquired a system of equations of Kolmogorov [90], which allows to determine all necessary weighting coefficients of interpolation $g(x, x_p)$, which minimize errors of the field $\hat{f}(x)$ calculation.

A well-known property of optimal interpolation algorithm of Kolmogorov is the possibility to estimate the field reconstruction error $\varepsilon(x)$ at a specified point x, before measurements in a particular set of measurement points surrounding it are made, because the formula for this estimate does not include measurements

$$\varepsilon(x) = K(x, x) - \sum_{p=1}^{N} g(x, x_p) K(x, x_p) \quad (3.31)$$

The first term on the right in (3.31) represents the variance of the field $f(x)$ at the point of interpolation, which took place before the implementation of the interpolation. Therefore, the obtained formula shows the degree to which the initial uncertainty of our knowledge of the field at point x diminished after execution of the interpolation at this point with optimal weighting coefficients.

Mapping Fields by Optimal Interpolation. The formula (3.28) and (3.30) of the algorithm for optimal interpolation was used to reconstruct the fields of chlorophyll and sea surface temperatures in the nodes of a square grid with a spacing of 1 km, covering the central part of the north-west Black Sea region. The choice of this step of interpolation grid was determined by the available remote sensing data of the fields described above. The analysis of statistical structure of random components of chlorophyll and sea surface temperatures field showed us that, to compare the results of interpolation with the available observational data, it is expedient to build fields in the grid with the spacing of 1 km, using grid measurements with the spacing of 10 km. In this case, there were two ways to calculate the contours of each of the fields: to use a grid measurement with the spacing of 10 km, or to use the spacing of 1 km. The results of optimal interpolation of chlorophyll concentration fields and sea surface temperature in the square region of the *NWBSS* are as follows: 1 degree of latitude and 1 degree of longitude are shown in Fig. 3.14 *a, d.*

The average values of absolute and relative interpolation error gave fairly good results: for the concentration of chlorophyll, they amounted to 0.56 mg·m^{-3}, and 23% ; for the sea surface temperature they were – 0,29 ^{o}C and 1,2 %, respectively.

In the further study, the computational experiment selected for the optimal interpolation region was expanded to the size of the whole *NWBSS* area. The results of the calculations are illustrated by Fig. 3.15. It shows contours of chlorophyll concentration C in mg·m^{-3} obtained on a grid with a spacing of 1 km: in Fig. 3.15, *a* – calculated, in Fig. 3.15, *b* – built on the field data. Comparison of these figures shows that the visual difference between the calculated data and the field data, is minimal. This was also confirmed by estimation of the error of these calculations.

The relative error of chlorophyll (Fig. 3.16, *a*) in the southern and the eastern parts of the *NWBSS* area has a minimum value of about 4%, while in the western and the northern parts, where the concentrations C in mg·m^{-3} have been dramatically increased to the highest values, the relative error increased up to 50%.

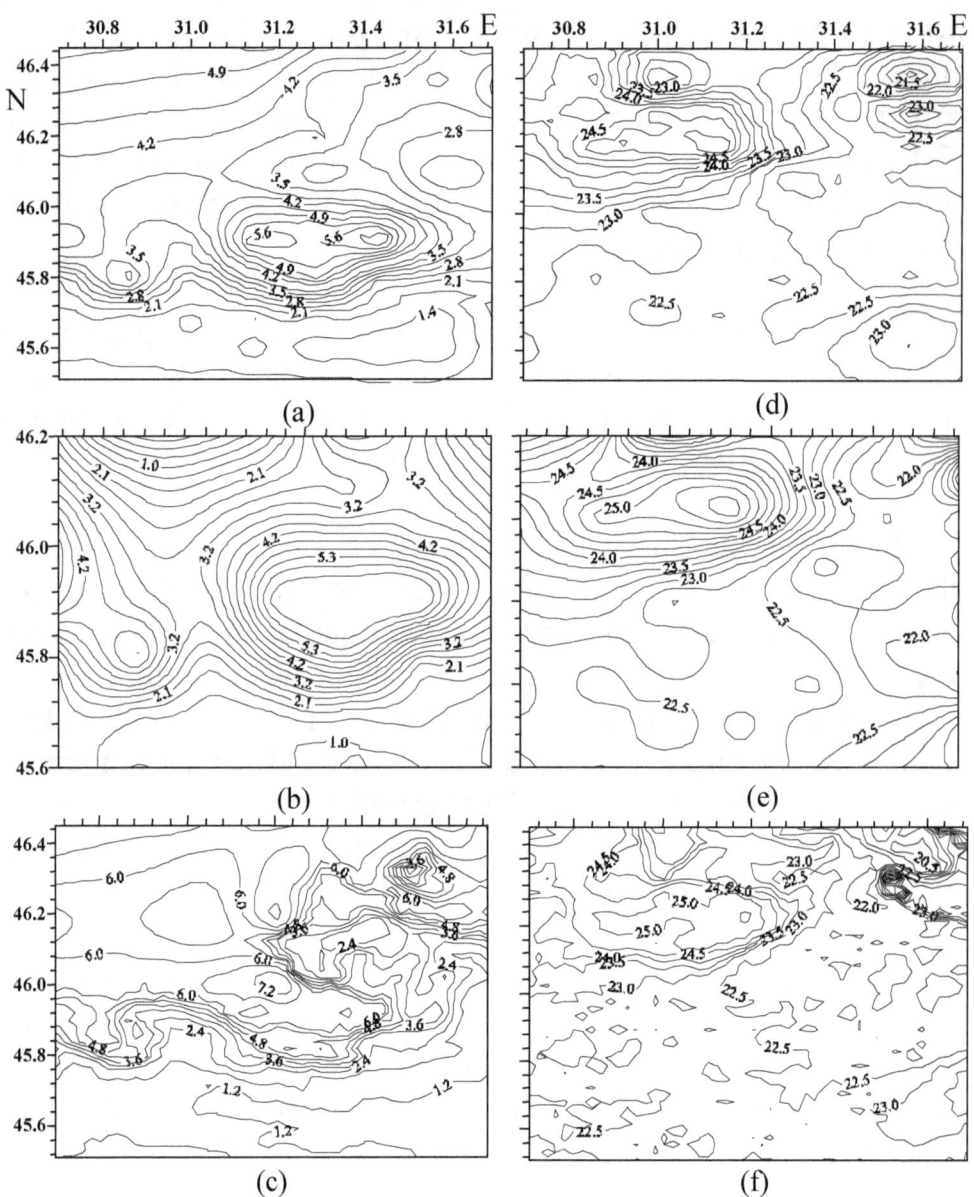

Fig. 3.14. Distributions of chlorophyll (*a, b, c*) and sea surface temperature

(*d, e, f*) fields in the center the *NWBSS* area: calculated (*a, d*) according to   prefiltered data on the grid with steps of 1 km, reconstructed according to   initial observations data on the grids with steps of 10 km (*b, e*) and

1km (*c, f*)

212

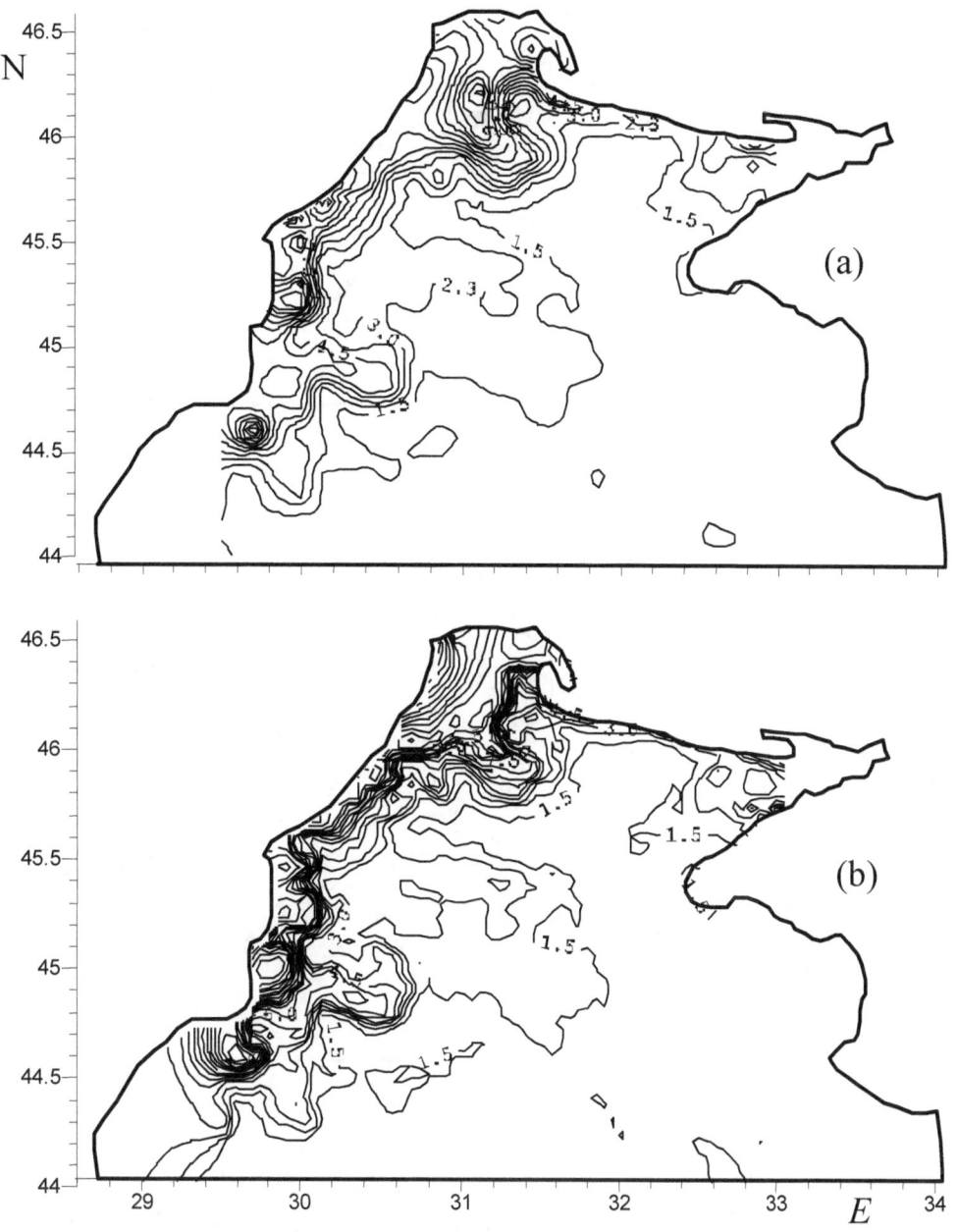

Fig. 3.15. Lines of the chlorophyll concentration constant values reconstructed from the grid with steps of 1km: (*a*) – pre-filtered values in nodes; (*b*) – initial observational data

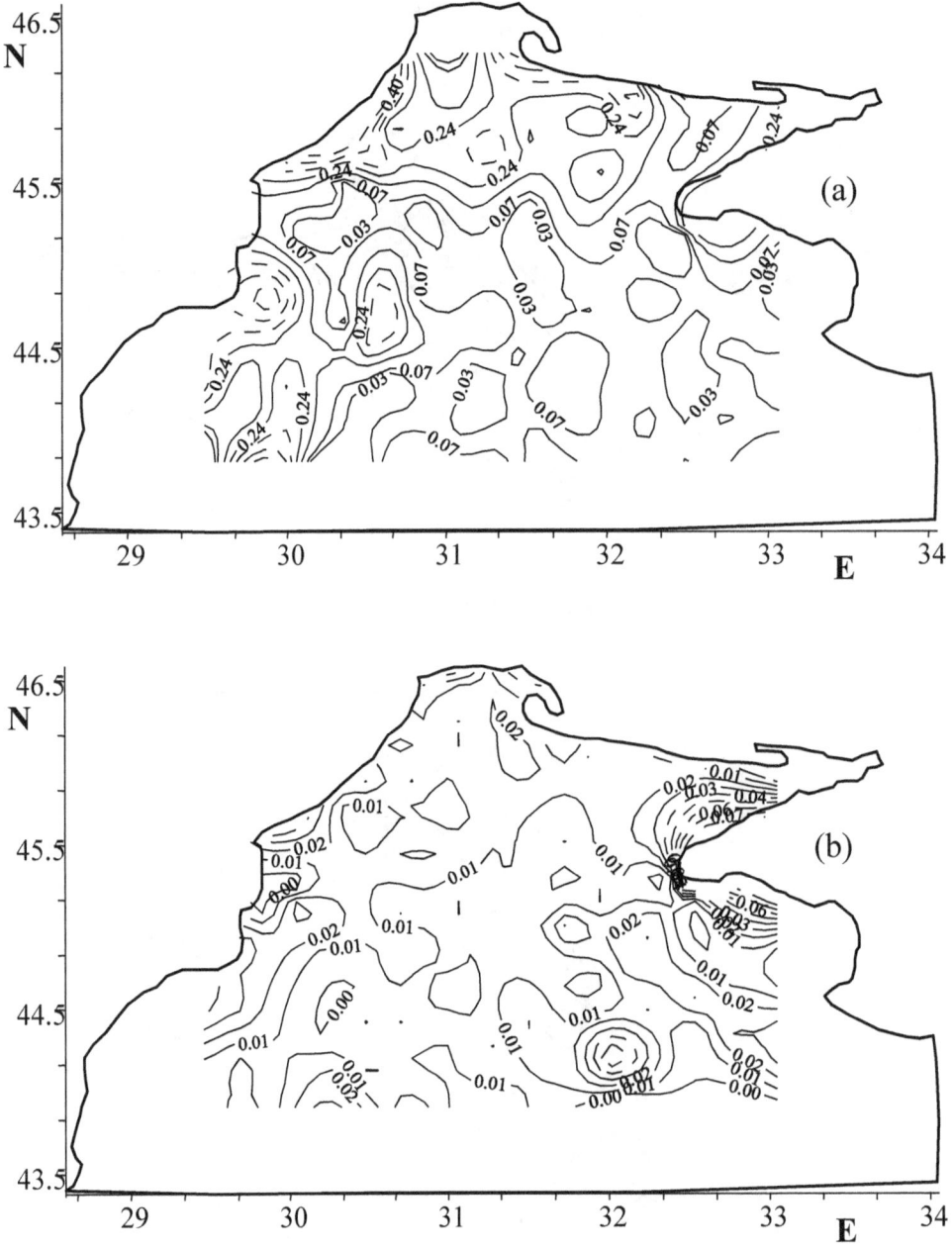

Fig. 3.16. Distribution about the *NWBSS* area of relative accuracy ($^\Delta f/f$) of the   optimal interpolation scheme applied to the fields of chlorophyll

concentrations, (*a*) and temperature, (*b*) from the grid with steps of 32 km

However, for the predominant area of water, the relative error of optimal interpolation of chlorophyll concentration was 20%.

The relative errors of optimum interpolation of temperature with minimal values of surface area occupied the dominant part of the *NWBSS* area. This is its western, central and southern parts, where the relative interpolation error was about 1%. Some comparatively large relative errors in temperature, reaching 5%, were observed in some areas of the west coast of the Crimea and Karkinitskiy Bay, where they occupied a small area (see Fig. 3.16, *b*).

Investigation of the statistical structure of random components of remote sensing data of the sea surface in the *NWBSS* area has shown the possibility of mapping the fields of surface temperature and chlorophyll concentration by the method of optimal interpolation in conditions of controlled data precision. For specific data sets of satellite observations, the optimal smoothing scales of the fields can be found in order to divide them into deterministic and random components. This can be dome by moving averaging of initial data with respect to their dependence on spatial coordinates. Distributions of relative errors ($\Delta f/f$) of the optimal interpolation of chlorophyll and temperature are shown in Fig. 3.15.

The appropriate optimality criterion is the minimum spread of spatial correlation coefficients of the random field components relative to their mean values calculated for different areas of the mapped area of the field, and for different space orientations of segments joining points of the field. In accordance with these criteria, correlation and spectral functions of random components of the surface temperature and chlorophyll concentration fields have been built for the *NWBSS* area. Correlation radiuses and optimal grid steps for a given accuracy of the optimal interpolation mapping have been found.

The method of optimal interpolation allows us to calculate field maps using only a part of the initial information contained in the remote sensing data, with virtually no significant loss of accuracy of the representation of fields. This means that the development of a probabilistic model of the field and analysis of the statistical structure of the array of remote sensing data can significantly reduce the amount of observational data to be archived and stored for later use. In the above example, due to observations of surface temperature and chlorophyll concentrations in water of the *NWBSS* area, it was possible to increase the spacing between grid nodes representing these fields, from 1 to 10 km. Calculations performed on a reduced array of data for the polygon area in the central part of the *NWBSS* and for the entire area of this region, showed that the mean relative error of mapping the chlorophyll concentration had the order of 0.2 of the variance of the field, but for the surface temperature, it had the order of 0.1 of the variance. With this accuracy in mapping, the main features of spatial distributions of fields were well traced in the maps.

3.4. Resource-Oriented Processes in Marine Ecosystems

A simplified model of the north-western Black Sea shelf ecosystem which we considered above, did not take into account the processes of respiration of marine organisms, which, taken along with the food chain: phytoplankton – zooplankton – bioresources, plays an important role in the ecosystem dynamics. In this section, we will focus our attention on the resources providing for living organisms, assuming as before, that each link of the food chain takes into account integrally the biomasses of all living organisms of the type in a specific volume of the marine environment.

We assume that for each of the three trophic levels one characteristic kind of aquatic living species that has specific resource requirements which could ensure the conditions of his existence, can be identified. For the purpose of modeling, we will choose a description of the state of dynamic balance of processes in the ecosystem, seeking to adapt to changing environmental conditions. Let us define the most important processes incurred in the achievement of this goal, with understanding that, among all processes, the priority must be given to the processes of respiration and nutrition, which provide the living creatures with energy and minerals.

Let us choose a day as a unit of time, and normalize this unit timeline. We agree to consider the average daily value of development processes in the ecosystem and to introduce the following notation for them:

OX – concentration of dissolved oxygen,
CD – concentration of dissolved carbon dioxide,
PP – concentration of phytoplankton,
ZP – concentration of zooplankton,
BR – concentration of bioresources,
BG – concentration of nutrients,
SR – light at sea surface.

Now, we will take a look at mutual relationship between the selected processes. As it was done above, we assume that one process has a positive influence on the other, when their trends are the same: with a decrease in the first process, the second one is also decreased. For example, phytoplankton has positive influence on oxygen, and with the increase of phytoplankton concentration, the concentration of dissolved oxygen increases. As zooplankton negatively influences on phytoplankton concentration, as it uses phytoplankton for food. We will introduce arrows to indicate the directions of the influences. Then, the selected processes can form a diagram of causal relationships, as it is depicted in Fig. 3.17, which we had agreed above to call a conceptual model of the ecosystem.

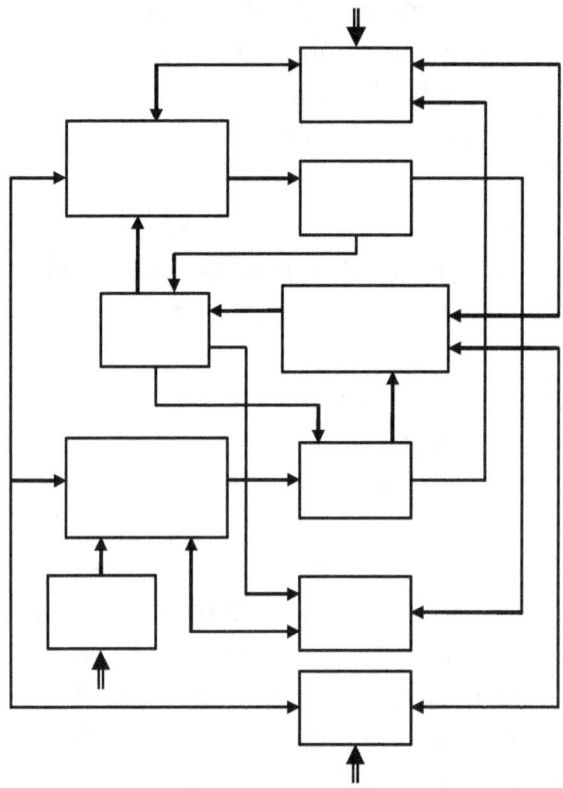

Fig. 3.17. Scheme of influences and management agents in the structure of
  conceptual model of the marine ecosystem

It is necessary to note that the inanimate objects of the conceptual model are directly related to each other, while the phytoplankton, zooplankton and biomass resources are influenced by management agents $A_{PP}(BG, SR, CD)$, $A_{ZP}(OX, PP, BG)$ and $A_{BR}(OX, ZP, BG)$. The difference in forms of interactions lies in the fact that the changes in the concentrations of inanimate objects (except for chemical reactions which are not considered here) can be represented as a weighted sum of concentrations increase of the other objects that influence them. What concerns the living objects, the changes in their concentrations must be subordinated to limiting operators, which retain only the influence of the process (at a given moment in time), which demonstrate minimal concentration compared with the other processes.

218

The operations of management agents represents changes in concentrations of aquatic species, which is necessary as an indicator of changes in resources' supply. Therefore, the processes affecting concentrations of living organisms, must be normalized to the portions (quanta) of resources support connected between them.

For example, the amount of dissolved oxygen in a given volume of the marine environment, which is a necessary living condition for some conventional aquatic organisms' population per time unit, represents oxygen provision for the population. Concentration of oxygen OX required for resource support of some conventional aquatic organisms, with the concentration ZP, can be represented as the product $OX = y_{ZP/OX} ZP$, in which $y_{ZP/OX}$ denotes the quantum – daily oxygen demand of representative of this type of aquatic organisms.

In order to simplify our study, we will measure concentration of environmental resources in such units. Therefore, the functions of management agents will be to find out the minimum of resources supply in each moment of time. To simulate the increments of phytoplankton concentrations PP we need to drive a comparison between available numbers of photons of solar radiation SR, concentrations of dissolved carbon dioxide CD, and nutrients BG, for increments of zooplankton ZP – concentrations of phytoplankton PP, oxygen OX and nutrients BG, for increments of bioresources BR – concentrations of zooplankton ZP, OX oxygen and nutrients BG.

In summary, we can formulate the following principle of modeling of resources-oriented development processes: the increment of resources-oriented process in a complex system should be determined by the type of development resources, which at this moment in time demonstrates minimal amount.

Formalization of the Conceptual Model Using Adaptive Balance of Causes. We will apply the method of adaptive balance of causes [181] in order to present the conceptual model of the ecosystem shown in Fig. 3.17, in the form of mathematical equations. The choice of this modeling approach can be justified by the stability

219

of solutions of equations with respect to the actions of management agents, which should produce changing of influencing functions in the right hand parts of equations.

Having said that, for formalization of the conceptual model in Fig. 3.17, it is sufficient to take standard dynamic equations of the *ABC*-model (2.16), and substitute x_i and $a_{ij}x_j$ in them with appropriate designations of processes in the ecosystem

$$\frac{dOX}{dt} = OX\{1 - 2[OX + a_{OX/BR}BR + a_{OX/ZP}ZP - a_{OX/PP}PP - F(WF)]\},$$

$$\frac{dCD}{dt} = CD[1 - 2(CD - a_{CD/BR}BR - a_{CD/ZP}ZP + a_{CD/PP}PP)],$$

$$\frac{dPP}{dt} = PP\{1 - 2[PP + a_{PP/ZP}ZP - A_{PP}(BG, SR, CD)]\},$$

$$\frac{dZP}{dt} = ZP\{1 - 2[ZP + a_{ZP/BR}BR - A_{ZP}(OX, PP, BG)]\}, \qquad (3.32)$$

$$\frac{dBR}{dt} = BR\{1 - 2[BR - A_{BR}(OX, ZP, BG)]\},$$

$$\frac{dBG}{dt} = BG[1 - 2(BG + a_{BG/BR}BR + a_{BG/ZP}ZP + a_{BG/PP}PP - BG_0)],$$

$$A_{PP}(BG, SR, CD) = a_{PP/BG}BG(t)A_{PP/BG}(t) + a_{PP/SR}SR(t)A_{PP/SR}(t)$$
$$+ a_{PP/CD}CD(t)A_{PP/CD}(t) \qquad (3.33)$$

$$a_{PP/BG}BG(t) = IF[M_{PP}(t) = BG(t); a_{PP/BG}BG(t); 0],$$

$$A_{PP/BG}(t) = IF[BG(t) < BG_c; -C_{PP/BG}(1 - e^{-\alpha_1 \tau_1}); C_{PP/BG}(1 - e^{-\alpha_2 \tau_2})],$$

$$a_{PP/SR}SR(t) = IF[M_{PP}(t) = SR(t); a_{PP/SR}SR(t); 0], \qquad (3.34)$$

$$A_{PP/SR}(t) = IF[SR(t) < SR_c; -C_{PP/SR}(1 - e^{-\alpha_3 \tau_3}); C_{PP/SR}(1 - e^{-\alpha_4 \tau_4})],$$

$$a_{PP/CD}CD(t) = IF[M_{PP}(t) = CD(t); a_{PP/CD}CD(t); 0]$$,

$$A_{PP/CD}(t) = IF[CD(t) < CD_c; -C_{PP/CD}(1 - e^{-\alpha_5 \tau_5}); C_{PP/CD}(1 - e^{-\alpha_6 \tau_6})]$$,

$$M_{PP} = \arg\min\{BG(t); SR(t); CD(t)\}$$, \hfill (3.35)

$$A_{ZP}(PP, OX, BG) = a_{ZP/PP}PP(t)A_{ZP/PP}(t) +$$
$$a_{ZP/OX}OX(t)A_{ZP/OX}(t) + a_{ZP/BG}BG(t)A_{ZP/BG}(t)$$, \hfill (3.36)

$$a_{ZP/PP}PP(t) = IF[M_{ZP}(t) = PP(t); a_{ZP/PP}PP(t); 0]$$,

$$A_{ZP/PP}(t) = IF[PP(t) < PP_c; -C_{ZP/PP}(1 - e^{-\alpha_7 \tau_7}); C_{ZP/PP}(1 - e^{-\alpha_8 \tau_8})]$$,

$$a_{ZP/OX}OX(t) = IF[M_{ZP}(t) = OX(t); a_{ZP/OX}OX(t); 0]$$,

$$A_{ZP/OX}(t) = IF[OX(t) < OX_c; -C_{ZP/OX}(1 - e^{-\alpha_9 \tau_9});$$
$$C_{ZP/OX}(1 - e^{-\alpha_{10} \tau_{10}})]$$, \hfill (3.37)

$$a_{ZP/BG}BG(t) = \arg\{IF[M_{ZP}(t) = BG(t); a_{ZP/BG}BG(t); 0]\}$$,

$$A_{ZP/BG}(t) = IF[BG(t) < BG_c; -C_{ZP/BG}(1 - e^{-\alpha_{12} \tau_{12}});$$
$$C_{ZP/BG}(1 - e^{-\alpha_{13} \tau_{13}})]$$,

$$M_{ZP} = \arg\min\{PP(t); OX(t); BG(t)\}$$, \hfill (3.38)

$$A_{BR}(ZP, OX, BG) = a_{BR/ZP}ZP(t)A_{BR/ZP}(t)$$
$$+ a_{BR/OX}OX(t)A_{BR/OX}(t) + a_{BR/BG}BG(t)A_{BR/BG}(t)$$, \hfill (3.39)

$$a_{BR/ZP}ZP(t) = IF[M_{BR}(t) = ZP(t); a_{BR/ZP}ZP(t); 0]$$,

$$A_{BR/ZP}(t) = IF[ZP(t) < ZP_c; -C_{BR/ZP}(1-e^{-\alpha_{14}\tau_{14}});$$
$$C_{BR/ZP}(1-e^{-\alpha_{15}\tau_{15}})]$$

$$a_{BR/OX}OX(t) = IF[M_{BR}(t) = OX(t); a_{BR/OX}OX(t); 0], \qquad (3.40)$$

$$A_{BR/OX}(t) = IF[OX(t) < OX_c; -C_{BR/OX}(1-e^{-\alpha_{16}\tau_{16}});$$
$$C_{BR/OX}(1-e^{-\alpha_{17}\tau_{17}})]$$

$$a_{BR/BG}BG(t) = IF[M_{BR}(t) = BG(t); a_{BR/BG}BG(t); 0],$$

$$A_{BR/BG}(t) = IF[BG(t) < BG_c; -C_{BR/BG}(1-e^{-\alpha_{18}\tau_{18}});$$
$$C_{BR/BG}(1-e^{-\alpha_{19}\tau_{19}})]$$

$$M_{BR} = \arg\min\{ZP(t); OX(t); BG(t)\}. \qquad (3.41)$$

The system of equations (3.32) is the *ABC*-model of dynamic processes in the integrated marine ecosystem, as shown in Fig. 3.17. The equations (3.33), (3.36) and (3.39) contain a combination of functions carried out by the control agents. These functions are expressed by logical operators (3.34) for phytoplankton, by operators (3.37) for zooplankton and by operators (3.40) for bioresources. The principle of resource-oriented development is executed by operators (3.35) for phytoplankton, (3.38) for zooplankton and (3.41) for bioresources. The parameters which are critical for development of organisms in the ecosystem were denoted as: SR_c, CD_c, PP_c, ZP_c, OX_c, BG_c.

To simplify the task of modeling, we used dimensionless (reduced) values that are associated with corresponding actual dimensional quantities A' in the following the dependence (2.22)

$$A = 5 A' [M(A')]^{-1},$$

where $[M(A')]$ is the average of the variability interval of the corresponding dimension value A'. This transformation restricts the variability of model scenarios of processes taking place in the marine ecosystem by the segment of values [0,10]. If necessary, we can easily carry our a return to initial (dimensional) values by the formula

$$A' = 0{,}2[\mathrm{M}\,(A')]A.$$

The model equations were solved numerically by 370 steps of the dimensionless time (a day was taken as a unit of time). To identify external influences on the model, we used function $SR(t)$, which imitated the annual course of daily averaged illumination of the sea surface; function $F(WF)$ imitated changes in concentration of dissolved oxygen due to external factors; and function BG_0, imitated the constant influx of nutrients. To identify the main external influence, the impact of the oxygen content scenario in the sea water was used, as shown in Fig. 3.18, *c*. The function $F(WF)$ was chosen in the form of harmonics, which had a large amplitude, in order to present more clearly the ecosystem's response to it, and to have the ability to analyze the impact of management agents on the development processes.

In experiments conducted with the ecosystem model under these external influences, two modes of management scenarios were investigated: behavior of the ecosystem near the border zones of homeostasis without regard to the resources' orientation of aquatic species, and formation of the resource-oriented scenarios, regardless of frontiers of the homeostasis zones. By dividing it into two separate modes we could study the effects of management by scenarios in each of the cases, separately.

Modeling the Variability of Processes Near the Boundaries of the Living Organisms' Homeostasis Zones. In the first series of experiments the influence of the homeostasis zones' boundaries was examined.

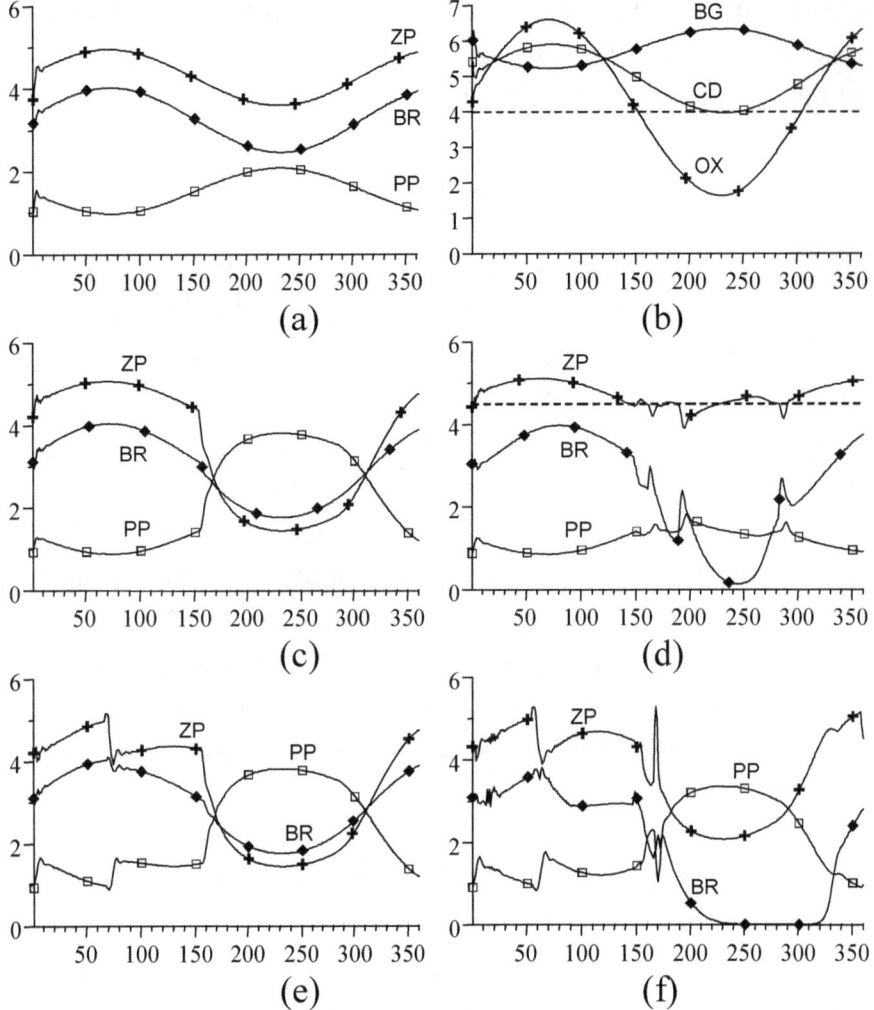

Fig. 3.18. The influence of the homeostasis zones' boundaries on the scenarios of marine organisms concentrations: (*a*) and (*c*) the original scenarios of processes without taking into account the conditions of extinction of

marine organisms; (*b*) extinction of zooplankton at the border zone of homeostasis on oxygen at the level of $OX_c = 4$; (*d*) extinction of

zooplankton at the border zone of homeostasis on phytoplankton at the level of $PP_c = 1$ and homeostasis on oxygen at the level of $OX_c = 4$;

(*e*) – extinction of bioresources at the border zone of homeostasis on zooplankton at the level of $ZP_c = 4,5$; (*f*) extinction of bioresources at the border zone of homeostasis on

224

zooplankton at the level of $ZP_c = 4{,}5$ and homeostasis on oxygen at the level of $OX_c = 4$. (CD – carbon

dioxide, OX – oxygen, BG – nutrients, PP – phytoplankton, ZP –

zooplankton, BR – bioresources)

Scenarios in Fig. 3.18, a and 3.18, c show the concentrations of living objects, carbon dioxide and nutrients, when the boundaries of homeostasis zones are not established. As it can be seen from the figures, the scenarios of zooplankton and bioresources smoothly follow the dynamics of the oxygen scenario, and the scenarios of phytoplankton, carbon dioxide and nutrients demonstrate smooth changes in agreement with the dynamics of their absorption by phytoplankton, zooplankton and bioresources.

The Fig. 3.18, b shows the same scenarios, taking into account the limits of the homeostasis on oxygen for a family of zooplankton, which was set at $OX_c = 4$. Other boundaries in this experiment were not set. With decreasing oxygen concentration to a critically low level, the agent $A_{ZP/OX}(t)$ from the group of formulas (3.36) included its function zooplankton extinction at a rate determined by parameters $C_{ZP/OX} = 5$ and $\alpha_9 = 0.7$ (50% for 10 days). The recovery rate of zooplankton concentration at the entrance to the zone of homeostasis was set up by choice of parameter $\alpha_{10} = 0.07$ (50% for 100 days).

The inclusion of a management agent led to significant change of scenarios, compared with Fig. 3.18, a. The zooplankton scenario obtained a further complication in the next experiment, when the border of homeostasis on phytoplankton at the level of $PP_c = 1$ was added to the border of homeostasis on oxygen at the level of $OX_c = 4$. The graphs, illustrating this case, are shown in Figs. 3.18, d.

Similar changes were observed in the experiments aimed at accounting of homeostasis zones for bioresources. Fig. 3.18, e shows the process of bioresources extinction at the border zone of homeostasis on zooplankton at the level of $ZP_c = 4{,}5$. Extinction speed control and renewal of bioresources concentration was conducted in the agent $A_{BR/ZP}(t)$ with the same values of parameters that were used in

the agent $A_{ZP/OX}(t)$ for zooplankton. Small fluctuations were observed in the scenario of bioresources when it was crossing the border zone of homeostasis, because reducing the concentration of bioresources restored the concentration of zooplankton for some period of time, to the values exceeding critical level of $ZP_c = 4,5$.

Another experiment was conducted under the condition that extinction of bioresources occurs at the boundary of the homeostasis zone, at the level of $ZP_c = 4,5$ for zooplankton. The next condition was set to provide that extinction of zooplankton, which also entails reduction in population of bioresources, occurs at the boundary of the zone of homeostasis on oxygen at the level of $OX_c = 4$. The results presented in Fig. 3.18, f, show a sharp decrease in bioresources concentration, which began to recover with delay, compared to the scenario of zooplankton concentration.

Application of Management Agents to Accounting and Management of Resources Constraints of Marine Organisms Development. The second series of experiments was conducted with the same input influences on the ecosystem, as the first one, except for the concentration of phytoplankton, the mean value of which was increased by 2 dimensionless units. The given scenario of oxygen (curve 2) and light (curve 4) are shown in Fig. 3.19, a. Initially, the operations of resources constraints (3.34), (3.37) and (3.40) were stopped. The model reproduced the development processes' adaptation to the external influences and to each other, as depicted in Fig. 3.19, a and 3.19, b.

As scenario of oxygen concentration was used as the dominant influence on concentrations of zooplankton and bioresources, ZP and BR curves in Fig. 3.19, b are smoothly tracking this scenario. In the absence of logical conditions (3.34), the scenario of phytoplankton concentration is formed as a weighted sum of the effects for concentrations of carbon dioxide, nutrients and luminosity level of the sea (curve CD, BG, and SR in Fig. 3.19, a). Since resource constraints were turned off, simultaneous account of all influences led to formation of maximum phytoplankton concentration in the vicinity of 230 steps of calculations.

Then the logical conditions (3.34), (3.37) and (3.40) were turned on, and the model reproduced the resource-oriented scenarios of phytoplankton, zooplankton and bioresources' concentrations. The effect of resource limitation becomes noticeable if we compare the respective graphs shown in Fig. 3.19, *b* and 3.19, *d*. Comparing phytoplankton scenarios with each other (*PP* curves in Fig. 3.19, *b* and 3.19, *d*, we can see that resources limitations, carried out by the agents $a_{PP/BG}BG(t)$, $a_{PP/SR}SR(t)$ and $a_{PP/CD}CD(t)$, led to a change from maximum phytoplankton concentration to its minimum in the neighborhood of 230 steps of calculations.

Significant changes in the *PP*-graph occurred in the first half-time of computing. In the neighborhood of step 70, the *PP*-graph ceased to grow (see *PP* curve in Fig. 3.19, *d*). Previously, it had been a consequence of oxygen scenario domination. During this interval of time, the management agents $a_{PP/SR}SR(t)$ and $a_{PP/BG}BG(t)$ targeted variability of phytoplankton concentration, first at the illumination minimum (see the curve of *SR* in Fig. 3.19, *a*), and then at the minimum of nutrients (see *BG* curve in Fig. 3.19, *a*) because exactly they were limiting sorts of resources for phytoplankton. Almost constant values took the graphs of zooplankton and bioresources concentrations (see the curves *ZP* and *BG* in Fig. 3.19, *d*). At the same time, the management agents $a_{ZP/BG}BG(t)$ and $a_{BR/BG}BG(t)$ directed their variability to the minimum of nutrient.

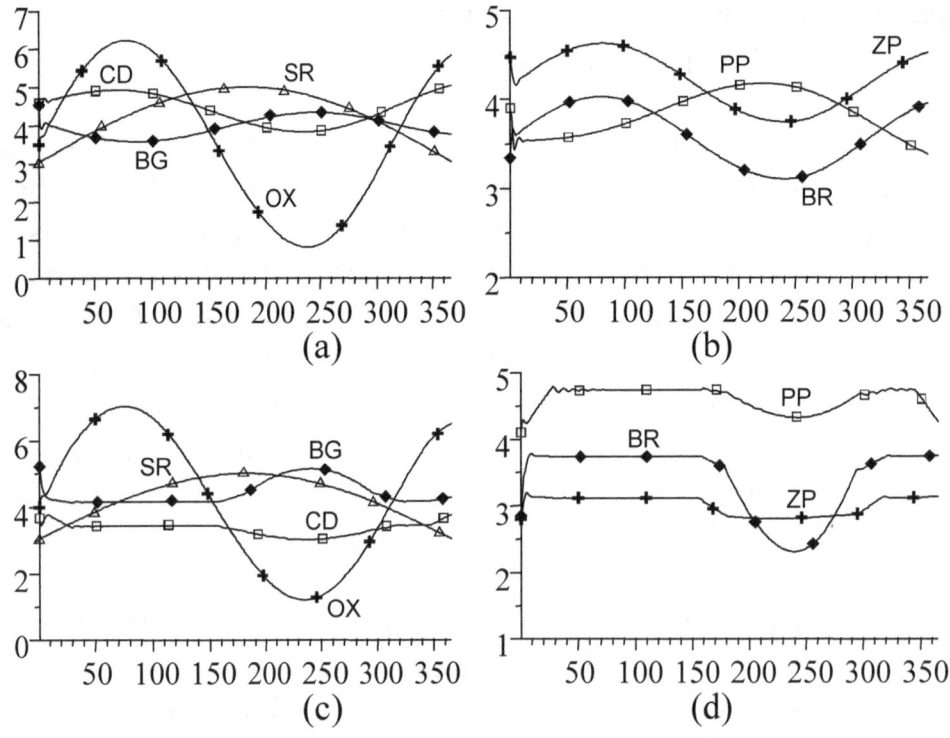

Fig. 3.19. Formation of scenarios of marine organisms' concentrations under the influence of resources limitation of development processes implemented by management agents: (*a*) and (*b*) scenarios of processes in the eco system without resources limitation; (*c*) and (*d*) resources limitation of phytoplankton concentration by the agent $A_{PP}(CD,SR,BG)$, zooplankton concentration by the agent $A_{ZP}(PP,OX,BG)$; (*d*) resources limitation of bioresources concentration by the agent $A_{BR}(ZP,OX,BG)$. (*SR* – the sea surface illumination, *CD* – carbon dioxide, *OX* – oxygen, *BG* – nutrients, *PP* – phytoplankton, *ZP* – zooplankton, *BR* – bioresources)

Results of computational experiments confirmed the conclusion that in models of marine ecosystems it is essential to take into account the border zones of homeostasis and the resource limitation of concentrations of living organisms. The proposed formal algorithms of the resources oriented scenarios represent logical control operators that switch on influence functions in the right-hand parts of the ecosystem's model equations, tracking the volume of available resources and the homeostasis zones boundaries of living objects.

It should be noted that, despite substantial nonlinearity, which is imposed on the model by such management functions, the structure of the equations used in the method of adaptive balance of causes is capable to ensure the sustainability of solutions. The presence of large number of parameters in the formulas of management agents makes it possible to take into account the rate of extinction and recovery of populations, which in turn, allows to conduct experiments with model ecosystems under different conditions of existence of marine organisms.

3.5. Building Scenarios of Processes Considering Resources Limitations

The processes of resources limitation can affect not only the concentrations of living organisms in marine ecosystems, but also the formation of new substances derived during chemical reactions. As a part of the integral description of these phenomena, it makes sense to consider formation of nutrients during the oxidation of detritus, since the dying off organisms produce detritus and living organisms need nutrients for their existence. Oxidation detritus' processes reduce the concentration of oxygen in sea water, which in turn significantly affects the conditions of existence for zooplankton and bioresources, as noted in the previous section. Therefore, taking into account the limitation of nutrients, should lead to more complex scenarios of development in marine ecosystems. To predict these scenarios we will use a conceptual model of the ecosystem shown in Fig. 3.20.

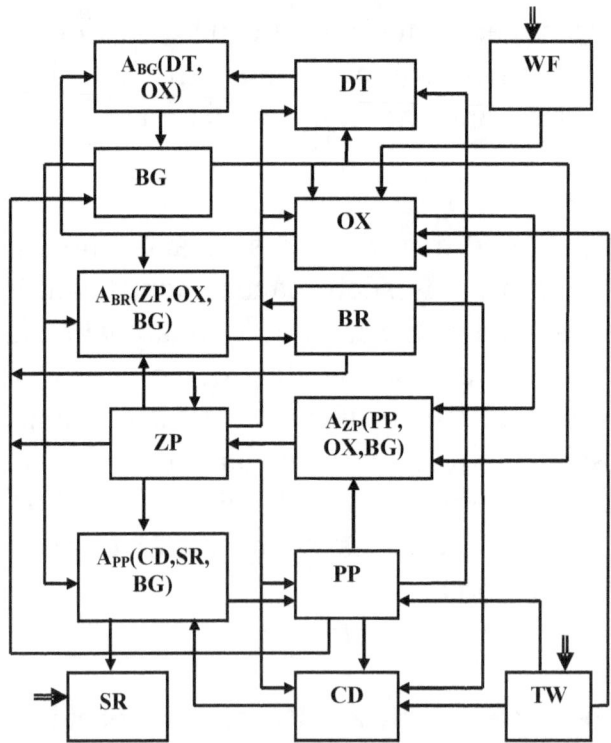

Fig. 3.20. Conceptual model of ecosystem, taking into account the limitation on the concentration of nutrients

This model is constructed by adding to the model shown in Fig. 3.17, *a* parameter of detritus *DT* concentration and a management agent, which limit the concentration of biogenes by oxygen and detritus. In addition to this, the new model takes into account the effect of marine water temperature *TW* on the concentration of phytoplankton, as well as on the concentration of dissolved oxygen and carbon dioxide. It is known that with increase of temperature, concentration of dissolved gases in sea water decreases.

Using the *ABC*-method, we can obtain the following dynamic equations of the ecosystem's model

$$\frac{dOX}{dt} = OX\{1 - 2[OX + a_{OX/BR}BR + a_{OX/ZP}ZP -$$
$$a_{OX/PP}PP + a_{OX/BG}BG + a_{OX/BG}TW - F_{OX}(WF)]\}$$

$$\frac{dCD}{dt} = CD[1 - 2(CD - a_{CD/BR}BR - a_{CD/ZP}ZP + a_{CD/PP}PP + a_{CD/PP}TW)]$$

$$\frac{dPP}{dt} = PP\{1 - 2[PP + a_{PP/ZP}ZP - A_{PP}(BG,SR,CD) - A_{PP}(TW)]\},$$

$$\frac{dZP}{dt} = ZP\{1 - 2[ZP + a_{ZP/BR}BR - A_{ZP}(OX,PP,BG)]\},$$

(3.42)

$$\frac{dBR}{dt} = BR\{1 - 2[BR - A_{BR}(OX,ZP,BG)]\},$$

$$\frac{dBG}{dt} = BG\{1 - 2[BG - A_{BG}(OX,DT) + a_{BG/PP}PP + a_{BG/ZP}ZP + a_{BG/BR}BR)]\}$$

$$\frac{dDT}{dt} = DT[1 - 2(DT - a_{DT/BR}BR - a_{DT/ZP}ZP - a_{DT/PP}PP + a_{DT/BG}BG)]$$

$$A_{BG}(OX,DT) = IF\{M_{BG}(t) = a_{BG/OX}OX(t); IF\{OX(t) < OX_{BG}*;$$

$$-OX_{BG}[1 - \exp(-\alpha_{BG}\tau)]; OX_{BG}[1 - \exp(-\beta_{BG}\tau)]\}; 0\} +$$

(3.43)

$$IF\{M_{BG}(t) = a_{BG/DT}DT(t); IF\{DT(t) < DT_{BG}*;$$

$$-DT_{BG}[1 - \exp(-\alpha_{DT}\tau)]; DT_{BG}[1 - \exp(-\beta_{DT}\tau)]\}; 0\}$$

$$M_{BG}(t) = \arg\min[a_{BG/OX}OX(t); a_{BG/DT}DT(t)],$$

$$A_{PP}(BG, SR, CD) = IF\{M_{PP}(t) = a_{PP/BG}BG(t); IF\{BG(t) < BG_{PP}^*;$$

$$-BG_{PP}[1 - \exp(-\alpha_{PP}\tau)]; BG_{PP}[1 - \exp(-\beta_{PP}\tau)]\}; 0\}$$
$$+$$
$$IF\{M_{PP}(t) = a_{PP/SR}SR(t); IF\{SR(t) < SR_{PP}^*;$$

$$-SR_{PP}[1 - \exp(-\alpha_{SR}\tau)]; SR_{PP}[1 - \exp(-\beta_{SR}\tau)]\}; 0\} +$$
(3.44)

$$IF\{M_{PP}(t) = a_{PP/CD}CD(t); IF\{CD(t) < CD_{PP}^*;$$

$$-CD_{PP}[1 - \exp(-\alpha_{CD}\tau)]; CD_{PP}[1 - \exp(-\beta_{CD}\tau)]\}; 0\}$$

$$M_{PP}(t) = \arg\min[a_{PP/BG}BG(t); a_{PP/SR}SR(t); a_{PP/CD}CD(t)],$$

$$A_{ZP}(OX, PP, BG) = IF\{M_{ZP}(t) = a_{ZP/OX}OX(t); IF\{OX(t) < OX_{ZP}^*;$$

$$-OX_{ZP}[1 - \exp(-\alpha_{ZP}\tau)]; OX_{ZP}[1 - \exp(-\beta_{ZP}\tau)]\}; 0\} +$$

$$IF\{M_{ZP}(t) = a_{ZP/PP}PP(t); IF\{PP(t) < PP_{ZP}^*;$$

$$-PP_{ZP}[1 - \exp(-\alpha_{PP}\tau)]; PP_{ZP}[1 - \exp(-\beta_{PP}\tau)]\}; 0\} +$$

$$IF\{M_{ZP}(t) = a_{ZP/BG}BG(t); IF\{BG(t) < BG_{ZP}^*;$$

$$-BG_{ZP}[1 - \exp(-\alpha_{BG}\tau)]; BG_{ZP}[1 - \exp(-\beta_{BG}\tau)]\}; 0\}$$

(3.45)

$$M_{ZP}(t) = \arg\min[a_{ZP/OX}OX(t); a_{ZP/PP}PP(t); a_{ZP/BG}BG(t)],$$

$$A_{BR}(ZP,OX,BG) = IF\{M_{BR}(t) = a_{BR/ZP}ZP(t); IF\{ZP(t) < ZP_{BR}*;$$

$$-ZP_{BR}[1-\exp(-\alpha_{ZP}\tau)]; ZP_{BR}[1-\exp(-\beta_{ZP}\tau)]\}; 0\} +$$

$$IF\{M_{BR}(t) = a_{BR/OX}OX(t); IF\{OX(t) < OX_{BR}*;$$

$$-OX_{BR}[1-\exp(-\alpha_{OX}\tau)]; OX_{BR}[1-\exp(-\beta_{OX}\tau)]\}; 0\} +$$

$$IF\{M_{BR}(t) = a_{BR/BG}BG(t); IF\{BG(t) < BG_{BR}*;$$

$$-BG_{BR}[1-\exp(-\alpha_{BG}\tau)]; BG_{BR}[1-\exp(-\beta_{BG}\tau)]\}; 0\}$$
$$M_{BR}(t) = \arg\min[a_{BR/OX}OX(t); a_{BR/ZP}ZP(t); a_{BR/BG}BG(t)],$$

$$A_{PP}(TW) = \exp[-\chi_{PP}(T-T_{PP})^2],$$

(3.46)

The way to describe limitation operations and accounting of homeostasis zones which was proposed above, includes time delays (the time needed for the growth of living organisms). The settings of the ecosystem's parameters which are critical to the development of living organisms, are identified as: $SR*$, $CD*$, $PP*$, $ZP*$, $OX*$, $BG*$, $DT*$.

The results of two computational experiments with the model are shown in Fig. 3.21 and 3.22. To construct scenarios of processes in the ecosystem, the external influences (their graphs are shown in Fig. 3.21, a) have been applied to it. Calculations were made for 370 time steps (days). The following external influences were simulated: the annual temperature variation of the sea upper layer (curve TW), the illumination of its surface (curve SR) and the modulus of wind velocity at the sea surface (curve WF). As the graph of TW the average long-term temperature variations were simulated, whereas for the graphs of illumination and wind speed, random variations of these parameters around the long-term annual cycles were used, which helped imitate the effects of current weather conditions.

All simulated processes were brought to a common dimensionless scale of their variability [0, 10] by the formulas (2.22).

233

Influence coefficients in the dynamic model (3.42) were chosen within the segment values [0,1; 0,5]. Rates of change for concentrations were taken equal for all the ecosystem's parameters: $\alpha = 0,1$ and $\beta = 1,1$.

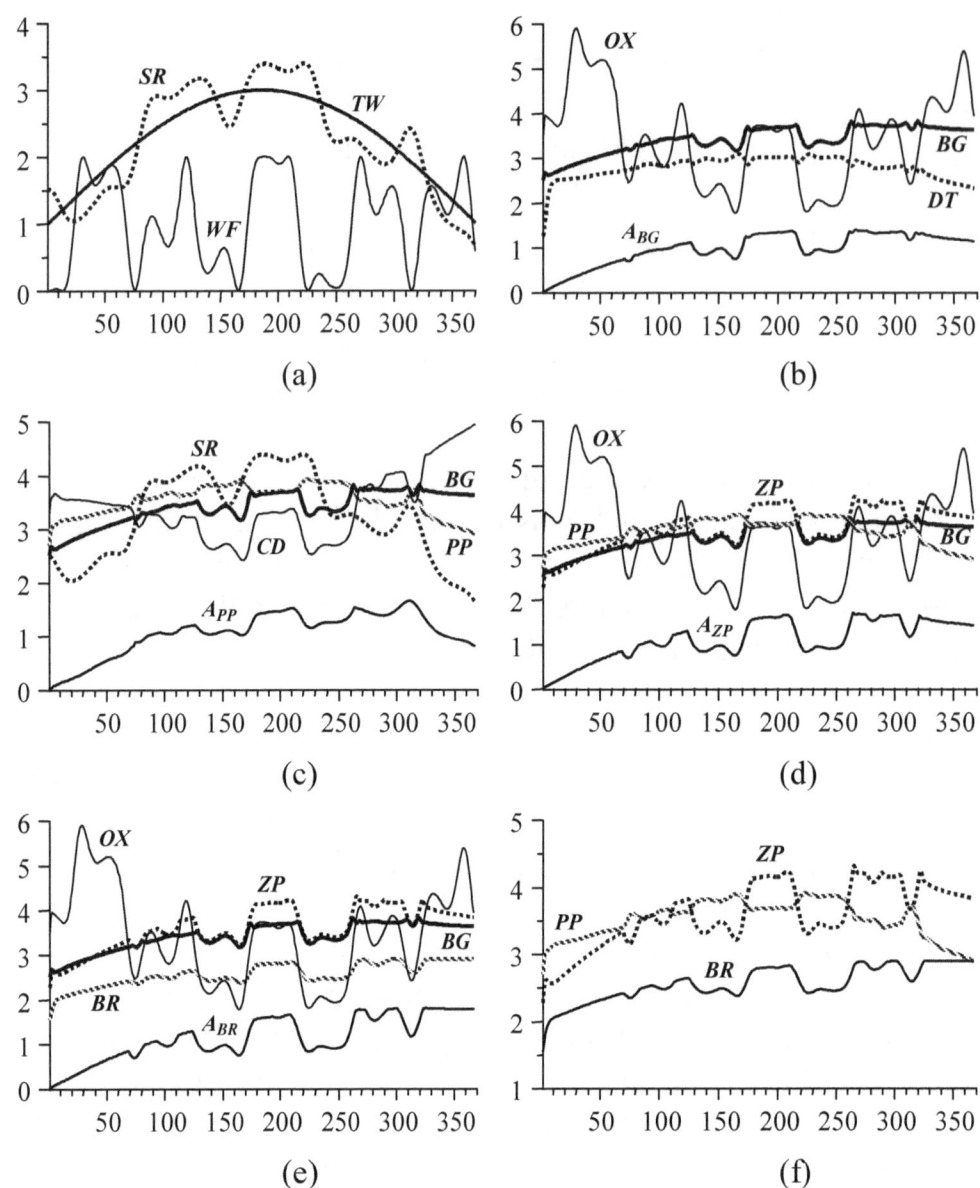

Fig. 3.21. Scenarios of processes in the marine ecosystem, obtained with

234

accounting of resources' constraints for development of marine
organisms.

In the first series of experiments (the results are shown in Fig. 3.21), the impact of resource limitation living organisms was analyzed, as well as the conditions of nutrients formation from detritus were studied. The influence of homeostasis zones' boundaries was excluded by equating the critical values (lower bounds) of the ecosystem's parameters to zero: $SR^* = CD^* = PP^* = ZP^* = OX^* = BG^* = DT^* = 0$. Under this condition, all parameters in equations for agents (3.43) - (3.46) surpass by values the lower boundaries of the zone of homeostasis.

Limiting operations performed by agents (3.43) - (3.46), are demonstrated with graphs in Fig.3.21. In particular, Fig. 3.21, b clearly shows that the scenario of nutrients concentration is formed by the graph of the management agent A_{BG} , which chooses the minimum values of the scenario of oxygen OX and detritus DT. In its turn, as it follows from Fig. 3.21, e, a scenario of nutrients concentration affects the graph of the management agent A_{BR}, which controls concentration of bioresources. It is noticeable, for example, at the end of the A_{BR} graph, when nutrients concentration was lower then oxygen concentration OX and zooplankton concentration ZP. The tendency of growth over time which is common for majority of the scenarios, is due to the setting of large parameter of development delay $\beta = 1,1$.

In the second series of calculations, the scenarios of processes in the ecosystem's model (3.42) were calculated in view of the lower boundaries of the homeostasis zones for living organisms, as well as for the chemical and bacterial reactions of nutrients formation from detritus. The results of calculations are shown in Fig. 3.22 and Fig. 3.23.

Calculations were performed with the same values of influence coefficients and model parameters as in the first series, but the following values of the lower boundaries of homeostasis zones were additionally set: $OX_{BG}{}^* = 2,5$ in the agent $A_{BG}(OX, DT)$ that managed nutrients concentration on oxygen, $BG_{PP}{}^* = 2,5$ in the agent $A_{PP}(BG, SR, CD)$, that managed phytoplankton concentration on nutrients, $OX_{ZP}{}^* = 2,0$ in the agent $A_{ZP}(OX, PP, BG)$ that managed zooplankton concentration on oxygen and $OX_{BR}{}^* = 2,5$ in the agent $A_{BR}(ZP, OX, BG)$ that managed bioresources concentration on oxygen.

The Inclusion in calculations of the homeostasis zones significantly changed the scenarios of processes. This became evident by comparison of graphs of management agents in nutrients (Fig. 3.14, b and Fig. 3.23, a), management agents in phytoplankton (Fig.3.21, c and Fig. 3.22, b), management agents in zooplankton (Fig. 3.21, d and Fig. 3.22, c) and management agents in bioresources (Fig. 3.21, e and Fig. 3.23, d).

Some significant changes also occurred in scenarios of phytoplankton and zooplankton concentrations. A comparison of Fig. 3.21, f with Fig. 3.23, f shows that the amplitude of these concentrations increased in 2 – 3 times. Variability of bioresource scenarios also markedly increased.

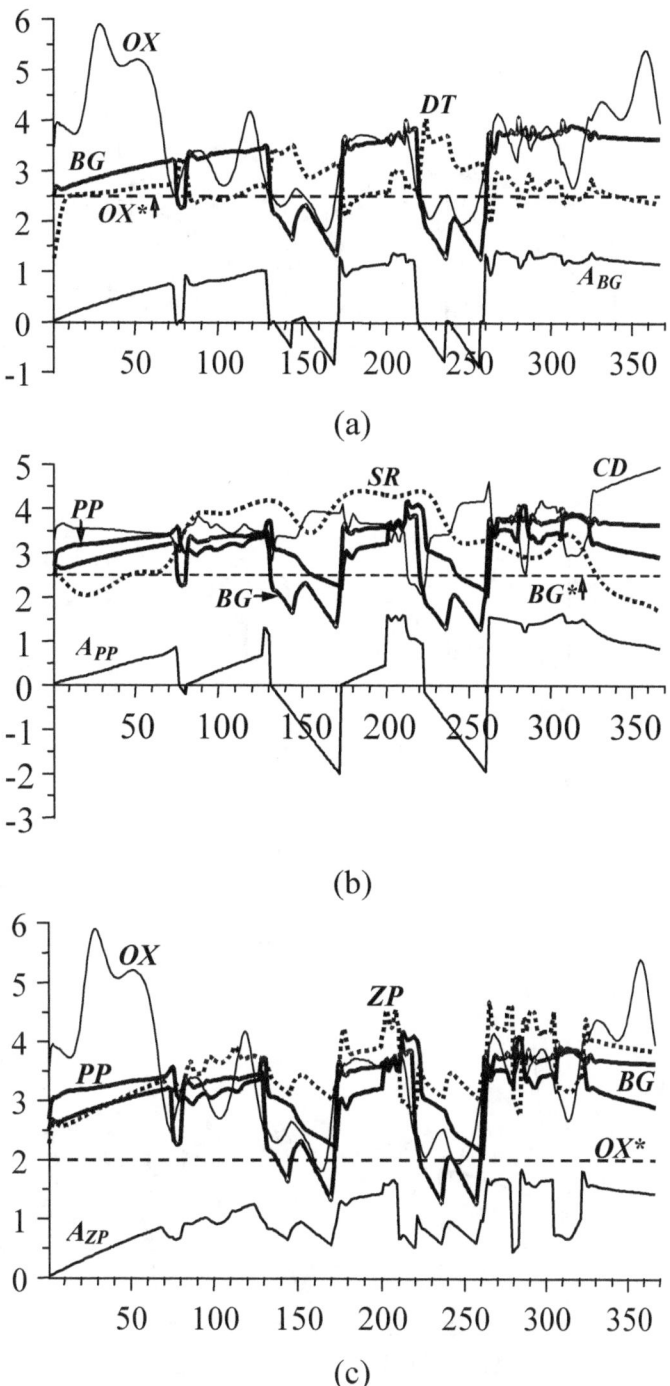

Fig. 3.22. The increasing complexity of scenarios for more explicit consideration   of marine organisms' homeostasis borders.

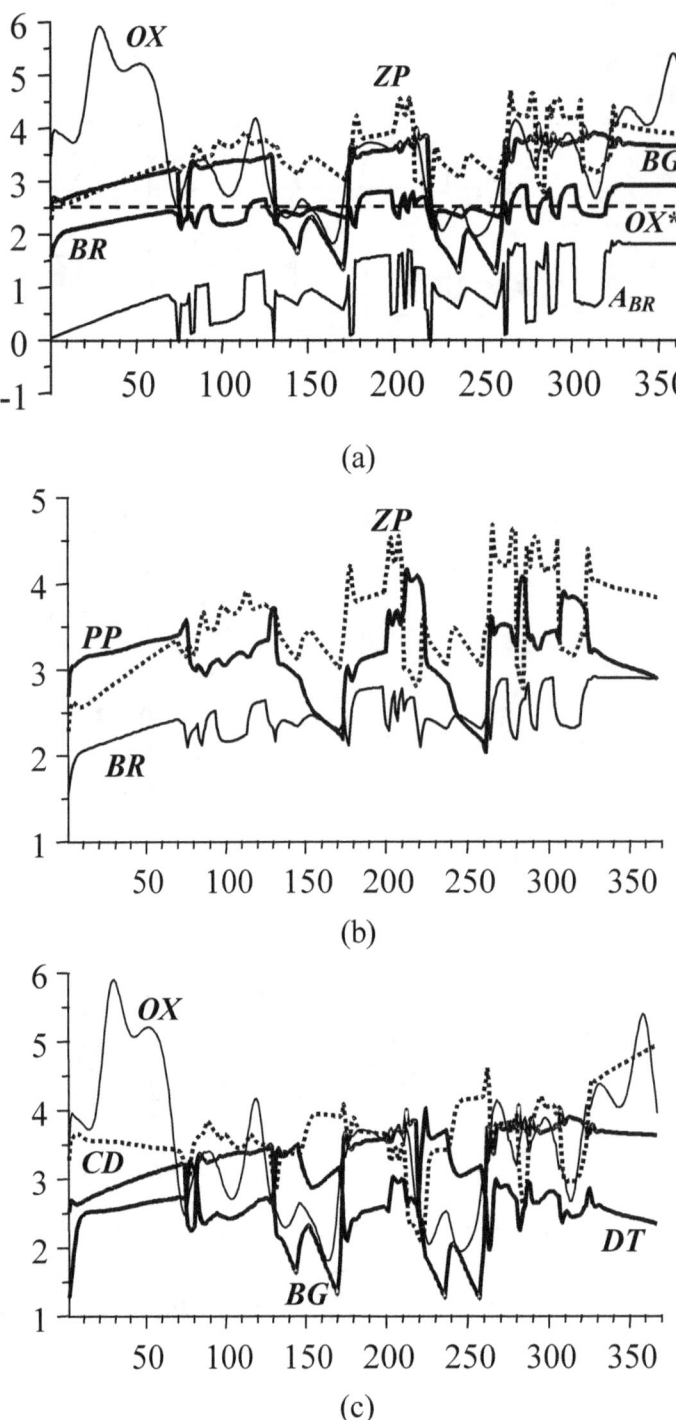

Fig. 3.23. The further increase of scenarios complexity for more explicit

consideration of resources limitations for chemical and bacterial reactions of nutrients and detritus formation

After comparing of the obtained scenarios between each other, we can make a conclusion that taking into account resources constraints and areas of homeostasis in models of the ecosystem leads to radical changes in development of these scenarios. Therefore, to ensure balance of consumption and reproduction of marine bioresources, it is necessary to use management agents in ecological models of the coastal zone area.

3.6. Modeling of Adaptive Balance of Processes in Coastal Zone Marine Ecosystem

The purpose of this section is to perform modeling of adaptive balance of development processes in a generalized model of a marine ecosystem, which includes control agents in its set of equations. Similarly to the earlier examples of the *ABC*-models, the chemical-biological processes in the marine environment will be presented in an aggregated form, i.e. without going into detailed study of living objects and their life cycles. However, the basic processes of coexistence and transformation of the living and nonliving matter will be considered in their interdependence, which is determined by logical rules implemented by the agents.

Frequent switching of the model coefficients from one value to another, which should be done by the management agents, complicates the solution of traditional systems of dynamic equations. Bringing stability to these systems has always been one of the most difficult problems of mathematical biology of the sea [107, 117]. As it was shown by computer experiments which we discussed above, the *ABC*-method of ecosystem models construction ensures stability of solutions, due to special modular construction of the method's equations.

Based on the goals set for this part of the research, let us select a few basic hydrochemical and hydrobiological processes that should be included in the model of marine ecosystem. All the living organisms naturally inhabiting the marine environment will be characterized by three general classes: phytoplankton PP, zooplankton ZP and bioresources BR. For the PP process representing phytoplankton, we will consider changes in its biomass per unit volume, as well as the process ZP will be considered as the changes in zooplankton biomass. With regard to bioresources BR, its value will represent the biomass of all other living organisms, including fish and other commercial species which justify the name "bioresources".

For further use, let us introduce processes that ensure the existence of these classes of living objects. For phytoplankton they are solar radiation SR and content of chemical compounds, necessary for photosynthesis and for representing products of decomposition of organic matter. Let us define the following processes, which are vital for existence of phytoplankton concentrations – biogenic elements: inorganic nitrogen (NO_2, NO_3, NH_4) and phosphorus PO_4. We need to add carbon dioxide CO_2 to these compounds, as one of the most important elements of the carbon cycle.

Conditions of existence of zooplankton are mainly determined by concentrations of phytoplankton and dissolved oxygen OX. For biological resources zooplankton, phytoplankton and oxygen are needed. However, in conditions of aggregation of the model's components, we will only consider the classical food chain: phytoplankton ☐ zooplankton ☐ bioresources.

We will also take into account traditional scheme of organic matter transformation, in which the dying organisms become sources of weighted particulate organic nitrogen PON, carbon POC and phosphorus POP. Particulate phase partly transfers into dissolved phase, forming the concentrations of dissolved organic nitrogen DON, carbon DOC and phosphorus DOP. Then, at the final stage of organic matter transformation mineral (inorganic) forms of nitrogen NH_4, NO_2, NO_3 and phosphorus PO_4^3 are formed in the stoichiometric ratio $N{:}P = 16{:}1$, according to the scheme of Horn [69].

$$(CH_2O)_{106}(NH_3)_{16}H_3PO_4 + 138O_2 \rightarrow 106CO_2 + 122H_2O + 16HNO_3 + H_3PO_4$$

These processes are subject to experience active impact of external influence factors. Besides solar radiation which we have introduced above, the driving sea-surface wind WF also affects the ecosystem. It increases the content of oxygen in the upper mixed layer. Temperature TW also has a noticeable effect on majority of the processes in the sea water ecosystem. Finally, freshwater runoff RF serves as an additional supplier of organic matter and nutrients NH_4, NO_2, NO_3, PO_4^3. A related inflow of pollutants PL into the sea affects their concentration in the marine environment. In general, freshwater runoff RF affects salinity of the sea water SA, and the content of nutrients, which (in a given volume of the marine environment) is also influenced by vertical displacement of water- masses VM. The latter can either deprive of, or enrich the whole volume of the researched marine environment with nutrients. The same phenomenon occurs at the horizontal water-masses exchange, which, however, we do not take into account in order to simplify the model.

A conceptual model of the ecosystem is a scheme of mutual influences of selected processes, which is shown in Fig. 3.24.

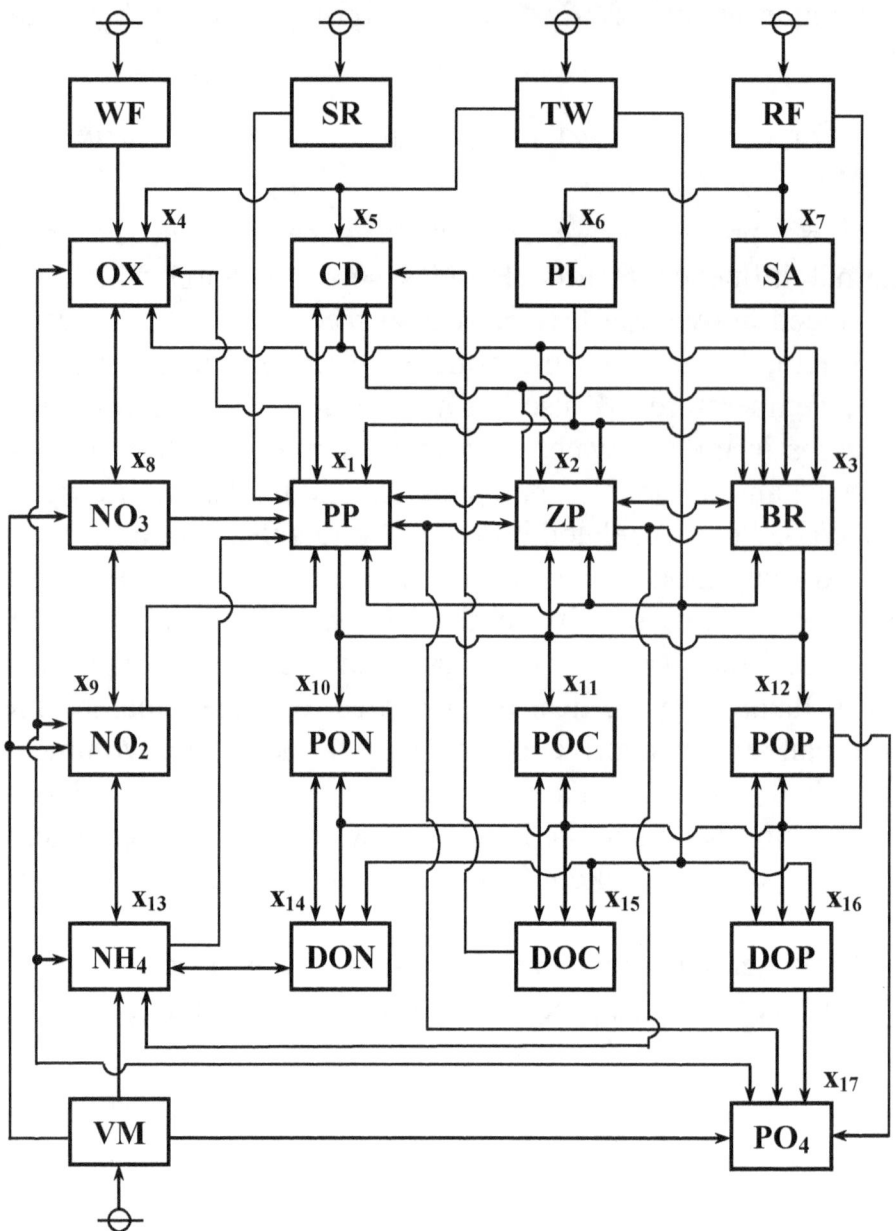

Fig. 3.24. Conceptual model of the marine ecosystem

At the moment of designing the scheme, expert information about modeling of natural phenomena should be used. This is information about the processes usually occurring in the selected volume of marine environment, that corresponds with the established scales of averaging or aggregation of the processes. Some directions of arrows on this scheme correspond to the directions of influences.

Therefore, an arrow with two possible directions, specifies mutual influences of processes on each other. Among the lines in the scheme, there are some (e.g., going from the block water temperature *TW*), that depict simultaneous impact of a process on a large number of other processes. For simplification of the scheme, such effects are shown in the form of main lines having branches.

The scheme of the conceptual model does not contain signs indicating positive or negative influences of the processes on each other. They are omitted in the diagram of conceptual model, since it is not difficult to establish the nature and sings of these influences. As we did above, we will agree to consider an influence as a positive one if an increase (decrease) in value of the cause gives a corresponding increase (decrease) in value of the effect. In the opposite case, we assume a negative impact. Examples of positive influences on the scheme are the cause-effect relationships *WF − OX* and *SR − PP*, and negative influences are represented by relationships *BR − ZP* and *ZP − PP*.

The conceptual model of the ecosystem, built with understanding of mutual influences, reflects the overall balance of the development processes' trends in the system. Such balance in the models of complex systems is supported by chains of positive and negative feedback. For example, a positive feedback which seeks to increase concentration of zooplankton in the marine environment is provided with the closed chain:

$$ZP - [PON, POC, POP] - [DON, DOC, DOP] - [NH_4, NO_3, NO_2, PO_4] - PP - ZP$$

This tendency is balanced by a closed chain of negative feedback: *ZP − PP − ZP*, which provides adaptation of zooplankton concentration to the concentration of phytoplankton.

As it was noted above, the principle is the fact that the growth and decrease in concentrations of biological objects depend on a number of factors at the same time. In order to take into account the limiting factors in a model of the ecosystem, as it was known from the previous section, it must be provided with some logic tracking operations for quantities of resources needed for vital functions of biological objects. Therefore, along with the equations describing the dynamics of phytoplankton, zooplankton, and several other processes, some intelligent agents that control growth and decay of proper concentrations should be included into the model.

The conceptual model of the ecosystem (Fig. 3.27) was designed with proper attention to the considerations made above. It was then formalized by the method of adaptive balance of causes (*ABC*-method) [181]

$$
\frac{dx_1}{dt} = x_1 \{ 1 - AD_{15} - AD_{18} - AD_{19} - AD_{1/13} - AD_{1/17} - AD_{1/SR} - \\
2[x_1 - AQ(x_5, x_8, x_9, x_{13}, x_{17}, SR) + a_{12}x_2 + a_{16}x_6 - a_{1/TW}TW]\},
$$

$$
\frac{dx_2}{dt} = x_2 \{ 1 - AD_{21} - AD_{24} - 2[x_2 - AQ(x_1, x_4) + a_{23}x_3 + a_{26}x_6 - a_{2/TW}TW]\},
$$

$$
\frac{dx_3}{dt} = x_3 \{ 1 - AD_{32} - AD_{34} - 2[x_3 - AQ(x_2, x_4) + a_{37}x_7 + a_{36}x_6 - a_{3/TW}TW]\},
$$

$$
\frac{dx_4}{dt} = x_4 \{ 1 - AD_{4/TW} - 2[x_4 - a_{41}x_1 + a_{42}x_2 + a_{43}x_3 + a_{48}x_8 + \\
a_{49}x_9 + a_{45}x_5 + a_{4/17}x_{17} - a_{4/WF}WF + a_{4/TW}TW]\},
$$

$$
\frac{dx_5}{dt} = x_5 \{ 1 - 2[x_5 - AP(x_4, x_{15}) + a_{51}x_1 - a_{52}x_2 - a_{53}x_3 - a_{5/TW}TW]\},
$$

$$
\frac{dx_6}{dt} = x_6 [1 - 2(x_6 - a_{6/RF}RF)],
$$

$$
\frac{dx_7}{dt} = x_7 [1 - 2(x_7 - a_{7/RF}RF)],
$$

$$
\frac{dx_8}{dt} = x_8 \{ 1 - 2[x_8 - AP(x_4, x_9) + a_{81}x_1 - a_{8/VM}VM]\},
$$

$$\frac{dx_9}{dt} = x_9\{1 - 2[x_9 - AP(x_4, x_{13}) + a_{91}x_1 - a_{9/VM}VM]\},$$

$$\frac{dx_{10}}{dt} = x_{10}[1 - 2(x_{10} - a_{10/1}x_1 - a_{10/2}x_2 - a_{10/3}x_3 + a_{10/14}x_{14} - a_{10/RF}RF)], \quad (3.47)$$

$$\frac{dx_{11}}{dt} = x_{11}[1 - 2(x_{11} - a_{11/1}x_1 - a_{11/2}x_2 - a_{11/3}x_3 + a_{11/15}x_{15} - a_{11/RF}RF)],$$

$$\frac{dx_{12}}{dt} = x_{12}[1 - 2(x_{12} - a_{12/1}x_1 - a_{12/2}x_2 - a_{12/3}x_3 + a_{12/12}x_{12} - a_{12/RF}RF)],$$

$$\frac{dx_{13}}{dt} = x_{13}\{1 - 2[x_{13} - AP(x_4, x_{14}) + a_{13/1}x_1 - a_{13/2}x_2 - a_{13/3}x_3 - a_{13/VM}VM)],$$

$$\frac{dx_{14}}{dt} = x_{14}[1 - 2(x_{14} - a_{14/10}x_{10} - a_{14/RF}RF - a_{14/TW}TW)],$$

$$\frac{dx_{15}}{dt} = x_{15}[1 - 2(x_{15} - a_{15/11}x_{11} - a_{15/RF}RF - a_{15/TW}TW)],$$

$$\frac{dx_{16}}{dt} = x_{16}[1 - 2(x_{16} - a_{16/12}x_{12} - a_{16/RF}RF - a_{16/TW}TW)],$$

$$\frac{dx_{17}}{dt} = x_{17}\{1 - 2[x_{17} - AP(x_4, x_{16}) - a_{16/RF}RF - a_{17/VM}VM)].$$

These equations for state-vector components of the ecosystem contain the following notations: x_1 – concentration of phytoplankton PP, x_2 – concentration of zooplankton ZP, x_3 – concentration of bioresources BR, x_4 – dissolved oxygen OX, x_5 – carbon dioxide CD, x_6 – concentration of pollutants PL, x_7 – marine salinity SA, x_8 – content of inorganic nitrogen in the form of NO_3, x_9 – substance of inorganic nitrogen in the form of NO_2, x_{10} – content of particulate organic nitrogen PON, x_{11} – content of particulate organic carbon POC, x_{12} – content of suspended organic phosphorus POP, x_{13} – content of ammonium NH_4, x_{14} – concentration of dissolved organic nitrogen DON, x_{15} – concentration of dissolved organic carbon DOC,

245

x_{16} – concentration of dissolved organic phosphorus *DOP*, x_{17} – content of inorganic phosphorus in form of PO_4.

The terms of the form *AD* in the right-hand sides of the equations (3.47) represent the first group of intelligent management agents. They have the following expressions:

$$AD_{15} = IF\{x_5 > x_5{}^*; 0; C_{15}[1 - \exp(-\alpha_{15}\tau)]\},$$

$$AD_{18} = IF\{x_8 > x_8{}^*; 0; C_{18}[1 - \exp(-\alpha_{18}\tau)]\}, \qquad (3.48)$$

$$AD_{19} = IF\{x_9 > x_9{}^*; 0; C_{19}[1 - \exp(-\alpha_{19}\tau)]\},$$

$$AD_{1/13} = IF\{x_{13} > x_{13}{}^*; 0; C_{1/13}[1 - \exp(-\alpha_{1/13}\tau)]\},$$

$$AD_{1/17} = IF\{x_{17} > x_{17}{}^*; 0; C_{1/17}[1 - \exp(-\alpha_{1/17}\tau)]\},$$

$$AD_{1/SR} = IF\{SR > SR^*; 0; C_{1/SR}[1 - \exp(-\alpha_{1/SR}\tau)]\},$$

$$AD_{21} = IF\{x_1 > x_1{}^*; 0; C_{21}[1 - \exp(-\alpha_{21}\tau)]\},$$

$$AD_{24} = IF\{x_4 > x_4{}^*; 0; C_{24}[1 - \exp(-\alpha_{24}\tau)]\},$$

$$AD_{32} = IF\{x_2 > x_2{}^*; 0; C_{32}[1 - \exp(-\alpha_{32}\tau)]\},$$

$$AD_{34} = IF\{x_4 > x_4{}^*; 0; C_{34}[1 - \exp(-\alpha_{34}\tau)]\},$$

$$AD_{4/TW} = IF\{TW < TW^*; 0; C_{4/TW}[1 - \exp(-\alpha_{4/TW}\tau)]\}.$$

The management agents (3.48) control behavior of the ecosystem near the borders of its homeostasis, i.e. within the boundaries of permissible values of the parameters of the system.

The homeostasis boundaries for the ecosystem were determined by setting of allowable values of phytoplankton $x_1{}^*$, zooplankton $x_2{}^*$, oxygen x_4^*, forms of nitrogen: x_8^*, x_9^* and x_8^*, x_9^*, as well as solar radiation SR^*. Similar conditions can be easily introduced for carbon and phosphorus. Rates of the ecosystem's degradation are regulated by the choice of parameters C_{ij} and $\square_{ij}$.

The second group of intelligent agents in the model (3.47), named as AP, controls the dependences of biomasses PP, ZP and BR on availability of resources they need. The simplest logical management operations are related to transformation of substances in chemical reactions and bacterial decomposition. We assume that this transformation takes place almost without any time delay. Therefore, the agents of the form AR control only the limiting factors of transformations. In our model, these agents are represented by the following conditions:

$$AP(x_1, x_4) = \min(a_{21}x_1, a_{24}x_4),$$

$$AP(x_2, x_4) = \min(a_{32}x_2, a_{34}x_4),$$

$$AP(x_4, x_{15}) = \min(a_{54}x_4, a_{5/15}x_{15}),$$

$$AP(x_4, x_9) = \min(a_{84}x_4, a_{89}x_9),$$

$$AP(x_4, x_{13}) = \min(a_{94}x_4, a_{9/13}x_{13}),$$ \hfill (3.49)

$$AP(x_4, x_{14}) = \min(a_{13/4}x_4, a_{13/14}x_{14}),$$

$$AP(x_4, x_{16}) = \min(a_{17/4}x_4, a_{17/16}x_{16})$$

Each of these agents continuously monitors the limiting factors and switches the relevant influences with coefficients assigned to them.

The third group of agents performs a few missions – it continuously monitors the limiting factors and sets the proper coefficients. It also takes into account the delays in development of the processes, in relation to the reasons which caused them. The inertia of the growth process of biomass *PP*, *ZP* and *BR* is due to natural delays associated with natural finite growth time of living organisms. We will characterize these times by respective constants $\square_{FP}$, $\square_{zp}$ and $\square_{BR}$. Assume also that the reduction in the number of organisms occurs without delays, as it may be caused with a momentary increase in the percentage of dying organisms. Then, the intelligent agents that are included into the equations of the model (3.47) can be represented in the following form

$$AQ(x_5, x_8, x_9, x_{13}, x_{17}, SR) = IF\{M_1(t) - M_1(t - \Delta t) < 0; M_1(t);$$

$$M_1(t) + [M_1(t - \Delta t) - M_1(t)]\exp(-\alpha_{M_1}\tau)\}$$

$$M_1(t) = \min[a_{15}x_5(t), a_{18}x_8(t), a_{19}x_9(t), a_{1/13}x_{13}(t), a_{1/17}x_{17}(t), SR(t)], \quad (3.50)$$

$$AQ(x_1, x_4) = IF\{M_2(t) - M_2(t - \Delta t) < 0; M_2(t);$$

$$M_2(t) + [M_2(t - \Delta t) - M_2(t)]\exp(-\alpha_{M_2}\tau)\}$$

$$M_2(t) = \min[a_{21}x_1(t), a_{24}x_4(t)],$$

$$AQ(x_2, x_4) = IF\{M_3(t) - M_3(t - \Delta t) < 0; M_3(t);$$

$$M_3(t) + [M_3(t - \Delta t) - M_3(t)]\exp(-\alpha_{M_3}\tau)\}$$

$$M_3(t) = \min[a_{32}x_2(t), a_{34}x_4(t)].$$

Parameter $\square$ in (3.50) can be regarded as a step of calculations in time. The agents analyze trends in the changes of limiting factors' values. If the tendency is negative, the usual mechanism of negative influence is switched on. If the same trend is positive, the increase of concentrations of *PP*, *ZP* and *BR* will be detained for some time, due to the expressions in brackets, in which the parameters $\square_{FP}, \square_{ZP}, \square_{BR}$ determine the growth of biological objects.

3.7. Simulation of Development Scenarios in the Coastal Zone Marine
Ecosystem

The choice of coefficients $a_{\square n}$ in the equations of the model determines the degree of influence of some processes in the marine ecosystem on each other. The values of the coefficients depend on conditions of development processes in a particular ecosystem. Therefore, to assess them we need additional information about the ecosystem's dynamics, of a particular or a general character. When the *ABC*-method is used for constructing of ecosystem models, it is possible to implement the statistical method of estimating coefficients, the essence of which was described above in section 2.8.

The general method for estimating coefficients is commonly used in modeling of marine ecosystems [107]. The method applies well-known conformities (which are known from theory or from laboratory experiments) to the laws of biochemical processes' dynamics. The most clearly defined is the relationship between concentrations of substances in the marine environment when chemical reactions are considered. Thus, for example, relationship between concentrations of NO_3 and NO_2 can be established by using stoichiometric-parameter relationships

$$2NO_2 + O_2 = 2NO_3$$

Considering concentrations of O_2 and NO_2 as the resources required for formation of NO_3, it is possible to determine what the amount of NO_3 will match the resources available in a particular volume of marine environment. From this equation of chemical reaction follows, that the relationship between coefficients of influences in the control agent $AP(x_4, x_9)$ should have the following form: $a_{89} = 0,5a_{84}$. Such quantitative relationships between the processes in the ecosystem permit direct identification for a part of the coefficients in its model.

To conduct simulation experiments, the model of ecosystem (3.47) was represented in the form of finite-difference equations of the *ABC*-method. All the modeled processes were reduced to dimensionless form and the total interval of variability (0,10) by the linear transformation (2.22). The influence coefficients in equations and expressions for the management agents were chosen within the range from 0,05 to 0,5. Calculations were performed on 370 time steps with an interval of one day. A large number of variables in the model ecosystems led to a variety of computational experiments. Let us take a brief look only at two of them.

In the first of these experiments, in accordance with the selected scales of variability processes, the annual courses of solar radiation *SR*, water temperature *TW*, (they are shown in Fig. 3.25), and oxygen *OX* (Fig. 3.26) were taken as external influences on the ecosystem. We considered the case when the ecosystem of the upper sea layer comes out at the border of its homeostasis zone on oxygen, due to increased sea temperatures in the absence of vertical water mass movements *VM* and driving wind force *WF*. The contents of zooplankton *ZP* was also fixed on the minimal content of phytoplankton and oxygen, which alternated with each other in different periods of time. This can be assumed from the Fig. 3.26, because the script of zooplankton's variability repeats the one of the two depicted graphs of phytoplankton and oxygen, which is lower than the other. Let us note that the setting of parameter $\square_{ZP,}$ taken in this experiment, had caused a 10-day delay for growth, which is related to the reason that caused this growth. This function was also carried out by the agent $AQ(x_1, x_4)$, as it could be confirmed by the lag of the *ZP* graph with respect to the graphs of *PP* and *OX* in Fig. 3.26.

The resulting scenarios *PP*, *ZP* and *BR* were obtained by using of agent $AQ(x_2, x_4)$, which in the course of bioresources management (in accordance with the equation for x_3) conducted the operation of the massive bioresources extinction caused with a sharp drop in oxygen concentration in the period from 150 to 230 steps of calculations. The same agent provided a lag in the 20 days of the restoring process of bioresources concentration to the previous level.

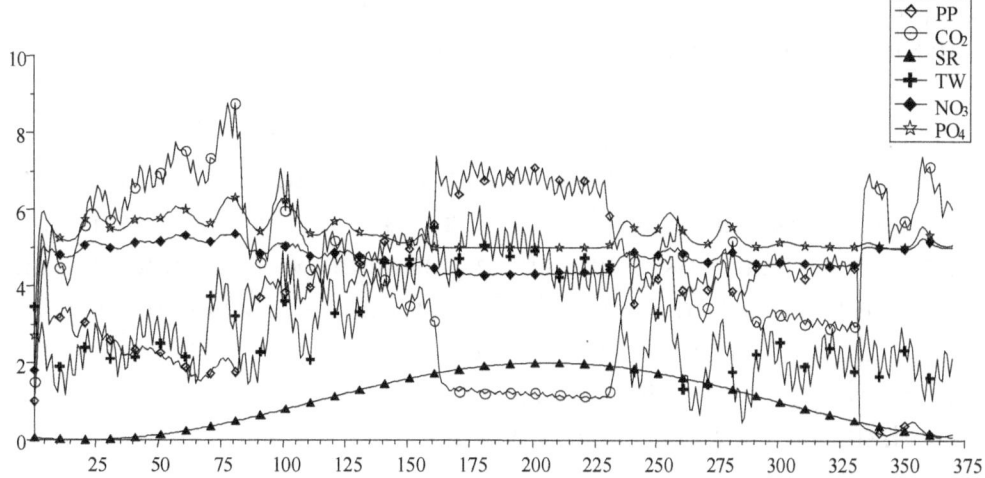

Fig. 3.25. Scenarios for the processes at external influences on the system:

PP – phytoplankton, CO_2 – carbon dioxide CD, SR – solar radiation,

TW – temperature of the sea water, NO_3 – nitrate concentrations,

PO_4 – phosphates concentration

We know that with increasing temperature, the oxygen content in the upper layer of the sea goes down as a result of decrease in its' solubility and releasing it into the atmosphere. As water temperature is a critical factor to the ecosystem, we provided it with the value of $TW* = 5$. This meant that the agent $AD_{4/TW}$ in the equation of oxygen dynamics from the system of model equations (3.47), had to switch on the regime of oxygen concentration decrease as soon as the temperature of the sea water began to exceed 5 dimensionless units. In the performed calculations, this mode took place between steps 150 and 230 of computing time (see chart TW in Fig. 3.25). A sharp drop in oxygen content in this period is clearly seen in Fig. 3.26.

The Fig. 3.26 shows the scenario of zoo and phytoplankton in relation to the dynamics of oxygen content in the sea water. As a result of the agent $AQ(x_1, x_4)$ actions, included in the dynamic equation for x_2, the development of zooplankton was also focused on the minimal values of phytoplankton and oxygen.

The second computational experiment covered the situation when the reduction in oxygen concentration was due to the release of high amount of organic matter and nutrients made by the river runoff *RF*, into the marine ecosystem. The experiments were planned in time for the summer period, in a calm sea conditions, when the sea barrier layer forms and prevents inflow of oxygen from the atmosphere. The increased need in the oxygen for the process of organic matter oxidizing, as well as on oxidizing of substances formed within the ecosystem due to the additional inflow of nutrients from the river, could be one of the reasons of the abovementioned massive bioresources extinction.

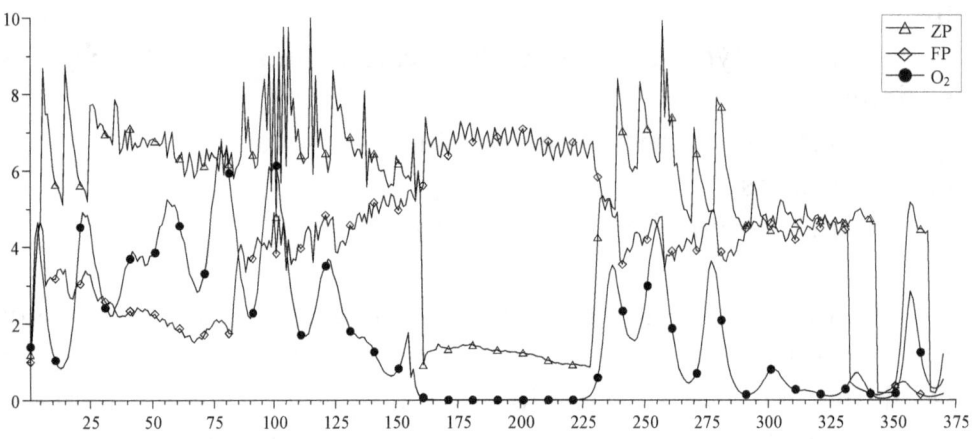

Fig. 3.26. Effect of dissolved oxygen concentration on the dynamics of zoo and phytoplankton: *ZP* – zooplankton, *PP* – phytoplankton, O_2 – oxygen

The time period of the organic matter coming into the ecosystem from the outside, was 100 days (see chart *RF* in Fig. 3.28), and the maximum of its income coincided with the maximum of solar radiation *SR* (185 days). Fig. 3.27 shows the scenarios of development processes in the ecosystem, presenting the growth and subsequent reductions in the concentrations of inorganic substances, which are used in the formation of phytoplankton, on the decomposition of which the dissolved oxygen was wasted.

Fig. 3.28 presents dynamics of biological objects in the ecosystem. It is determined with the graphs of changes in nutrient contents (see Fig. 3.27) and changes in concentration of dissolved oxygen. The changes of OX, PP, ZP and BR were observed in this connection. The agents controlling operations in the model, were placed into the conditions which imitated expansion of the ecosystem beyond the boundaries of homeostasis on oxygen concentration for zooplankton, and on zooplankton concentration for bioresources.

To this end, in the expressions for the agents AD_{24} and AD_{32} (see formulas (3.48)) were set the follows parameters' values $\square_{24} = \square_{24} = 0,8$.

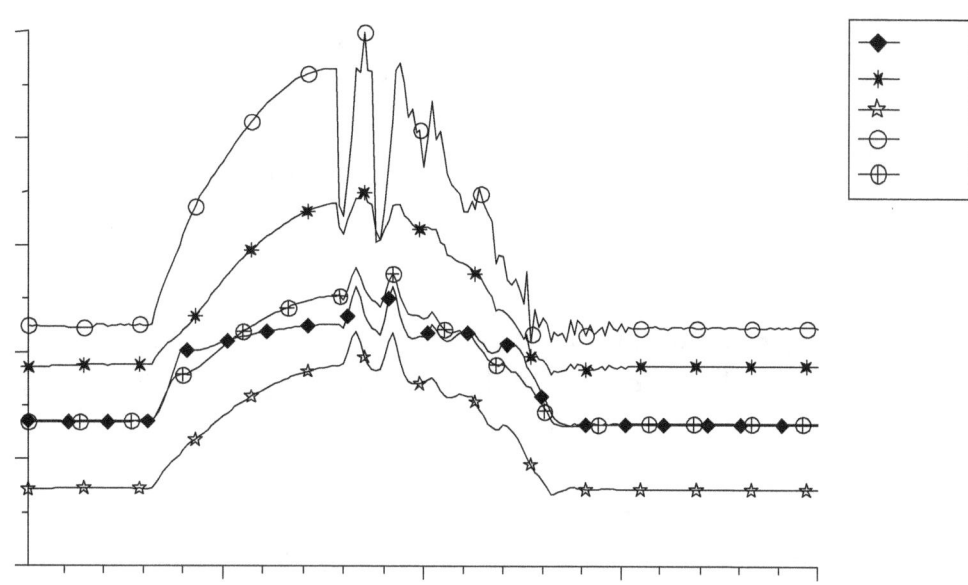

Fig. 3.27. Simulated income of nutrients from the river runoff: NO_3 – nitrates, NH_4 – ammonia, PO_4 – phosphates, CO_2 – carbon dioxide,
        NO_2 – nitrites

The oxygen reached the minimum of its allowable concentrations twice during the incoming of organic matter and nutrients with river flow: at 125 and 135 steps of calculations. At these steps, concentration of zooplankton and bioresources also fell sharply. It provided a reason for a slight decrease in oxygen consumption, resulting in subsequent increase of its concentration and return of the ecosystem back into the borders of its homeostasis zone.

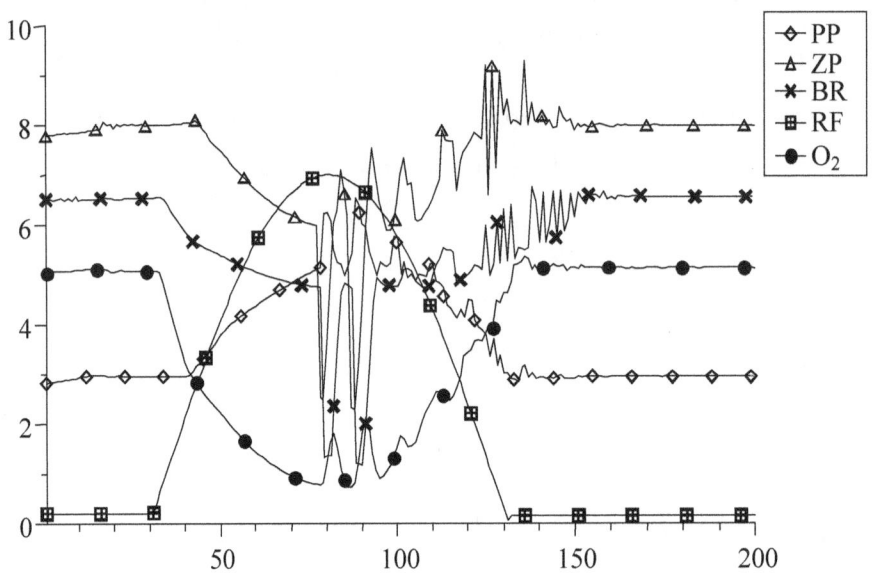

Fig. 3.28. Reaction of the ecosystem on income of nutrients from the river runoff: *PP* – phytoplankton, *ZP* – zooplankton, *BR* – bioresources, *RF* – river inflow, O_2 – oxygen

The computational experiments which we conducted showed us that the use of intellectual management agents in the equations of marine ecosystems' models can produce complex scenarios of chemical and biological processes that occur in real ecosystems due to nonlinear interactions and effects of delay of the ecosystem's responses on external influences.

Considered in this section, the *ABC*-model of marine ecosystem had sustainable solutions were working in conditions of sharp switching of its coefficients, carried out by agents of governance, and retardations in growth of zooplankton and phytoplankton concentrations. This gives grounds for concluding that the method of adaptive balance of causes, coupled with the management agents, has a good potential application in the development of information management technologies for marine ecosystems.

Chapter IV. MODELING OF COASTAL ZONE ECONOMIC SYSTEMS

4.1. Conditional Potentials of Coastal Zone Resources

The economic value of natural resources of a territory or a sea-water area is determined by goods and services' production technologies which provide everything necessary for development of the social ecological-economic system associated with this area. Among the key production technologies of this system, we will consider the industrial technologies presented in equations of the input-output economic balance system of the researched region. Natural, economic, environmental and other types of coastal zone resources can be used in different technologies of goods and services production (hereinafter – products). Each local terrestrial area and marine area of the *CZA* must be regarded as a source of resources for production, depending on economic benefits and environmental feasibility of consumption of these resources.

Conditional Resource Potential within General System of Nature Management. Among general conditional characteristics of resource properties of local *CZA* sites, the resource potentials of these sites should be studied specifically [177]. The main blocks of the conceptual model of conditional resource potential formation are shown in Fig 4.1.

To design scenarios of local resources' potential, a quantitative model is needed. It should provide connections between variable development processes in the social ecological-economic system that consumes natural resources. In accordance with the concepts of the systems approach, such model can be based on the expert analysis of cause-effect relationships between the processes of development. Let us consider the concept of the *CZA* resource potential in connection with technologies of utilizing of its natural resources.

Let us assume that in the researched *CZA* economic system there are n production technologies $T_i (i = 1, 2, \ldots , n)$, and the selected area of natural environment has m beneficial natural properties, which may be regarded as resources c_j , $(l = 1, 2, \ldots, m)$ utilized by a number of specific production technologies. Production of goods with the use of different technologies would have different economic effects. In order to evaluate the perspectives of using the resources of this area in relation to any of production technologies, we will research average indexes of production for some time interval (e.g., day, quarter or year).

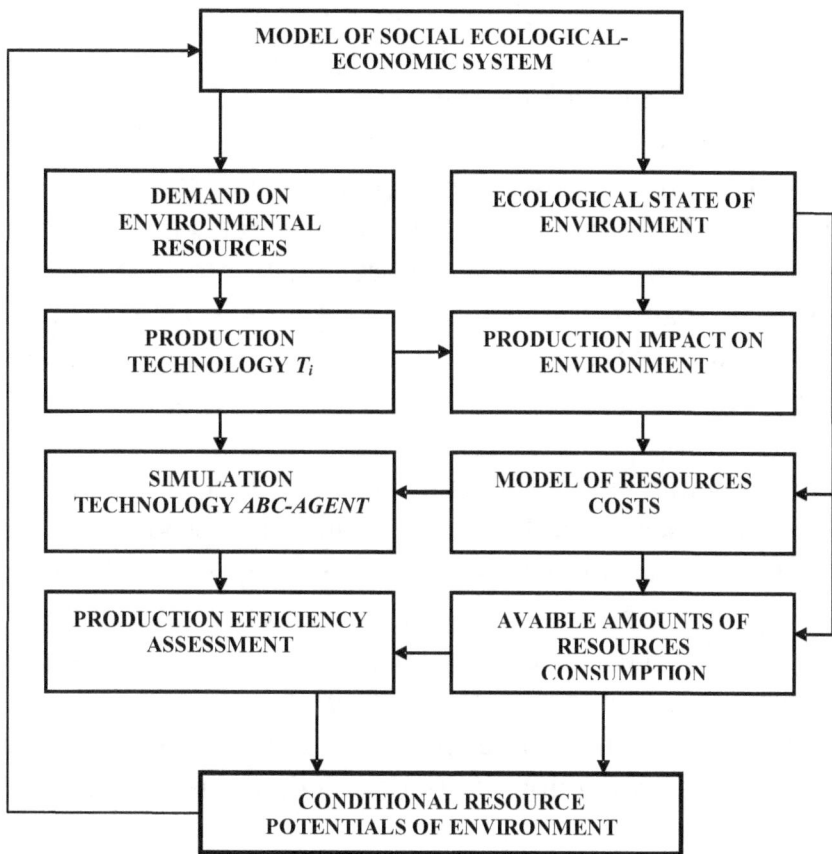

Fig. 4.1. Conditional resource potentials in general system of nature management

Then, it is possible to introduce conditional index of economic efficiency (cost-benefit) Φ_i for the selected area in relation to technology Ti

$$\Phi_i = \frac{p_i S_i}{\left[U_i + \left(U_i^0 - U_i\right)\exp\left(-\alpha_i \tau\right)\right]},$$

$$U_i = q_i V_i, \quad (i = 1, 2, \ldots, n).$$

(4.1)

In this formula the following notations are used: p_i – unit price of a product, manufactured by technology T_i, S_i – volume of sales of products manufactured by technology T_i, q_i – cost of production, V_i – volume of output, α_{1i} – a parameter, establishing the rate of repayment of initial investment (loan) used for creation and for bringing fixed production assets into operation, U_i^0 – volume of initial investment associated with organization of production process at this area, manufactured by technology T_i, U_i – operating costs associated with consumption of natural resources and production itself, τ – time.

The numerator in the formula (4.1) stands for current profit of the industry that uses natural resources, and also produces and markets its products. The higher the income, the more efficient is this technology for this area. Also, the investments assigned to maintain these resources, have a certain value. Current profit correlates with ongoing costs, which are provided in the denominator of the formula. They constitute initial investment of the industry, and the costs associated with production of goods and depreciation of fixed assets.

In course of time, the cost of creating necessary fixed assets U_i^0 is compensated with accumulated profit, and the operating costs take the value U_i, which is determined by the volume of output V_i and the cost of production. Thus, the cost-effectiveness of the natural environment area is related to the technology T_i as a variable function of time (according to the scenario).

Each local site of the *CZA* environment has its own significant properties, which are important for both, nature and society. To characterize ecological state of the environment on a site, experts commonly study values of maximum permissible concentrations of the chemicals that can have harmful effect on the living organisms populating the site. In addition to this, they take a look at another general criterion of the environment's ecological state. It is the level of biological diversity of the site, which can integrally determine the number of various forms of living matter, their total concentration in a single volume of the environment, and study variability of this concentration in space and time.

Typically, a hierarchy of three related levels of living matter' organization are used: genetic level, species-characterizing level and ecosystem level [137]. Each of the levels bears its specific properties, which can be used to estimate the value of biodiversity. The most promising one appears to be the method of expert assessment of biodiversity, which is based on establishing general tendencies in degradation or restoration of diversity: being unable to estimate the number of organisms, we can compare the observations of them at different times. Carrying out a comparison of observational data enables us to evaluate tendencies of the processes which are taking place on a site of a *CZA*.

To identify socially significant properties of *CZA's* natural environment, we can take a look at its basic properties that serve to promote health of human beings, their well-being and positive look on life. Pure and clean sea water, including the water of coastal recreational areas, is a good example of such properties. Later in Chapter VI, we will consider recreational properties of the *CZA* as an important social factor in the use of its resources.

Ecological properties of natural environment on a local site are mainly determined by economic activities of the people who inhabit this land. Economic and domestic consumption of natural resources usually leads to environmental degradation. In order to quantify this change, we will introduce the factor of environmental degradation of the site Ψ_i, caused by natural resources consumption resulting from production technologies development T_i. We will introduce this factor by putting together the following formula

259

$$\Psi_i = \sum_{l}^{h} \varpi_l \frac{\Delta c_{il}}{c_{il}}, \qquad (l = 1, 2, \ldots, h).$$

(4.2)

Formula (4.2) is a weighted (with some weights ϖ_l) sum of relative deviations of harmful chemicals' concentrations c_{il} found in the natural environment, from their maximum permissible values. Assuming that the deterioration of a site's ecological state by a company is a sequence of utilizing environmental resources with industrial technology T_i, a fine should be imposed on the company that implemented this technology. Sanctions may be imposed by taking a percent of the company's profit, proportionally to Ψ_i, which will logically cause an increase in operating costs to a value of $U_i^e = \gamma_i \Psi_i$.

In addition to the penalties, each producing company has to pay rent for the right to use natural resources. The value of the rent should also be linked to the general state of natural environment, if we bear in mind its' renewable living resources. Thus, we will assume that consumption of resources by using the technology T_i reduces initial values of resources concentration B_{ik}^0, and with time these concentrations are restored to the levels of B_{ik}. Then, summary relative change in the concentration of living organisms on the site (i.e. their number per unit volume of the environment) can be estimated by the formula

$$\rho_i = \sum_{k}^{e} \frac{B_{ik} + \left(B_{ik}^0 - B_{ik}\right)\exp\left(-\beta_{ik}\tau\right)}{B_{ik}}, \qquad (k = 1, 2, \ldots, e).$$

(4.3)

Restoring concentrations to the levels B_{ik} occurs with rates that are determined by the parameters β_{ik}. Let us note that self-purification of the natural environment against antropogenic pollution, i.e. gradual decrease in the values of Δc_{il} in the formula (4.2), should also be taken into account. We will introduce it in a form of an expression similar to formula (4.3). The resource rent should be proportional to magnitude ρ_i. Consequently, it increases the production cost to a quantity, which we will denote as $q_i^e = \delta_i \rho_i$.

As we discussed above, rational environmental management involves finding and preserving dynamic balance between economic benefits received from consumption of natural resources and maintaining ecological state of the environment within acceptable limits. The above factors Φ_i, Ψ_i and ρ_i provide an opportunity to characterize the environmental and economic potential of this environmental area with a projected profitability scenario of its resources use, with utilization of technology T_i.

The following expression can be called conditional with respect to technology T_i resource potential of the local area-

$$\varphi_i = \ln \Phi_i = \ln p_i S_i - \ln \left[U_i - \left(U_i^0 - U_i \right) \exp(-\alpha_i \tau) \right],$$
$$(i = 1,2, \ldots , n) \tag{4.4}$$

where $U_i = (q_i + q_i^e) V_i + U_i^e$.

The conditional resource potential depends on a large number of economic, environmental and generally speaking, of some social factors, which are presented by scenarios of socio-economic development. Positive values of φ_i correspond to economic viability based on consumption of the resources of the local environment' area. Negative values for the entire time of proposed exploitation of the local area environment, show an excess of expenditure over income because of unfavorable economic and (or) environmental situation.

The scenarios φ_i should save positive values for potential cost-effective areas of the environment in relation to the technology of production. This process is set to start at a certain moment in time τ_i^*, when current income from sales of products is compared in magnitude to current expenditures (including a part of the initial loan U_i^0, which has not been returned by this time). This process, being typical for the technology "time-for-investment-return " can be estimated by the formula

$$\tau_i^* = \alpha_i^{-1} [\ln(U_i^0 - U_i) - \ln(U_i - p_i S_i)].$$
(4.5)

The revenues received from the sale of produced goods, which are represented by the first term in the equation (4.4), depend on the prices p_i, affecting the sales volumes of S_i. In its turn, the price is determined by the cost of production q_i, which is affected by the size of the resource rent q_i^e. In addition, penalties for pollution of the environment U_i^e can significantly affect the second term of the formula (4.4). Therefore, conditional resource potential of local environmental areas is a set of complex functions of time from all three of factors: Φ_i, Ψ_i and ρ_i. In regards to marine (or terrestrial) areas, they are characterized with space-time fields of the values φ_i, determining the perspectives of application of the technology T_i in this coastal zone.

4.2. Dynamic Input–Output *ABC*-Model of a Coastal Zone Regional Economy

As a rule, each region containing a *CZA* as its part, produces a wide variety of goods and services, and utilizes a number of technologies. To design scenarios of local resources' potential, we need a quantitative model which would connect the variables in formulas (4.1) – (4.4) with the development processes in the social ecological-economic system of the region, that consumes the *CZA's* natural resources. The economic value of natural resources is determined by technologies of goods and services production. We will assume that the industries of the local economy which constitute the equations of input-output balance of the economic system of the *CZA* region, are the key production technologies in our system.

As it was noted above, the environmental management means finding and maintaining dynamic balance between economic benefits obtained from consumption of natural resources and maintaining ecological state of the environment within acceptable limits. To achieve and preserve such balance, systematic management of its resources is required, where a certain informational technology for support of management decisions must be applied.

Natural resources of each local site can be used in different technologies of production of goods and services. In the above discussion, when defining the term "resource potential", we decided that in our *CZA* economic system there are *n* production technologies T_i ($i = 1, 2, \ldots , n$), and our local site of natural environment has *m* useful natural properties, considered as its' economic resources, c_j ($l = 1, 2, \ldots , m$). The profitability of the product depends on the manufacturing cost and on customers' demand for the product.

Fig. 4.2 illustrates the input-output balance in the *CZA* region, showing how the demand for goods and services produced by different technologies, transforms into the demand for resources on local sites of the *CZA*. Ecological-economic model of this local site must ensure continuous assessment of economic viability of resources consumption, taking into account the overall system of input-output balance of the region, as well as the environmental restrictions on the use of resources in the local area. Comparing the evaluation of cost-effectiveness of resources on different production technologies, one can choose the most rational ways for managing the development of the *CZA*.

Dynamics Equations of Input-Output Balance in the *CZA* Region. Let us take a look at the ways of possible interaction between industries of a regional economic system. We will see how this interaction can be represented in case of utilizing the *ABC*-method for the purpose of input-output balance modeling in the researched region.

The conceptual model that takes into account the effects of mutual influences between the industries of the economy on each other, is shown in Fig. 4.3. Every industry is characterized by its output, which is identified with X_i in this figure. The purpose of each product type is to meet consumers' demand A_i and to satisfy technological demand TA_i. The ratio of prices for goods is of great importance for economic viability of each of the industries and services P_I ,

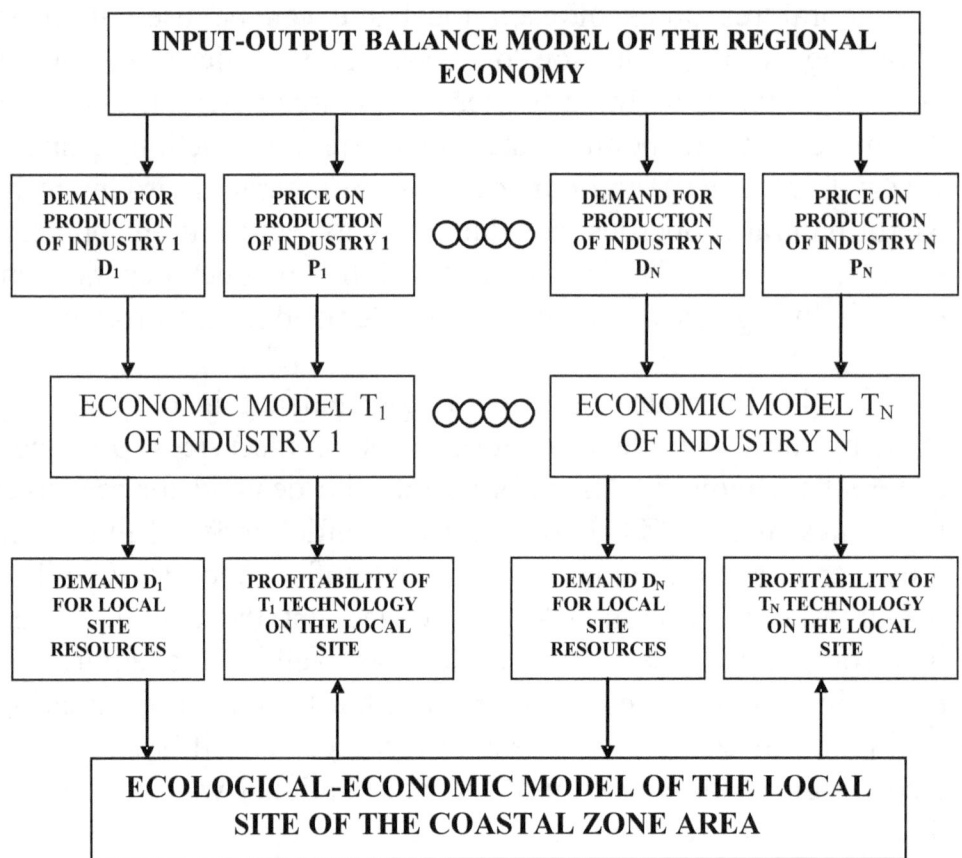

Fig. 4.2. Evaluation of profitability of resources consumption on a local site of the *CZA* environment, with utilizing various technologies

and the cost of production TR_i, which determines the size of the product's added value R_i. The figure of the conceptual model shows that all these characteristics are intimately linked with each other. Therefore, any modification of any parameter of the input-output system affects the value of all other parameters of the system.

In accordance with the concepts of systems modeling, the *ABC*-model of a complex system provides dynamic balance of influences between the processes that compose this system, and balance the external influences applied to the system. In the case of the input-output model, such external influences can be regarded as factors that can change values of consumers' and technological demand for products of a particular industry, as well as the factors that modify relative prices and production costs of each industry.

In this section we will use the method of adaptive balance of causes to build a dynamic *ABC*- model of input-output economic balance in the *CZA* region. Let us represent the output X_i of industries T_i in the region in a reduced form, by expressing them with dimensionless quantities v_i, which vary within the range of values (0, 1). The part of the industrial production T_i which is aimed to meet technological needs of the industry T_j, will be regarded as the value which

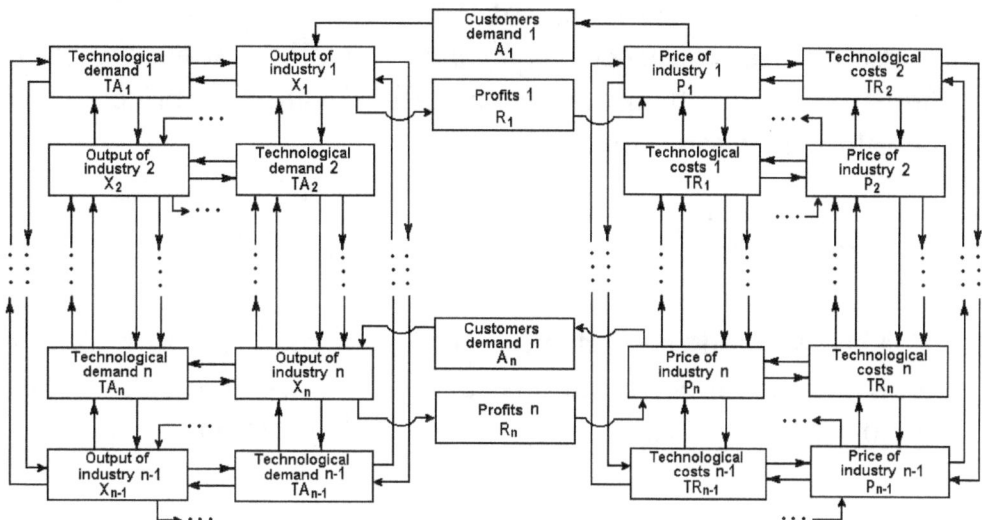

Fig. 4.3. Conceptual model of input-output balance of a regional economy, that takes into account the effects of influences produced by various
industries of the economy on each other

265

characterizes the degree of influence of the industry T_j on production output v_i. Then, the dimensionless production volume v_i can be assumed as a result of joint influences on the industry T_i, imposed by all other industries of the economy (technological demand TA_i), and by the consumers demand A_i for its products, as we can assume from the socio-economic system of the region (final consumption).

$$v_i = A_i + \sum_j a_{ij} v_j \qquad (i,j = 1,2, \dots , \text{n}), \; i \neq j \qquad (4.6)$$

In this expression, coefficients a_{ij} establish the degrees of influences of the economy's industries on each other.

Similar relations link the prices of products produced by different industries. All the technological costs associated with manufacturing of a production unit, represent summary effects on the prices formation for the products of other industries. If we denote the degree of these influences as b_{ij}, and normalized added values of the industry T_i as Q_i, then for the normalized values of prices we obtain the following expression:

$$p_i = Q_i + \sum_j b_{ij} p_j \qquad (i,j = 1,2, \dots , \text{n}) \; i \neq j. \qquad (4.7)$$

The Method of Adaptive Balance of Causes provides the following dynamic equations of the input-output model of the economic balance' state of a regional *CZA* system [181]

$$\frac{dv_i}{dt} = v_i \left[1 - 2\left(v_i - A_i - \sum_j a_{ij} v_j \right) \right],$$

$$\frac{dp_i}{dt} = p_i \left[1 - \left(p_i - Q_i - \sum_j b_{ij} p_j \right) \right],$$

$$\frac{dA_i}{dt} = A_i \left[1 - 2\left(A_i + a_{Api} p_i \right) \right],$$

(4.8)

$$\frac{dQ_i}{dt} = Q_i \left[1 - 2\left(Q_i + a_{Qvi} v_i \right) \right].$$

The input-output model (4.8) constitutes a system of relations imposed on the industries of the *CZA* region in accordance with the conceptual model shown in Fig. 4.3. They express the state of dynamic balance in the system, which is being controlled by external influences' factors. In what follows, we will not take into account possible changes in production technologies, which may occur as a result of scientific and technological progress. Then, as the main external factors of economic development in the region we will regard the variable consumer demand for products of all sectors of the economy A_i, and the added values Q_i.

Let us consider the basic components of production costs. The costs of manufacturing of industrial goods are defined by technological costs (the cost of labor, energy, transportation, etc.), the costs of environmental actions (resource rents, ecological fines, etc.) and costs of acquiring natural resources (cost of obtaining, shipping, handling, storage, etc.). Let us assume that the unit of production contains y_1 units of the first type of natural resources, y_2 units of the second type, etc. up to y_m units of the m-type of natural resources, inclusive. Then, we have the following expression for the production cost of the technology T_i

$$q_i = q_{Ti} + q_i^e + \sum_{l=1}^{m} r_l y_l,$$

(4.9)

in which q_{Ti} – the technological costs, q_i^e – the cost of environmental protection actions, and r_l – the prices of the relevant natural resources.

We should clarify the meaning of the phrase "cost of a unit of natural resources". We will take it as a summary of costs that are needed for acquisition of this particular resource unit and for keeping it on the industry's production plant, ready for use. The market value of each type of the resources can become a criterion to measure effectiveness of consumption of this type of resources on a local site of the environment. In other words, an enterprise of the industry can either buy the missing amount of natural resources by itself, or can start maintaining its development independently, in the areas where it is possible and profitable.

Taking decisions regarding this issue requires a certain amount of informational support, which can be provided by a dynamic model of the industry. To accomplish this, we can use a standard dynamic model of production. We will use the informational technology *ABC-AGENT*, which takes into account and connects all the processes taking place in the economic system of goods and services production, with each other. This technology will be considered in the next section.

4.3. Information Management Technology of Regional Economic Processes

The developers of the Method of Adaptive Balance of Causes (*ABC*-method [181]) proposed a universal informational technology for management of processes which occur in economic systems producing goods and services. The technology is called *ABC-AGENT* technology [183]. The principle of construction of an economic model of goods and services production, which was implemented in this technology, supposes applying to agents (logical control operators) in balance equations of economic processes, built by the *ABC*-method. The structure of *ABC AGENT* informational technology is shown in Fig. 4.4. The control agents presented in the figure, consistently perform basic operations inherent in any economic systems.

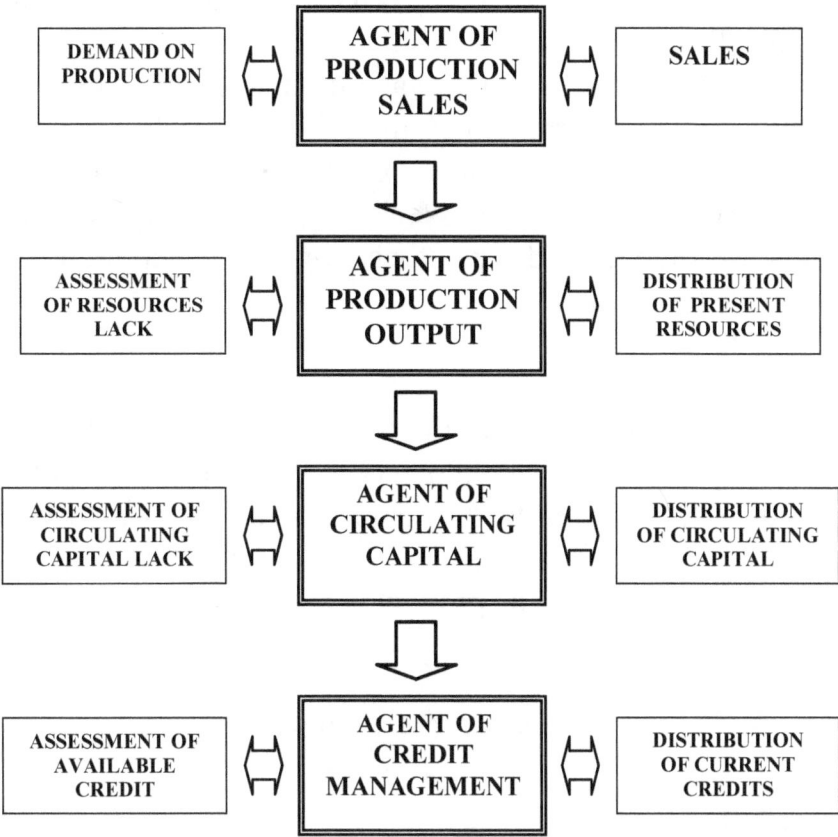

Fig.4.4. Operations of main control agents that run in the standard management   technology of an economic system

Let us give a thorough study to functions of management agents in the example of an industrial technology T of an economic system. The demand for products of the industry (company) can be satisfied by sales of finished products accumulated in its warehouse, as well as by production of additional amounts of goods. Let us denote the number of ready-to-sell products as H, produced by the industrial technology T. In the case of continuous sales, the balance equation of finished products for the function H can be represented by the following equation of the ABC-model

$$\frac{dH}{dT} = H\left[1 - 2\left(H - V + S\right)\right], \tag{4.10}$$

269

where V is the receipt of finished goods to the industry's warehouse, S is its sales.

The rate of profit N_i which was obtained from sales of a unit of production will be determined by the difference between its price and cost: $N_i = p_i - q_i$. Sales of products becomes unprofitable when $N_i < 0$. Therefore, in the *ABC AGENT* technology, in every two successive moments of time t_j and t_k, , these operations are controlled by production sales agents.

The functions of a production sales agent are to continuously collate current demand for products of the industry with availability of finished products in the industry's warehouse, and to estimate benefits of selling the products at current market prices (see Fig.4.5).

Thus, the sales of a product are subject to the following conditions:

$$S_k = IF(N_k < 0; 0; R),$$

$$R = IF(D_k < H_j; D_k; H_j),$$

where D_k is current demand for the product.

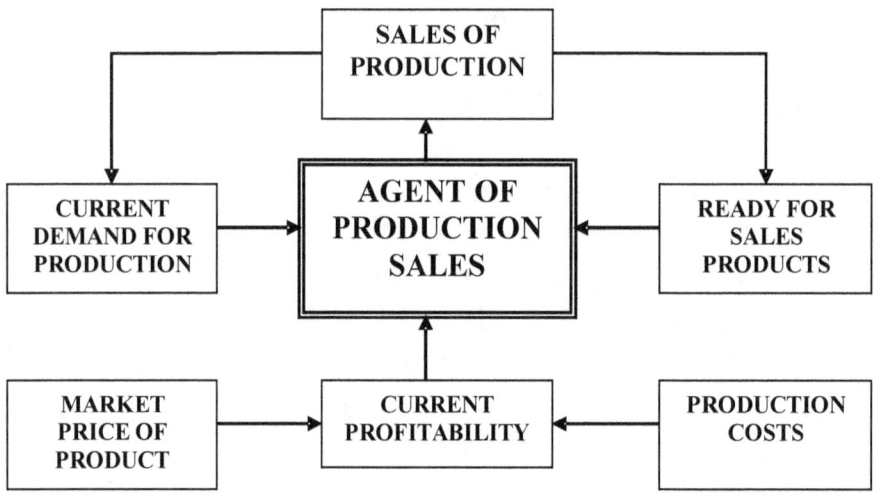

Fig.4.5. Operations performed by the agent of production sales

The agent of production output controls the release of products. Its main functions are shown in Fig. 4.6. The release should not be made when accumulated storage in the industry's warehouse exceeds the demand. If the amount of produced goods is insufficient, the production of the necessary amount depends on availability of resources. It is evident that the output volume V will be limited by those types of resources which are insufficient at the moment. If we denote as M_i the minimal amount of resources available for production under production technology T, the amount of released goods will be determined by the following logical operations performed by the agent of production release:

$$V_k = IF(D_k < H_j; 0; M_{1k}),$$

$$M_{1k} = IF(D_k - H_j < M_j; D_k - H_j; M_j),$$

$$M_j = \min(m_{1j}; m_{2j}; ...; m_{lj}),$$

where m_{ij} are the identified quantities of resources available for production:

$$m_{ij} = \frac{H_{ij}}{y_i}, \ (i = 1, 2, ..., l).$$

Operations carried out by the agent of production release are summarized in Fig. 4.6.

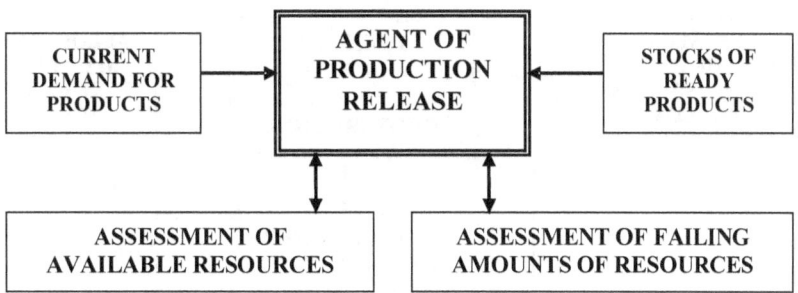

Fig.4.6. Operations carried out by the agent of production release

In order to increase production to the levels determined by current demand for it, the industry T can buy the missing resources by using its existing working (circulating) capital. Let us denote current volume of working capital of the industry as H_2. Its dynamics is determined by the amounts of current profits I, credits H_3 and the costs of acquired resources S_2:

$$\frac{dH_2}{dt} = H_2[1 - 2(H_2 - I - H_3 + S_2)] \qquad (4.11)$$

The functions of financial resources' control and allocation of the industry, which are run by the agent of working capital distribution, are presented in Fig. 4.7. We will assume that the available working capital H_2 is allocated for acquisition of missing resources in the same proportion in which each of the resources types is involved in the production. Let us now introduce proportionality coefficients of the form

$$\rho_i = \frac{r_i y_i}{r_1 y_1 + r_2 y_2 + \dots + r_l y_l}, \quad (i = 1, 2, \dots, l).$$

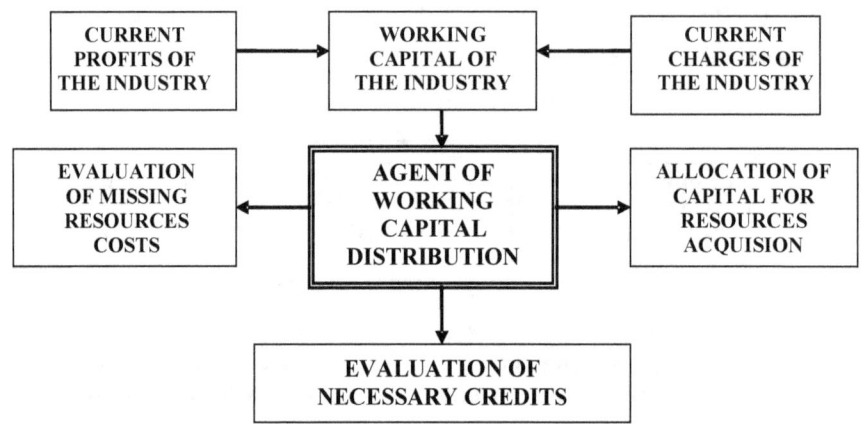

272

Fig. 4.7. Distribution of working capital of the industry and evaluation of necessary credits to acquire the missing resources

Then, for acquisition of i – type of resource, the industry has the $\square_i H_2$ part of their working capital. If this amount is not enough, the industry has an opportunity to acquire resources by taking a loan, provided that its total debt H_3 (accumulated to the current time of the credit) does not exceed a certain pre-established norm H_3^*. It is necessary to clarify that this "norm" can serve as a control level of the industry's development in cases when uncontrolled consumption of natural resources of the *CZA* begins to threaten the ecological state of the surrounding area. In such cases, by setting up a certain value of maximum permissible loan means that the administrative body managing the *CZA* prohibits consumption of natural resources in amounts exceeding some established level of rental payment. Functions of the agent of the accumulated credit' management are demonstrated in Fig. 4.8.

In a case when an industry takes a credit to purchase scarce or rare resources, additional terms S_3 appear in circle brackets of equations (4.11); they denote the cost of the loan repayment. Taking into account the percent of the credit repayment $\square$, logical conditions for S_3 take the form

$$S_{3k} = IF[H_{3j}\theta < H_{2j}; H_{3j}\theta; H_{2j}].$$

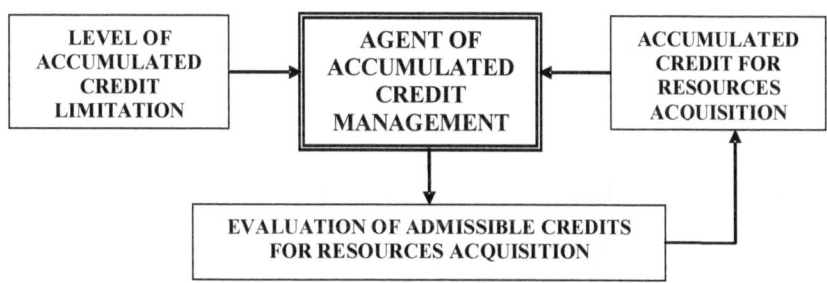

273

Fig. 4.8. Evaluation of permissible credits for resources acquisition by the agent of accumulated credit management

Let us denote the volumes of resources purchased in credit as V_{11}, V_{12}, ... , V_{1l}. Then the total cost of the current credit will be

$$V_3 = r_1 V_{11} + r_2 V_{12} + ... + r_l V_{1l}.$$

Current value of the accumulated credit can be expressed by the balance equation

$$\frac{dH_3}{dt} = H_3 \left[1 - 2 \left(H_3 - V_3 + S_3 \right) \right]. \tag{4.12}$$

Thus, the balance equations of the *ABC*-model (4.10) - (4.12) are essential for strategy planning of production processes, as shown in Fig. 4.3 – Fig. 4.7.

Now, it is time to turn our attention to the processes related to expenditure on acquisition of resources. We will first consider the amount of the resources which are available to the industrial technology T at its warehouses. Let us introduce the designation H_{1ik} to them, bearing in mind that the amount H_1 of the i means a type of resource at a time k on the storage of industry T. With continuous production regime, these quantities are expressed by balance relations:

$$\frac{dH_{1i}}{dt} = H_{1i} \left[1 - 2 \left(H_{1i} - V_{1i} + S_{1i} \right) \right].$$

In these equations, the expenditure of each type of resources is in proportional dependence to the number of released products $S_{1i} = V y_i$. In the case where the storage of resources is sufficient to produce the required volume of production, the purchase of resources is not performed. Otherwise, the industry has to purchase the amount of resources F_i

$$V_{1i} = IF[(D-H)y_i < H_{1i}; 0; F_i].$$

Function F_i restricts the acquisition of resources by the amount of available working capital of the industry. If the amount of available capital funds for the acquisition of i- kind of resource is $\square_i H_2$, and the cost of the missing i- kind of resource is equal to p_i ($y_i D - H_{1i}$), then the logical condition for this function takes the form

$$F_i = IF[r_i(y_i D - H_{1i}) < \rho_i H_2; y_i D - H_{1i}; R_i].$$

In the above expression, the function R_i restricts acquisition of the resource to the amount of funds equal to the value of the credit which the industry gained for acquisition of the resources; it approaches predetermined value H_3^*. As we noted before, the organization sponsoring the industry (or the state administrative body regulating consumption of natural resources) sets the requirement on the production industry to withdraw some of its gains in favor of the state. These operations are presented in Fig. 4.9.

In [181], a scheme of integrated management of the coastal zone resources' consumption was proposed. It was based on management of production output by setting values of H_3^*. This function has the meaning of maximum permissible level of rent payments for the use of resources that the industry has already taken from the natural environment in order to create some resource storage in its warehouses. By reducing the value of H_{3i}^*, the society restricts consumption of natural resources in order to protect the environment. We will use this scheme below to assess the resource potential of local areas of the environment.

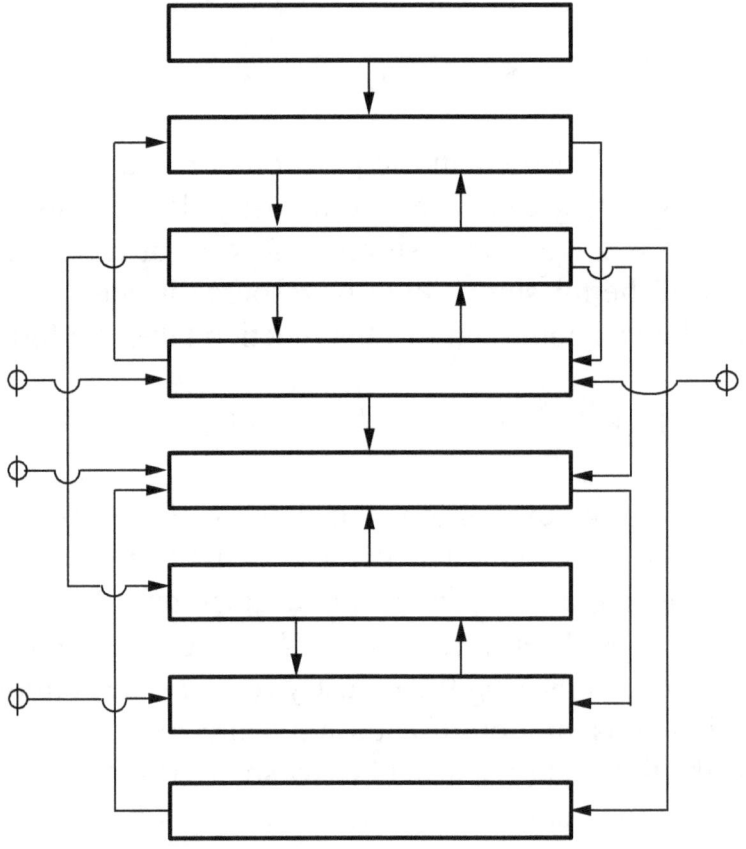

Fig.4.9. Basic operations in the *ABC AGENT* technology of a standard economic system

4.4. Integrated Management of Regional Economy Considering Consumption of the Coastal Zone Resources

Let us take a look at a case-study of volumes of resources consumption management in a local area of natural environment. To accomplish this, we can use a prototype model for an industry's production. We will use the model presented above, as it takes into account all basic economic processes of a technology that produces products and services. In this technology, a possibility to adjust the volume of output by changing the limit of allowable credit H_3^* for purchasing of natural resources of the *CZA* is implemented.

276

Let us suppose that, to produce a specific product, it is necessary to use three kinds of *CZA's* natural resources. We will assume that the prices for these kinds of resources, which include resource rents and environmental payments, may change over time. Then, the return from consumption of these resources will depend on the prices' dynamics. In cases when the prices rise, the production industry may be forced to purchase resources for a credit, thus accumulating the total amount of debt H_3. With deterioration of ecological situation in the local site of the *CZA* area, the limit of allowable credit H_3^* could make the industry to terminate consumption and depletion of resources at all.

As the main variables describing the state of the resources' market, we will take price of the resource r_j, quality of the resource u_j, demand for resource d_j and the volume of its supply c_j ($j = 1, 2, ...,$ l). Given the known causal relationships between the introduced variables, the dynamic equations of the *ABC*-market model of resources may take the following form

$$\frac{dr_j}{dt} = r_j \left[1 - 2\left(r_j + b_{rcj}c_j - b_{rdj}d_j - b_{ruj}u_j \right) \right],$$

$$\frac{dc_j}{dt} = c_j \left[1 - 2\left(c_j - b_{crj}r_j - b_{cdj}d_j + b_{cuj}u_j \right) \right],$$

(4.13)

$$\frac{dd_j}{dt} = d_j \left[1 - 2\left(d_j - b_{duj}u_j + b_{drj}r_j \right) \right].$$

$$\frac{du_j}{dt} = u_j \left[1 - 2\left(u_j - b_{urj}r_j + b_{udj}d_j \right) \right].$$

It is necessary to note that the magnitude of demand for this type of resource positively depends on its quality. Therefore, for the influence function $b_{duj}u_j$ the following expression can be used:

$$b_{duj}u_j = b_{duj}[1 - \exp(-\alpha_{duj}u_j)]$$

In addition to this, the quality of a kind of resources depends on the technology of its use. Therefore, quality restrictions for resources should be taken into account: the demand for a certain type of resource is terminated when the quality falls below a certain limit u_{ij}^{0}

$$d_j = IF[u_{ij} > u_{ij}^{0}; u_{ij}; 0)].$$

Consumption of resources in a local area of environment leads to development of local prices for resources and local supply functions of the local resources. Prior to using this type of resource in a certain local area, its price is equal to the market price p_j. As soon as the development and consumption of the resource begins, its price usually falls to the values of r_j, which should be considered as the local costs of resources

$$r_j = [p_j - (r_j - p_j)\exp(-\delta_j\tau)] \tag{4.14}$$

A significant reduction in the cost of local resources compared to the cost established in markets, is a major encouragement to their development and consumption. Coefficient δ_{lj} determines the speed of this process. Large-scale resources development with the purpose of meeting the needs of all sectors of the economy, could significantly affect the supply function for this type of resources in the market. If we denote local supply function as c^l_j, i.e. the volume of locally produced kind of resource j, in this case we obtain the dependence

$$c_j = [c^l_j + (c_j - c^l_j)\exp(-\gamma_{lj}\tau)] \tag{4.15}$$

For the convenience of simulation, we will use the dimensionless (reduced) values of A, which are connected with corresponding actual dimensional quantities A' by the following dependence

$$A = 5 A' [M (A')]^{-1},$$

where [M (A')] is the mean value of variability interval of corresponding dimensional value A'. This transformation reduces variability of scenarios to the set of values [0, 10]. As we mentioned above, a return to the original (dimensional) variables can be easily made by the formula

$$A' = 0,2[M(A')]A.$$

In the first experiment, we will consider scenarios of economic processes for a simulation model of a production industry, which uses three types of natural resources of the local CZA area. We will assume that the prices of these resources have constant values: $r_1 = 3,7;$ $r_2 = 7,0;$ $r_3 = 4,5$. Let us further suppose that, for manufacturing of a production unit, an industrial technology T is used, for which it is necessary to use the following amount of resources available in the local area: $y_1 = 1,3;$ $y_2 = 3,3;$ $y_3 = 2,0$.

Let us set up an anticipated price of a production unit $p_1 = 7,0$ and define constant volume of its realization $S_i = 42,0$. We will also assume that the initial investment into production is equal to $S_i^0 = 10,0$ and the percent of deductions from profit to cover these costs is equal to 1.

Now, we will introduce environmental restrictions on consumption of natural resources. To control the level of environment contamination, we will use the following scenario of ecological deductions from the profits of production industry. To simplify the description, we will further on define production industry with a term "company". We will set a condition that the company is unable to accumulate higher debt of payments for utilizing natural resources than the established reference values of H_3^*. Let the integrated management of the ecological-economic system be subject to the following scenario:

$$H_3^* = H_3^0 \exp\left(-0,01(\tau - 33)^2\right), \qquad (4.16)$$

where τ is the time of computations. Such a choice of control function means that the most favorable time for obtaining loans to purchase natural resources is the interval of time in the vicinity of the 33-rd step of the computation.

Calculations on the model were performed for 370 steps. Fig. 4.10 shows predicted scenarios of development processes in the industry that consumes resources of the *CZA*. Let us note that on this and on all subsequent figures, the scripts of processes are shown in the dimensionless units reduced to a single scale of values [0,10]. In the initial moment of time, the company had some reserve of the working capital for acquisition of resources, through which at the step 60 of calculation the product release was launched. However, to meet the demand, the industry was forced to borrow money and accumulate credit, and its volume increased linearly. At the step 190, the value of accumulated credits reached its critical limit, which at that moment of time was represented by the function (4.16).

The further lending and procurement of resources were discontinued, and production was halted. The balance of working capital was used for partial repayment of the accumulated credit.

Having scenarios of economic processes, it is easy to construct a temporal graph of T_1 resource potential of the local site of the *CZA* natural environment. Calculated by the formula (4.4), the values of φ_1 are shown in Fig. 4.10 – 4.12. As it can be seen from Figure 4.10b, application of the technology T_1 in this sector had proven to be unprofitable, as the resource potential φ_1 almost always had a negative value.

Now, let us consider an example of use of natural resources on the same local site of the *CZA* environment, but in terms of changing policy of accumulated credit limitation. We will assume that ecological conditions of the environment have improved, thereby expanding the time interval of consumption of resources, as shown in Fig. 4.11. Then, the predicted scenarios of economic processes shape much more favorably to the production (see Fig. 4.11b).

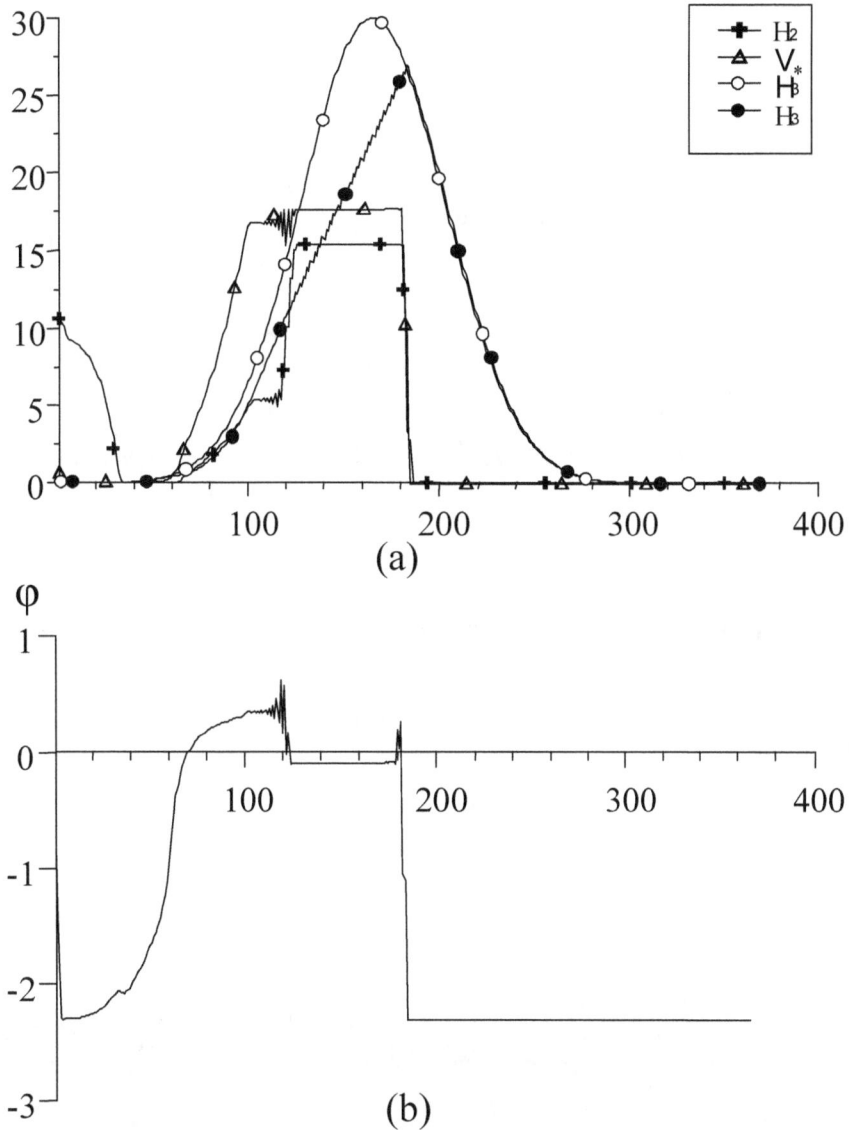

Fig. 4.10. Scenarios of economic processes in integrated management of resources consumption: (a) working capital H_2, production volume V_1, administration H_3^*, accumulated debt of environmental payments H_3; (b) profitability of production φ

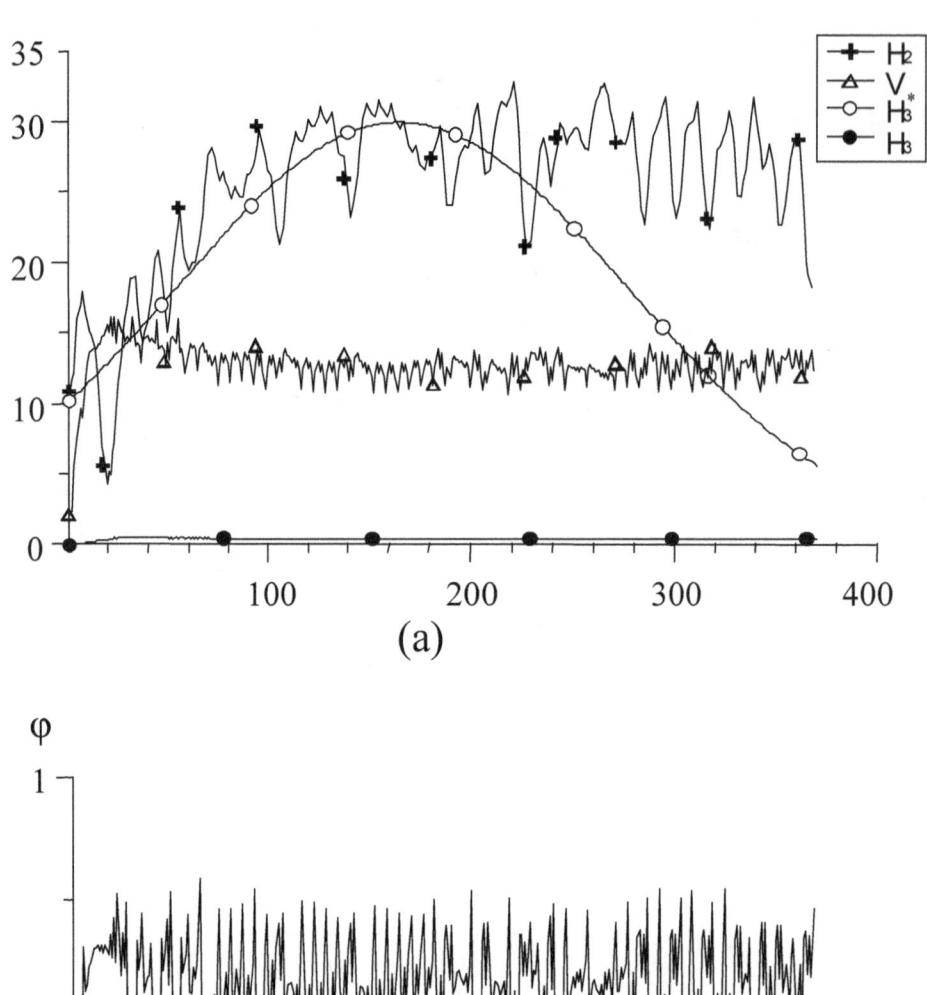

Fig. 4.11. Scenarios of economic processes with changing management of resources consumption: (a) working capital H_2,

production volume V_1, administration H_3^*, accumulated debts of environmental charges H_3;

(b) profitability of production φ

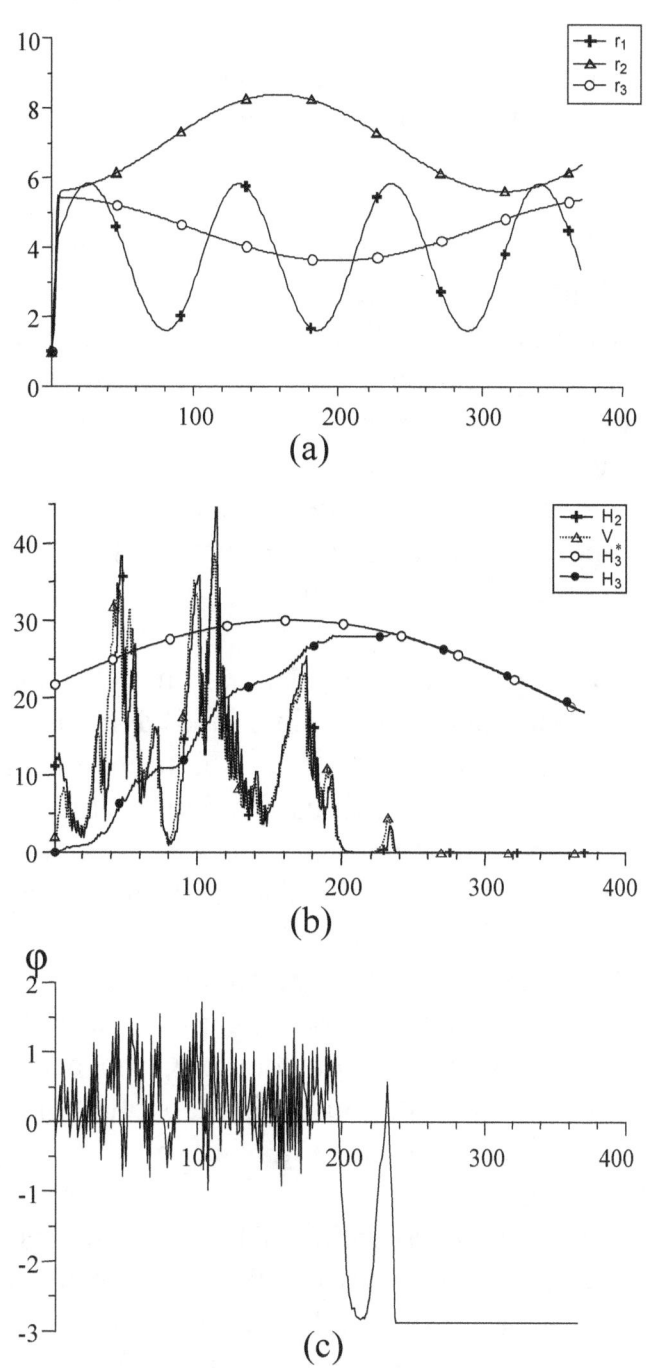

(a)

(b)

(c)

Fig. 4.12. Scenarios for economic processes with use of resources on technology T_2: (a) the dynamics of values of the three types of resources;

(b) working capital H_2, production volume V, administration H_3*,

accumulated debts of environmental charges H_3;

(c) profitability of production φ

As the demand for products was fully satisfied, this provided substantial profits and stable level of working capital. The company had a small accumulated credit that was repaid practically right after realization of the produced products. In the new conditions, the technology of production ensured low (compared to the maximum allowable quantities) accumulated environmental charges for use of resources. As a consequence, the scenario of conditional resource potential of the local *CZA* area, shown in Fig. 4.11, *b*, gives a forecast of cost-effective production.

In the following computational experiment, changes in costs of resources were simulated. The simulated dynamics of prices for resources is presented in Fig. 4.12, *a*. The terms of loans were more relaxed compared to the previous version of the calculations. To simulate real market of finished products, a random function of external influence was added into the equation for the magnitude of demand. Random variations of the demand and changing prices for resources led to emergence of more complex scenarios of economic processes. While the current working capital and volumes of output in these conditions changed considerably, the accumulated credit continued to increase monotonously till step 240, when it reached the maximum permissible value (see Fig. 4.12, *b*).

Conditional resource potential of the *CZA* site is shown in Fig. 4.12, *c*. As it can be seen from the figure, even in the condition of changing prices for resources, production continued to remain profitable throughout the whole period of the forecast.

According to summarized results of the computational experiments, we can conclude that it is possible to predict profitability of resources' consumption at local sites of the *CZA*, based on the specific ecological and economic conditions of its use. The profitability of production is dependent on the technology of the resources use, which, in its turn, is determined by the volume of purchased resources and by the size of environmental payment for their consumption.

4.5. Industrial Efficiency Evaluation Concurrent with Environmental Criteria Assessment

The possibility to predict processes in ecological-economic systems of *CZA* creates foundation for resources management is such areas. The companies competing for natural resources in conditions of market economy, pursue their economic interests, which may contradict to the interests of society. Administrative bodies are compelled to constantly monitor the state of natural environment and regulate consumption of natural resources in order to avoid negative impact on the nature. The rational consumption of resources supposes that two factors must be satisfied at the same time - the economic interests of the society, and the society's desire to maintain clean environment. Keeping this balance is one of the main concepts and tasks of sustainable ecological and economic development of natural-economic systems.

A number of recent studies proposed information management technologies for socio-economic systems which enable us to evaluate permissible limits of natural resources consumption by performing model experiments [181. 183]. Because of complexity and diversity of ecological and economic processes that need to be taken into account simultaneously with resolving the problems of sustainable development, these information management technologies often use integrated assessment of economic benefits and environmental "utility" of natural resources consumption. In particular, we can look at the example of *ABC AGENT* technology [181], which allows to predict scenarios of profit and takes into account environmental constraints of resources consumption process. In this section, we will consider application of the simulation model for management of development processes by different ecological and economic criteria, which we constructed above.

Earlier, in the algorithm of economic model of an industry, we used the maximum permissible value of accumulated loan H_3^*, which could be applied to the industry as a competent control factor for an economic production system. The scenario of the H_3^* value was set experimentally, based on some additional information about the state of the *CZA* environment. Now, we will introduce a scheme of resources' management feedback, which will automatically set the value of H_3^* depending on various factors of the observed state of natural resources. In addition to this, we will introduce another factor of control, the value of which should depend on the state of natural resources use: it is the resource rent. The management structure of *CZA* resources consumption processes is presented as a conceptual model in Fig. 4.13.

Assuming that industrial production has continuous negative impact on ecological state of *CZA* environment, we will suppose that an increase in production is going to cause increasing damage of the surrounding environment. Therefore, the influence of production output is introduced into the model with a positive sign. The problem of continuous control consists in limiting the production volume to a level at which the proper ecological state of natural environment will meet certain quality criteria.

To perform this task, two chains of negative feedback are envisaged in the management model. The first chain creates connection between the environmental damage caused by production and the maximum permissible debt accumulated by production company which purchases resources per credit. The higher the damage, the less this value should be. Reduction of credit availability will limit acquisition of natural resources and, consequently, the volume of production will be reduced.

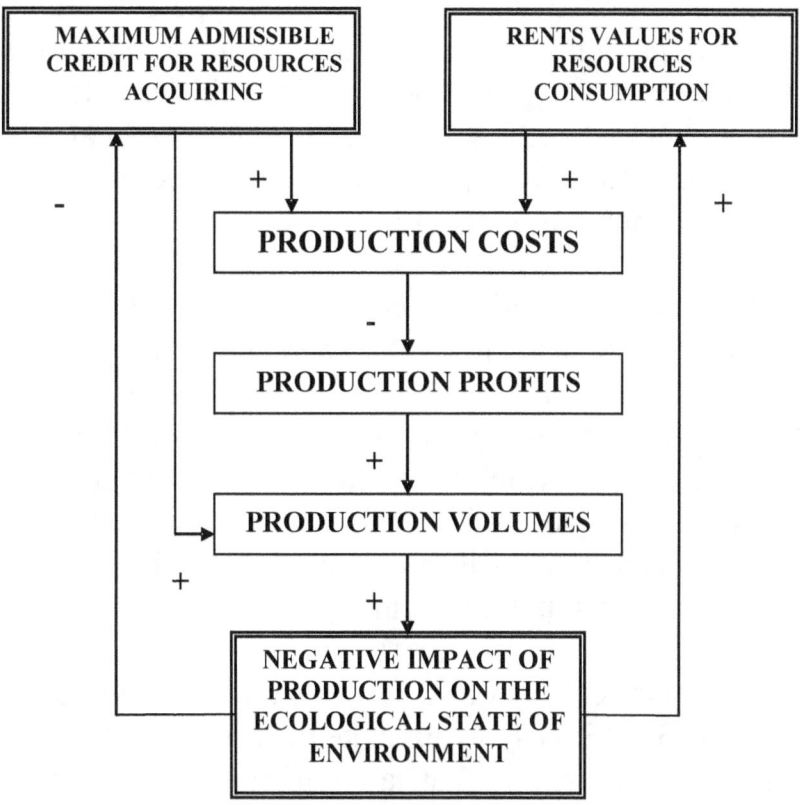

Fig. 4.13. Management structure of *CZA* resources consumption

In addition to this, credit limitation will force the production company to seek for a replacement of insufficient resources. The company will have to change the production technology, and very probably, it will perform other actions that will inevitably lead to increase in production costs and prices for the resulting products. This, in its turn, will reduce the demand for the products, and decrease production profitability. Thus, the debts of the production company can help regulate the intensity of production's impact on the environment.

The other major mechanism of resources management also links the ecological state of environment with the cost, and hence, with profitability of production. Consumption of natural resources constitutes a significant portion of the production cost. Therefore, if an available amount of this type of resources in *CZA's* natural environment is linked with price for this type of resource, then a reduction in natural reserve of this resource will automatically increase the cost of production. This will lead to decrease in production output and will relieve environmental damage. A corresponding closed feedback chain is contained in the right part of the conceptual model in Fig. 4.13.

Let us now consider some examples of applying these management mechanisms to *CZA* resources management. The results are presented in Fig. 4.14 and 4.15. As a goal of our first computer simulation, we will set a task to build a forecast for scenarios of economic processes, which take into account changes in storage of natural resources caused by their consumption. The scheme of resources management will meet the left chain of the feedback presented in a conceptual model in Fig. 4.13.

As the model of economic system contains equations of market resources (4.13), it is easy to simulate the observed dynamics of their variability. Let us assume that, a significant drop in storage of the first kind of the utilized resources takes place in the local *CZA*. The fluctuations in the resource storage of the second type, and a slight decrease in the resource storage of the third type, are shown in Fig. 4.14a.

The main parameters of production technology will remain the same as before (see the previous paragraph): the input prices are constant values: $r_1 = 3,7;$ $r_2 = 7,0;$ $r_3 = 4,5$, for manufacturing of a production unit the following amounts of resources of the local *CZA* area must be used: $y_1 = 1,3;$ $y_2 = 3,3;$ $y_3 = 2,0$. The expected price of a unit of production is defined as $p_1 = 7,0$. We will also assume that initial investment into production is $S_i^0 = 10,0$, the interest deductions from earnings to cover the expenditure is equal to unity, and the demand for the finished product suffers of random fluctuations.

As it was assumed in the calculations, the maximum allowable amount of accumulated credit is proportional to the offer of the first type of *CZA* resources $H_3^* = kc_1(\tau)$. This means that with decrease in

288

the volume of this type of resource in the *CZA* local area, the threshold for the allowable resource consumption credit is also automatically lowered, as it is shown in Fig. 4.14, *b*. Scenarios of working capital dynamics and production output dynamics, shown in the same figure, demonstrate that production was automatically stopped at that time interval when the storage of the first type of the *CZA* resources were minimal. The calculated scenario of production profitability (see Fig. 4.14, *c*) confirms this result.

As the goal of the second computational experiment, we set the task to perform   management of economic processes' scenarios, taking into account the increased costs for the resources, the reserves of which are shrinking. The scenarios obtained in this experiment are shown in Fig. 4.16.

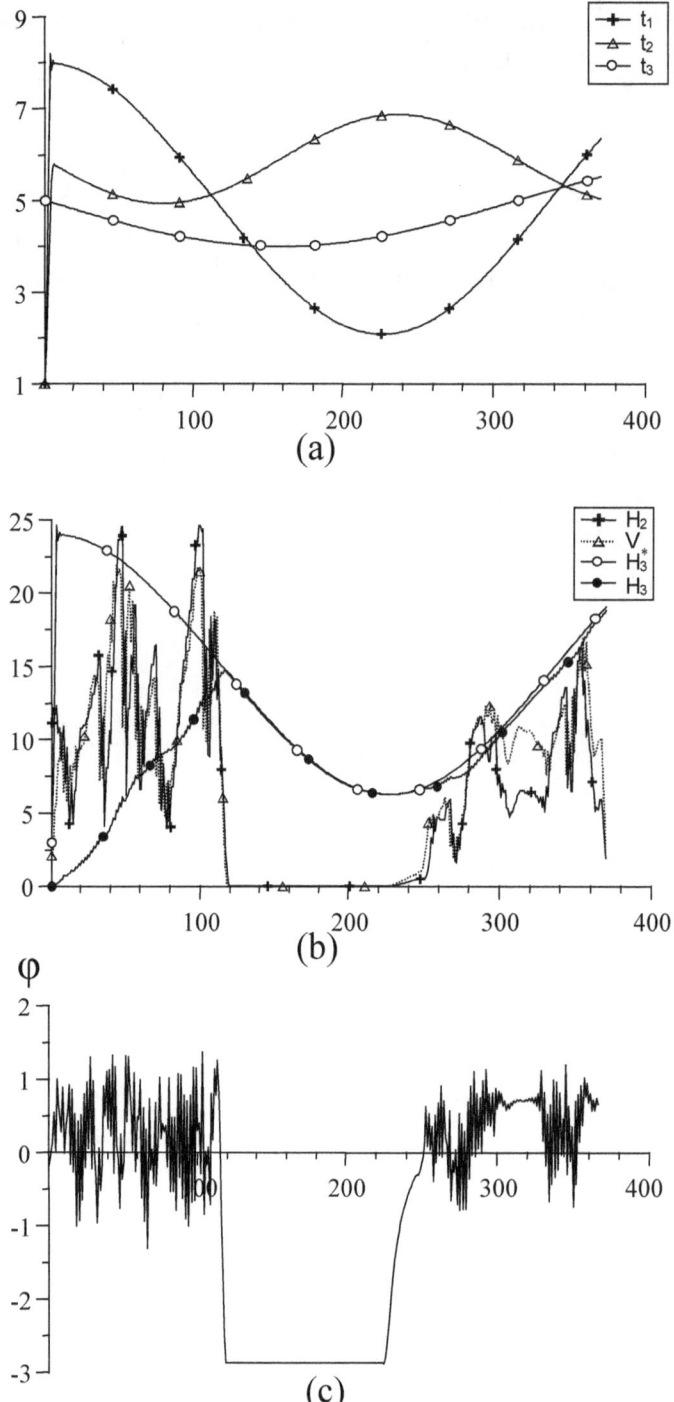

Fig. 4.14. Economic scenarios of management by the criterion of the first type of resources' supply: (a) the dynamics of storage of the three types of resources; (b) working capital H_2, production volume V, administration H_3^*, accumulated debts of environmental charges H_3; (c) profitability of production φ

Fig. 4.15, *a* illustrates scenarios of the amounts of each of the three types of resources obtained by the production company. Fig. 4.15, *b* shows volumes of credits allocated for purchase of the resources. The negative values of scenarios appeared in these figures at the time when the production process was stopped. This means that calculation of the required amounts of resources was performed, and the previously purchased and stored amounts of the resources were taken into account in the calculations by the production company. A similar explanation can be given to the negative values of accumulated credit: with termination of production, the model of economic system continued to count credits, which were required for purchasing resources and continuing the production process.

As it was noted above, it is possible to set up more stringent restrictions on consumption of *CZA* resources. This can be done when, along with utilizing maximum permissible credit for purchasing resources, a direct relationship between rental payments for the resources use and the state of natural environment in the *CZA* is established. In this case, the management process is carried out on all feedback circuits shown in Fig. 4.13.

Fig. 4.17 shows scenarios of purchasing of each kind of resources, accounting for the dynamics of demand for finished products and sales.

Model calculations were made based on the same parameters of the production technology, as in the previous example. However, in this example the prices for the resources have changed and become unstable. In accordance with the *ABC*-model of market resources (4.13), the prices were calculated based on the available storage of each type of the resources, as shown in Fig. 4.16, *a*. Thus, the opposite relationship was used: the smaller was the stored amount of a resource type, the higher was it's cost.

The dynamics of prices for resources is shown in Fig. 4.16, *b*. As we could expect, the greatest increase in prices was associated with the first type of resources, because the storage of this resource type decreased the most dramatically. Therefore, in the run of computational experiment, the value of resource rent was set proportional to the changes of prices for this type of resource.

Scenarios of resources' procurement are shown in Fig. 4.17. Fig. 4.18 depicts the dynamics of production capital inputs, its volumes and the values of accumulated credits.

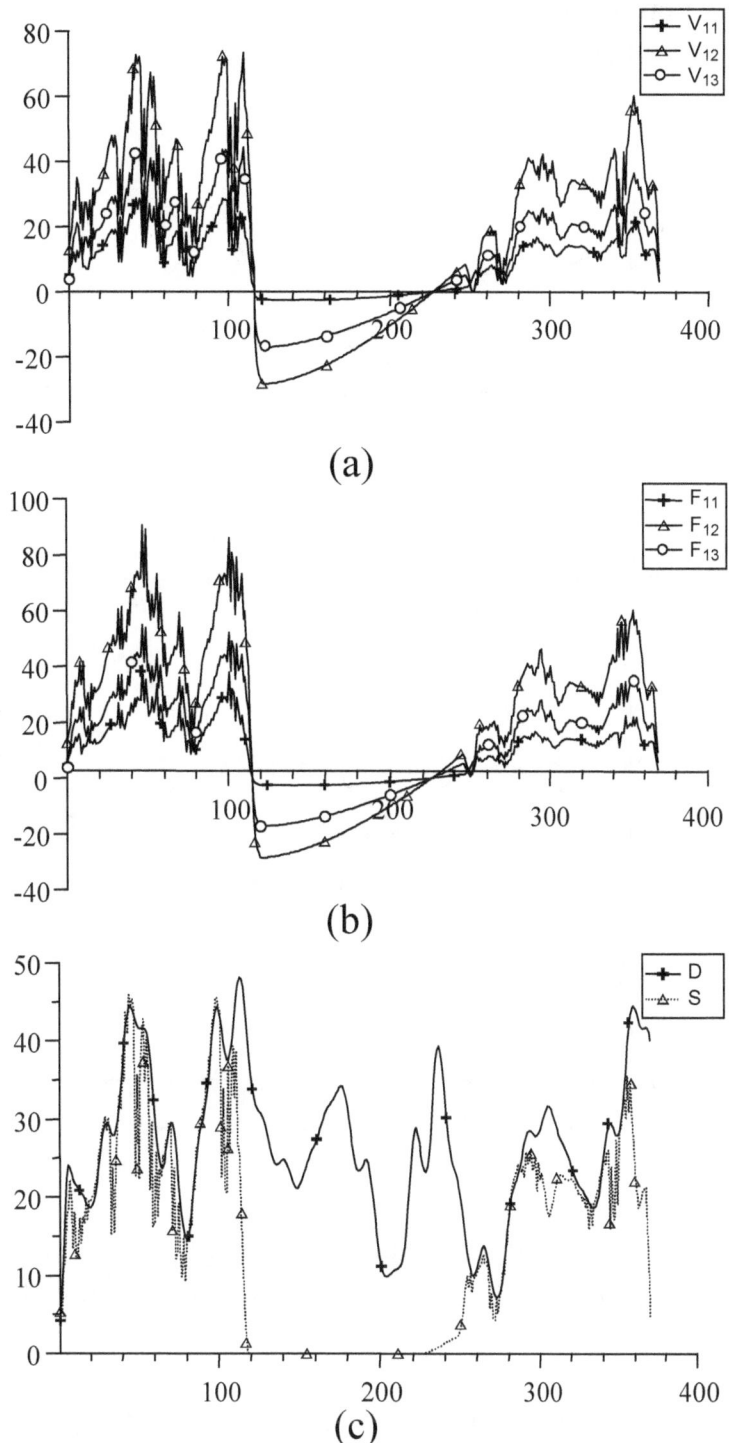

Fig. 4.15. Predicted scenarios: (*a*) volumes of purchased resources V_1, V_2, V_3 ;

293

(*b*) credits obtained to purchase of resources F_{11}, F_{12}, F_{13}; (*c*) demand D and release S of produced goods.

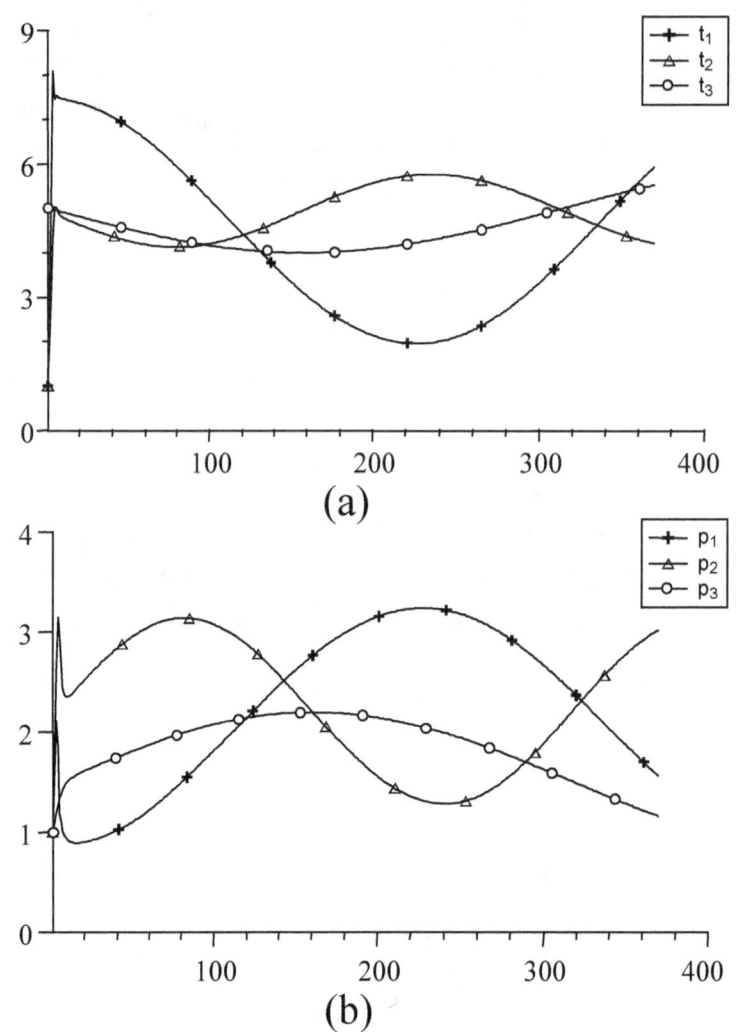

Fig. 4.16. Scenarios of available storage of resources and their costs: (*a*) kinds of resources t_1, t_2, t_3; (*b*) costs of the three types of resources p_1, p_2, p_3

The imitated reduction of storage of the first type of resources significantly affected the dynamics of all economic processes. In the period from step 110 to step 220 of computation, the price of this resource and the size of the resource rent increased so much that the management agent of the output was forced to halt the production.

Scenarios of resources' procurement (Figure 4.17, *a*) and assignment of loans (Figure 4.17, *b*) demonstrate this effect. The sales volumes of finished products in relation to the demand for it, are presented in Fig. 4.17, *c*. In cases when before the termination of production the demand was met virtually completely, the resumption of the production process at the step 220 required some time to accumulate resources and to restore the production output.

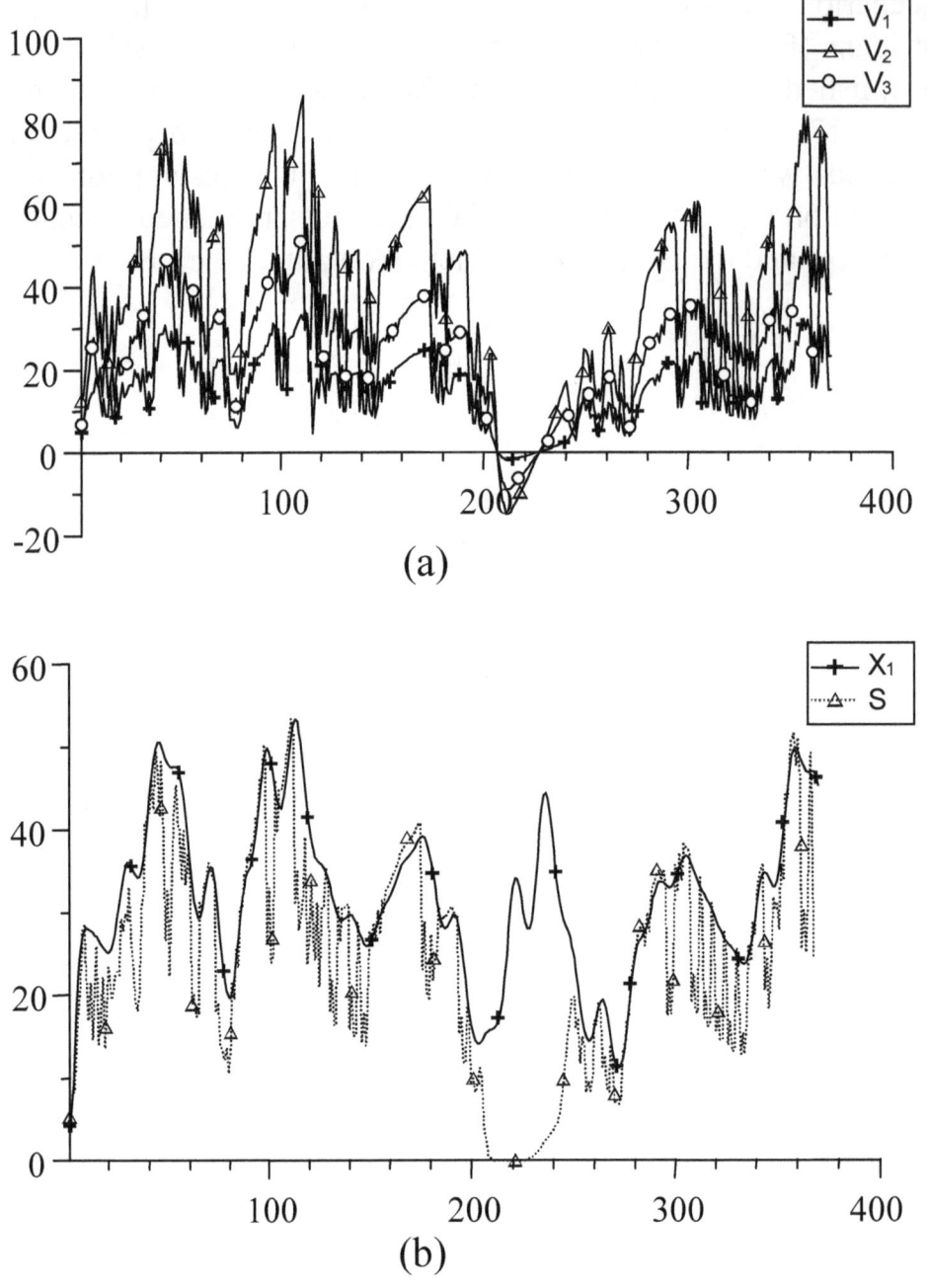

Fig. 4.17. Scenarios of economic processes: (a) volumes of purchased resources V_1, V_2, V_3; (b) demand x_1 and sales of products S

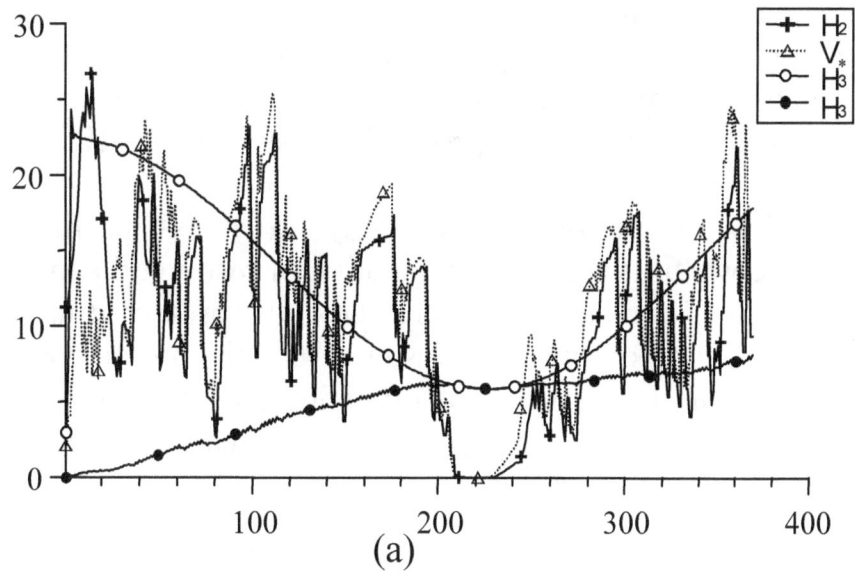

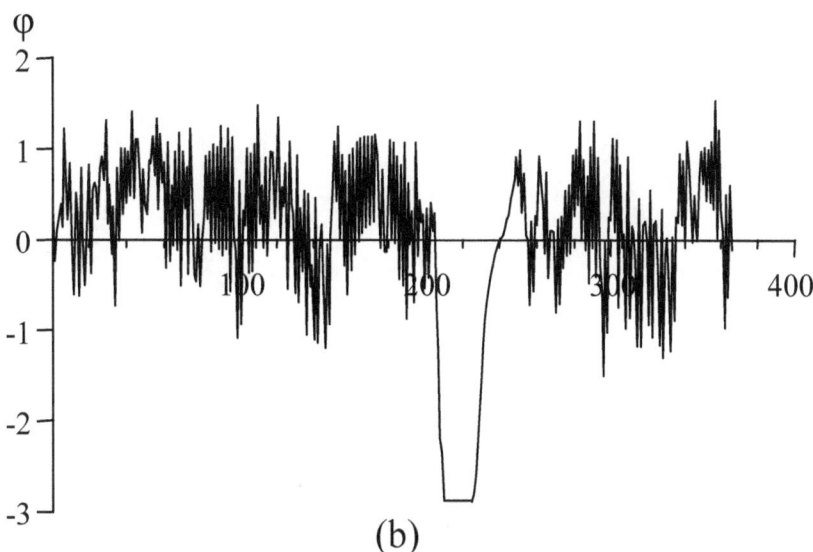

Fig. 4.18. Predicted scenarios of an economic system of production: (*a*) working capital H_2, production volume V, administration H_3^*, accumulated debts of environmental payments H_3, (*b*) profitability of production φ

It is easy to see that, in the initial period of time, the output followed the demand, which experienced random fluctuations. Profits from sales of products did not cover the cost of resources completely, because the cost of the first type of resource and the value of the resource rent related to it, had increased. The production company

297

was forced to acquire resources by credit. The value of accumulated loan increased monotonically up to step 110 of the calculations, when it reached its maximum permissible value. At that moment the credit was terminated and the production was suspended.

Fig. 4.18, *b* provides opportunity to assess profitability of production, which is subject to regulation of resources' consumption. The scenario of conditional resource potential presented in it, shows that in conditions of first-type resource deficit, production became unprofitable due to existence of limitation for maximum allowable credit for purchasing resources, and due to increased resource rent.

Chapter V. BALANCING CONSUMPTION AND CONSERVATION OF MARINE BIORESOURCES IN "SEA – LAND" SYSTEMS

5.1. Conceptual Model of Integrated Bioresources Content in a Marine Ecosystem

To study processes in an ecosystem of a coastal zone area, a computer technology of decision support is required. Such technology is needed to control marine biological resources. In this section, we will build a relatively simple model of integrated processes involved in formation and concentration of bioresources. The model contains 9 variables: phytoplankton, zooplankton, fish larvae, the fish, oxygen, carbon dioxide, nutrients and detritus, which experience influences of solar radiation, sea surface wind, sea temperature and chemicals' inflow with river runoff.

In accordance with fundamental principles of the systems approach [181], construction of a conceptual model of an ecosystem must begin with analysis of the goal of modeling and with evaluation of the processes which are involved into the overall process of the ecosystem's development in its natural environment. At this early stage of modeling, it is necessary to identify all processes which form closed chains of cause-effect relationships existing, as these chains form the basis of the system. This approach is commonly used in the method of system dynamics [50] and it allows to take into account the most important modeling goals of the processes, as well as to clarify the very formulation of the problem, so that the selected system of processes would be adequate to it. Therefore, it makes sense to develop ways of mutual adaptation between the goals of modeling and the structure of the future conceptual model.

We will select predicted scenarios of development processes in a marine ecosystem as our modeling goal. These scenarios are averaged for the period of a few days' concentrations of living organisms and chemicals in a certain volume of the marine environment, characterizing the coastal zone area. Among the living organisms, we will consider four categories of the modeling objects: phytoplankton (PP), zooplankton (ZP), fish larvae (LF) and fish (BR). Among other factors supporting existence of these organisms, the following factors will be used in the model: solar radiation (SR), carbon dioxide (CD), oxygen (OX), nutrients (BG), detritus (DT) and contribution of the river runoff in concentrations of nutrients and detritus in the sea.

As the main external influences on the ecosystem we will select the following factors: the amounts of nutrients and detritus coming into the ecosystem with river runoff (RF), the intensity of the driving wind force (WF), and the sea water temperature (TW).

This selection of processes provides a large number of closed feedback chains in the model of the ecosystem. In particular, phytoplankton releases oxygen consumed by zooplankton, fish larvae and fish. This causes some increase in concentration of carbon dioxide, consumed by phytoplankton. This chain gives an example of a positive feedback, which is stabilized by a negative feedback chain, because as a sequence of increase in zooplankton, fish larvae and fish concentrations, the concentration of oxygen decreases. The conceptual model of processes in the marine ecosystem built with these designations, is shown in Fig. 5.1.

In addition to the variables selected as components of the ecosystem's model, it is necessary to take into account the dependence of the ecosystem's living organisms on conditions of their existence in the marine environment. As it was mentioned above, there are minimal allowable concentrations of food chain elements, oxygen and carbon dioxide. If concentration of some elements reaches its minimum, a sharp reduction of amount of the living organism's populations occurs, due to their death. To describe the conditions of homeostasis in the ecosystem's model, we need to introduce logical operators or management agents G_j, which will be responsible for "watching" the processes, finding out such conditions in the very beginning, and switching on necessary operations to imitate processes of decreasing concentrations of living organisms in the model.

An important role in development of processes in the ecosystem must be given to reactions of the processes on the events (influences) that cause these reactions. For example, an increase in concentration of phytoplankton can't cause an increase in concentration of zooplankton, until the whole process is accompanied with simultaneous increase in concentration of oxygen in sea water. At any given time, and for any of the ecosystem's living organisms, there is a minimum amount of one type of resources necessary for their development. The influence of this resource, which can limit growth of a population of organisms, should be introduced into the model of the ecosystem with the help of another group of management agents, which we will denote as C_i.

The sea surface temperature significantly affects the rate of development of living organisms. Typically, there is a well-defined temperature range in witch the development conditions are formed most favorably. In order to take this into account, we will introduce management agents A_k into the structure of the

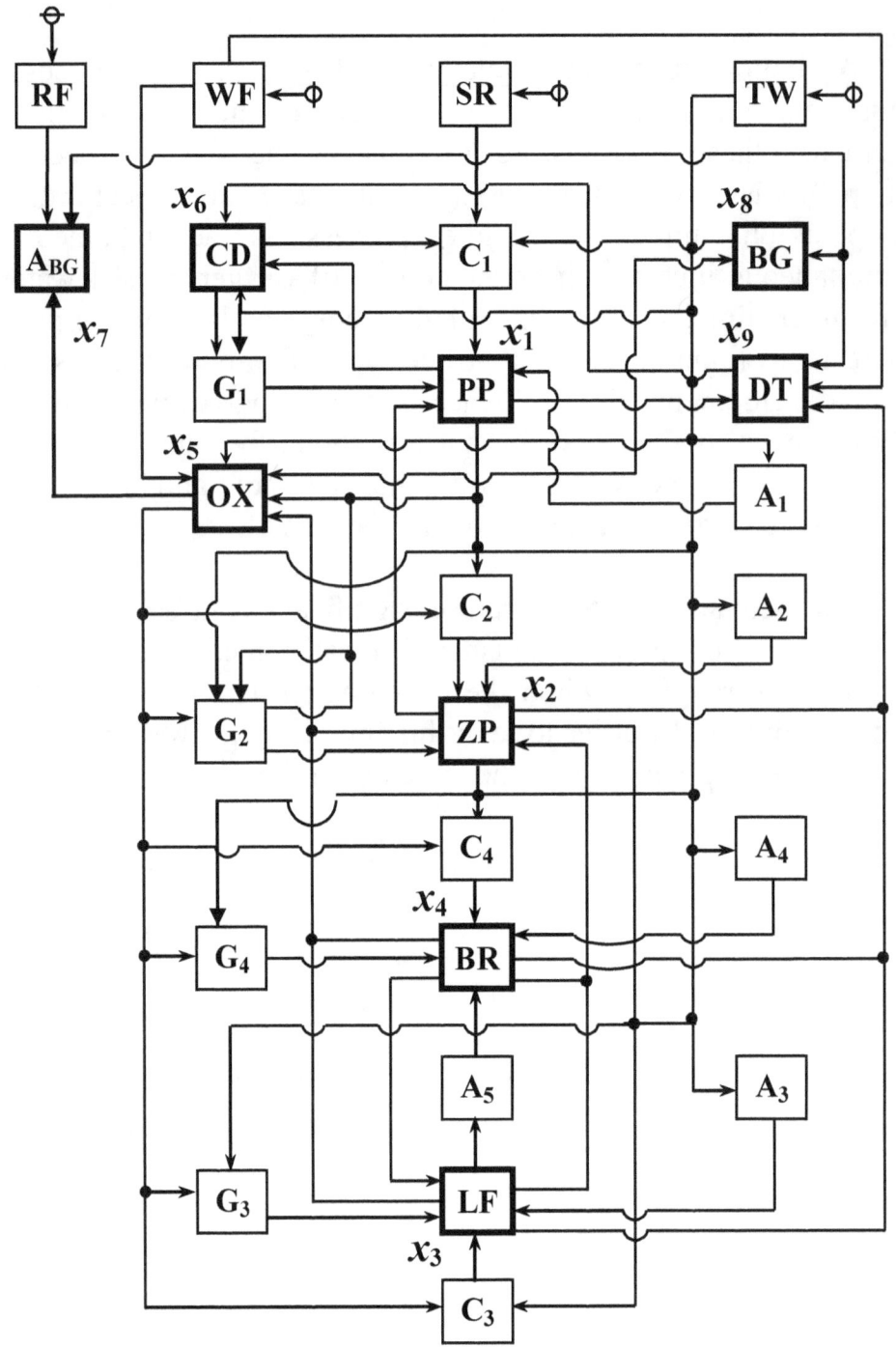

Fig.5.1. A conceptual model of bioresources' formation in marine ecosystem

ecosystem's model. They will enhance the growth processes of living organisms within the values of the sea temperature which are the most favorable for life of these organisms.

To carry out computational experiments with the model of the ecosystem, we will suppose that the river flow carries some amounts of production and domestic waste into the sea, and they contain nutrients and detritus. It is well-known that the sea temperature and solar radiation experience some annual variations, whereas the intensity of wind mixing of the upper sea layer has usually a random character. In order to simplify our model, the convective mixing and the associated seasonal increase in the concentration of nutrients will not be counted. Also, the processes associated with sinking of some amount of detritus to the bottom, will be excluded from our consideration.

5.2. Dynamic *ABC*-Model of Integrated Bioresources in a Marine Ecosystem

For presentation of the ecosystem's model in the form of a system of differential equations, the following notations for concentrations of substances and objects in the sea water were chosen: x_1 – phytoplankton, x_2 – zooplankton, x_3 – fish larvae, x_4 – fish, x_5 – oxygen. x_6 – carbon dioxide, x_7 – contribution from the river runoff in the concentration of nutrients and detritus, x_8 – concentration of nutrients in the marine environment, x_9 – concentration of detritus. The equations of the *ABC*- model of the ecosystem had the following form

$$\frac{dx_1}{dt} = x_1\{1 - 2[x_1 - AQ_1(x_6, x_8, SR) + G_1(x_6) + G_1(x_8) + a_{12}x_2 - A_1(TW)]\},$$

$$\frac{dx_2}{dt} = x_2\{1 - 2[x_2 - AQ_2(x_1, x_5, x_8) + G_2(x_5) +$$
$$G_2(x_1) + G_2(x_8) - a_{23}x_3 - A_2(TW)]\}$$

,

303

$$\frac{dx_3}{dt} = x_3\{1 - 2[x_3 - AQ_3(x_2, x_5, x_8) + G_3(x_2) +$$

$$G_3(x_5) + G_3(x_8) - a_{34}x_4 - A_3(TW)]\}$$

,

$$\frac{dx_4}{dt} = x_4\{1 - 2[x_4 - AQ_4(x_2, x_5, x_8) + G_4(x_2) + G_4(x_5)$$

$$+ G_4(x_8) - A_5(x_3) - A_4(TW)]\},$$

$$\frac{dx_5}{dt} = x_5\{1 - 2[x_5 - a_{51}x_1 + a_{52}x_2 + a_{53}x_3 + a_{54}x_4 +$$

$$a_{58}x_8 + a_{59}x_9 - a_{5/WF}WF]\},$$

(5.1)

$$\frac{dx_6}{dt} = x_6\{1 - 2[x_6 + a_{61}x_1 - a_{69}x_9 - a_{62}x_2 - a_{63}x_3 - a_{64}x_4 + a_{6/TW}TW]\},$$

$$\frac{dx_7}{dt} = x_7\{1 - 2[x_7 - A_{7/BG}(x_5, RF)]\},$$

$$\frac{dx_8}{dt} = x_8\{1 - 2[x_8 - a_{87}x_7 + a_{81}x_1 + a_{82}x_2 +$$

$$a_{83}x_3 + a_{84}x_4 - A_{8/BG}(x_5, x_9)]\},$$

$$\frac{dx_9}{dt} = x_9\{1 - 2[x_9 - a_{91}x_1 - a_{92}x_2 - a_{93}x_3 - a_{94}x_4 - a_{97}x_7 + a_{95}x_5]\}.$$

In addition to affecting functions of the form $a_{kl}x_{kl}$ in the right-hand side of the *ABC*-model equations (5.1), we included management agents which have designations G_i, A_i and C_i, that performed miscellaneous support operations. Below, we will focus on the structure of these terms and on the role which they play in models of the ecosystem.

Application of Management Agents in the Model of the Ecosystem. The living organisms in the sea have a limited living space, not only geographically, but also in the physical-chemical sense. For each kind of the organisms, there exists a typical area of necessary environmental parameters – the zone of homeostasis, in which the organism can survive and reproduce. The group of management agents, that control zones of homeostasis of living organisms, should be represented in the *ABC*-model (5.1) by the following formulas:

$$G_1(x_6) = IF\{x_6 > x_{16}*; 0; k_{16}[1 - \exp(-\alpha_{16}t)]\};$$

$$G_1(x_8) = IF\{x_8 > x_{18}*; 0; k_{18}[1 - \exp(-\alpha_{18}t)]\};$$

$$G_2(x_5) = IF\{x_5 > x_{25}*; 0; k_{25}[1 - \exp(-\alpha_{25}t)]\}; \qquad (5.2)$$

$$G_2(x_1) = IF\{x_1 > x_{21}*; 0; k_{21}[1 - \exp(-\alpha_{21}t)]\};$$

$$G_2(x_8) = IF\{x_8 > x_{28}*; 0; k_{28}[1 - \exp(-\alpha_{28}t)]\};$$

$$G_3(x_5) = IF\{x_5 > x_{35}*; 0; k_{35}[1 - \exp(-\alpha_{35}t)]\};$$

$$G_3(x_1) = IF\{x_1 > x_{31}*; 0; k_{31}[1 - \exp(-\alpha_{31}t)]\}; \qquad (5.3)$$

$$G_3(x_8) = IF\{x_8 > x_{38}*; 0; k_{38}[1 - \exp(-\alpha_{38}t)]\};$$

$$G_4(x_5) = IF\{x_5 > x_{45}*; 0; k_{45}[1 - \exp(-\alpha_{45}t)]\}; \qquad (5.4)$$

$$G_4(x_8) = IF\{x_8 > x_{48}*; 0; k_{48}[1 - \exp(-\alpha_{48}t)]\};$$

$$A_{BG} = \{A_{7/BG}; A_{8/BG}\},$$

$$A_{7/BG} = IF\{x_5 > x_5*; a_{7/RF}RF; -a_{7/RF}RF[1 - \exp(-\alpha_{7/RF}\tau)]\}, \qquad (5.5)$$

$$A_{8/BG} = \min(a_{85}x_5; a_{89}x_9)$$

In these formulas, the lower border zones of homeostasis marked as $x*$. Parameters k_i and α_{ij} establish the death rates of living organisms outside the zone of homeostasis.

To perform the operations limiting the growth of living organisms in dependence of the available resources, we included a group of restrictive agents into the equations of the model, which have the form

$$AQ_1(x_6, x_8, SR) = IF\{M_1(t) - M_1(t - \Delta t) < 0; M_1(t);$$

$$M_1(t - \Delta t) + [M_1(t) - M_1(t - \Delta t)][1 - \exp(-\alpha_{AQ_1}\tau)]\},$$

$$M_1(t) = \min[a_{16}x_6(t); a_{18}x_8(t); a_{1/SR}SR(t)];$$

$$AQ_2(x_1, x_5, x_8) = IF\{M_2(t) < M_2(t - \Delta t); M_2(t);$$

$$M_2(t - \Delta t) + [M_2(t) - M_2(t - \Delta t)][1 - \exp(-\alpha_{AQ_2}\tau)]\},$$

$$M_2(t) = \min[a_{21}x_1(t), a_{25}x_5(t), a_{28}x_8(t)]$$

$$AQ_3(x_2, x_5, x_8) = IF\{M_3(t) < M_3(t - \Delta t); M_3(t);$$

$$M_3(t - \Delta t) + [M_3(t) - M_3(t - \Delta t)][1 - \exp(-\alpha_{AQ_3}\tau)]\},$$

$$M_3(t) = \min[a_{32}x_2(t), a_{35}x_5(t), a_{38}x_8(t)], \qquad (5.6)$$

$$AQ_4(x_2, x_5, x_8) = IF\{M_4(t) - M_4(t - \Delta t) < 0; M_4(t);$$

$$M_4(t - \Delta t) + [M_4(t) - M_4(t - \Delta t)][1 - \exp(-\alpha_{AQ_4}\tau)]\},$$

$$M_4(t) = \min[a_{42}x_2(t), a_{45}x_5(t), a_{48}x_8(t)],$$

where Δt is a step of calculation algorithm in time.

This group of agents allows us to imitate instantaneous decrease in concentrations of living organisms, when the amount of the resource which is limiting their concentration, decreases. In an opposite case, when there is a growing amount of the limiting kind of resource, its impact on the concentration of living organisms is delayed for some time τ_k. The parameters α_{C_i} in expressions for management agents determine the growth rate of concentrations in the ecosystem in terms of the increase of resource potential in the marine environment. Thus, this group of agents combines functions of limitation and delays the growth of concentrations of living organisms.

Another group of agents in the equations of the ecosystem's model, which has the symbol A_i, can enhance the growth processes of living organisms, where the temperature conditions became the most favorable for their development.

Let us denote the most favorable values of temperatures and use the Gaussian functions for agents $A_1 - A_4$ in the following form [89]:

$$A_i(TW) = \exp[-\beta_i(TW - TW_i^*)^2], \quad (i = 1, 2, 3, 4). \tag{5.7}$$

Management agent $A_5(x_3)$ included in the equation of fish concentration x_4, provides an opportunity to simulate the effect of fish larvae on this variable. As the concentration of larvae grows, its influence on the concentration of fish appears with delay that takes

307

into account the finite time of the larvae growth to the size of the fish, and is determined by the choice of parameter α_{43}

$$A_5(x_3) = IF\{x_3(t) - x_3(t - \Delta t) < 0; 0;$$

$$a_{43}x_3(t - \Delta t) + a_{43}[x_3(t) - x_3(t - \Delta t)][1 - \exp(-\alpha_{43}\tau)]\}. \tag{5.8}$$

It is necessary to note that we excluded from consideration the early stage of the fish larvae development, when they feed exclusively with phytoplankton. In fact, this is not about the fish larvae but the juveniles, which have not yet reached reproductive age, but have switched to feeding with zooplankton and are continuing to grow, albeit more slowly than the larvae at an early stage of development. In this particular period of development, the agent $A_5(x_3)$ is introduced into the model and begins to accomplish its function. The formal model which we built here, contains a large number of coefficients, which allows us to simulate development processes in the marine ecosystem under different conditions.

We believe that the main influence on the concentration of Bioresources (fish) in the modeling marine ecosystem is produced with lack of oxygen caused by spending it on the oxidation of products, which are made with the river runoff. To maintain the necessary supply of fishery resources, the inflow of these products into the sea should be limited by making of the environment protection actions (the reduction of hazardous waste getting into the rivers from industrial and agricultural enterprises, the use of wastes treatment facilities, etc.). Implementation of these expensive operations requires informational support, which must be performed by the calculation of the model scenario processes in the marine ecosystem.

5.3. Reaction of the Coastal Zone Marine Ecosystem to the Impact of Nutrients and Detritus Coming in with River Flow

The above model allows us to connect concentrations of fish in the sea with the amount of nutrients and detritus x_7 coming into the ecosystem with the river runoff *RF*. To organize management of removal of these substances, it is necessary to make sure that the inverse relationship between concentration of oxygen in marine environment x_5 and the volume of incoming chemicals in it is established, which can be implemented by a control agent (5.5).

To develop scenarios of bioresources' concentrations, all the modeled processes were brought together into a common dimensionless scale of variables (0, 10). The coefficients of the equations of the ecosystem (5.1) were chosen in the range values from 0.1 to 0.8. We took step calculations Δt, as a unit of time measurement – each equal to one day. Calculations were made at 500 time steps. Fig.5.2 shows simulated scenarios of annual variations of external influences applied to the ecosystem.

To simulate temporal variability of sea water temperature, we used harmonic function with annual period; with the peak attributed to the month of August. A scenario of sea surface illumination was presented with a similar function with the peak in mid-June, and with a low intensity modulation of the annual harmonic amplitude. This helped us imitate random distribution of cloudy days during the year. Random fluctuations of average wind speed (on a unit of the area) were set, and then we simulated an incoming flow of nutrients and detritus (significant in the amplitude) during a period from day 250 to day 350, the maximum of which attributed to day 300 of computing.

The results of management by oxygen concentration are shown in Fig. 5.3, *a, b*.

The logical operator of control included in the equation for x_7 was to ensure that concentration of oxygen in the sea did not fall down below the allowable minimum value of $x_5{}^*$. In cases where this condition was not fulfilled, the agent began to reduce the amount of *RF*, compared with unregulated scenario. It continued the management operation until the condition $x_5 > x_5^*$ was restored. This way, we were imitated environment protection actions aimed at preserving a certain concentration of oxygen in the sea.

As it can be seen from Fig. 5.3, concentrations of oxygen in the sea significantly depended on concentrations of nutrients, because the changes in their calculated scenarios took place in opposite directions. The critical level of oxygen concentration was set as $x_5^* = 1$.

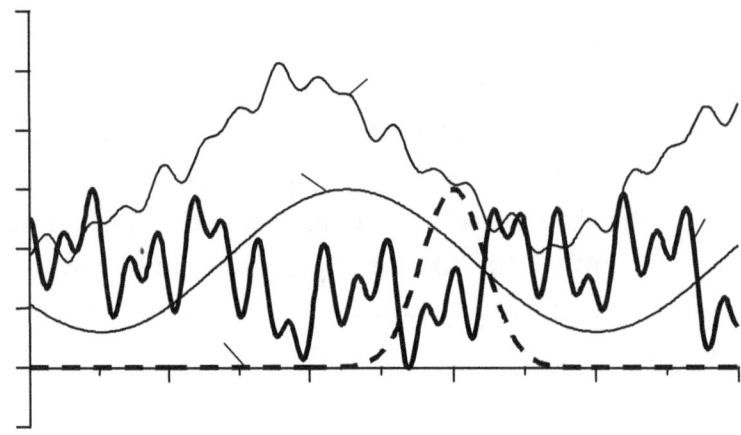

Fig. 5.2. Dimensionless scenarios of external influences on the ecosystem as
functions of calculation steps: 1 – *SR*, 2 – *TW*, 3 – *WF*, 4 – *RF*

Leaching of nutrients with the river runoff led to significant fall of the scenario x_5. In Fig. 5.3, *a* the scenario of oxygen concentration is shown in absence of management (curve *3*). We can conclude from it that the period of critically low values of x_5 in this case was about 100 days. The same figure shows the scenario x_5 after the management control (curve *4*). A comparison of these two scenarios shows that, as a result of the management, the period of low values for x_5 decreased by about a half.

The functioning of the management agent, which simulated environment protection actions, is shown in Fig. 5.3, *b*. This function allowed to vary the rate of the oxygen concentration recovery and, consequently, to vary the duration of environmental actions. In particular, the intensity of control of the nutrients inflow for the scenarios shown in Fig. 5.3, *b*, was set by selection of coefficient $\alpha_{7/RF} = 0.1$, which corresponded to reduction of nutrients inflow due to the river runoff, by 50% in 70 days.

Fig. 5.3, *a*, *b* shows the results of management of biological resources of the marine ecosystem. In order to assess how much the management of nutrients and detritus influenced them, we performed two computational experiments. In the first experiment (see Fig.5.3, *a*) management was absent. Scenarios of concentrations of all living species, which depend on oxygen concentration, show extinction of organisms as a result of hypoxia. In the second experiment, management was switched on (see Fig. 5.3, *b*). It noticeably weakened the value of decrease of the oxygen concentration, and shortened the duration of hypoxia for fishery resources.

The constructed scenarios allow us to estimate the influence of constraints that simulate environment protection actions aimed at nutrients and detritus volumes' reduction on bioresources concentration in the sea. By comparing each scenario of bioresources concentration before and after removal of management control, it is possible to obtain estimations of influence on increase of fish resources concentration in the sea.

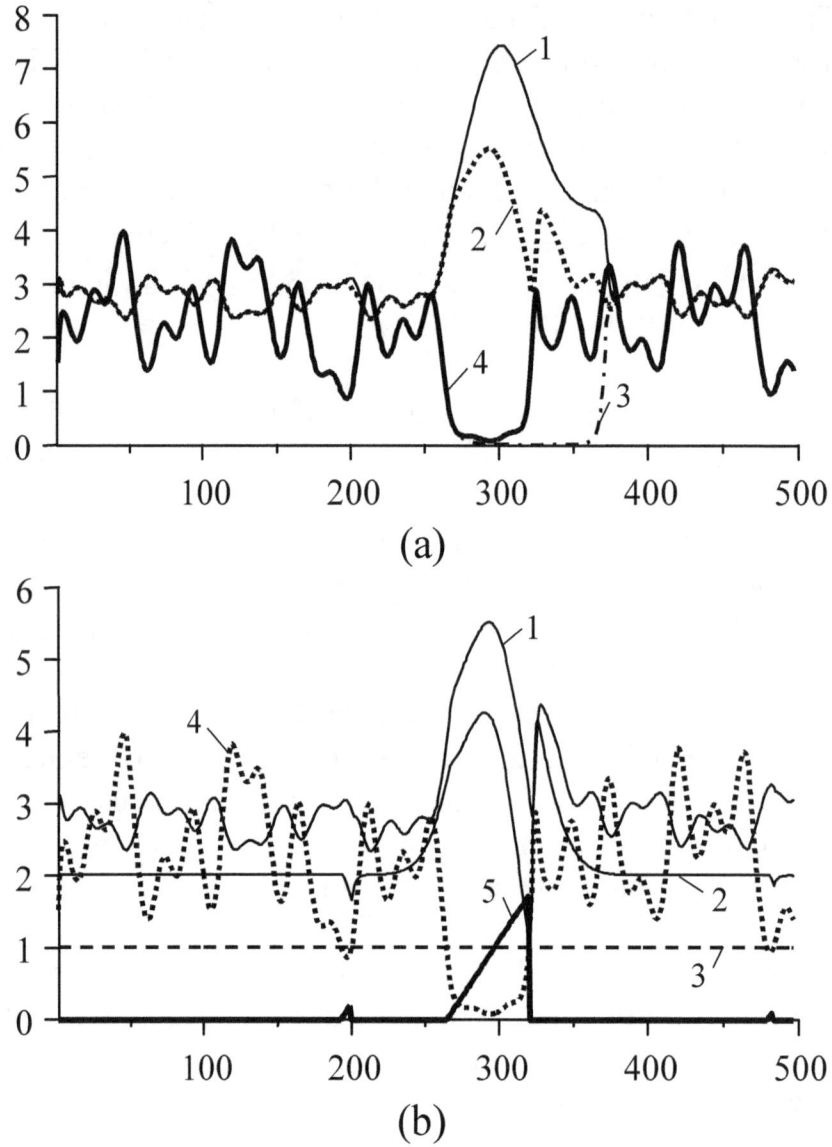

Fig. 5.3. Results of oxygen concentration management by regulation of the river inflow of nutrients and detritus in the sea: (*a*) 1 – concentration of nutrients in the sea *BG* before control, 2 – *BG* in the sea after control, 3 – concentration of oxygen *OX* before control, 4 – *OX* after control; (*b*) 1 – *BG* in the sea after control, 2 – removal of chemicals into the sea *RF* after control, 3 – lower limiting boundary of the Bioresources homeostasis x_5* on oxygen, 4 – *OX* after control, 5 – a possible scenario for management agent $A_{7/BG}$ (was not switched on)

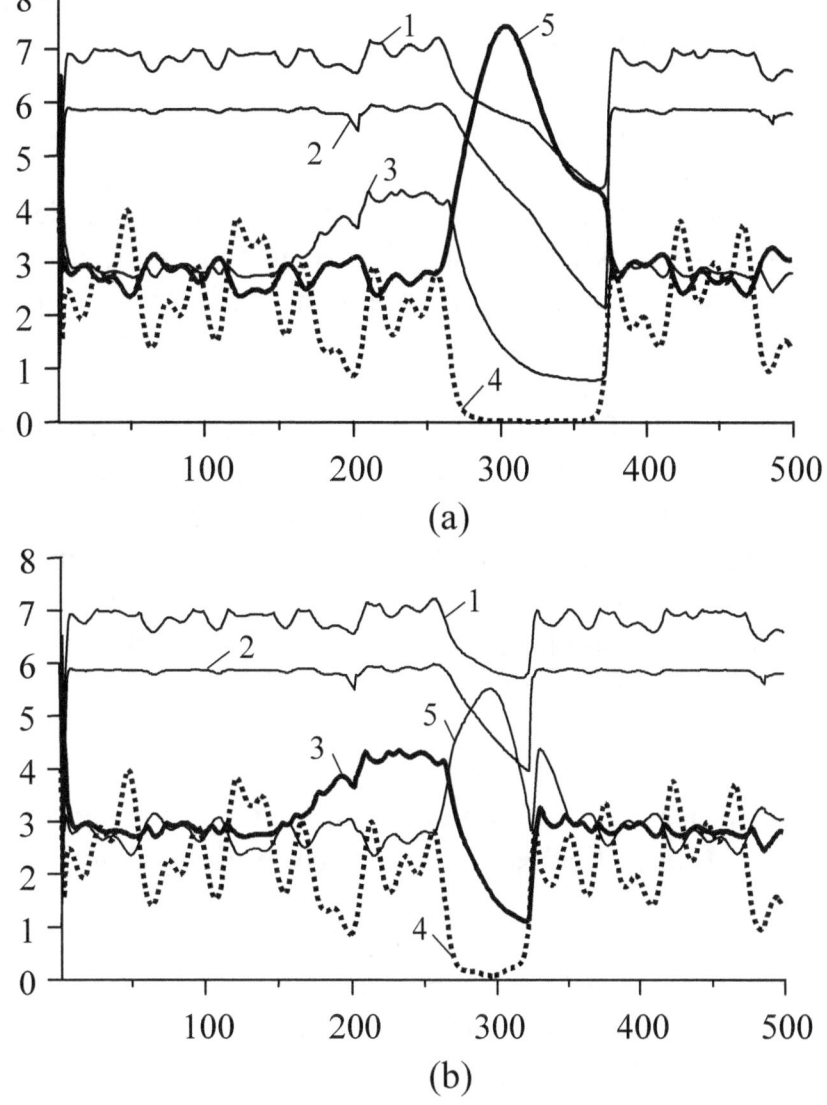

Fig. 5.4. Reaction of the ecosystem on the remove of nutrients and detritus with river runoff: (*a*) concentrations in the absence of control on the amount of oxygen: 1 – Bioresources *BR*, 2 – fish larvae *LF*, 3 – zooplankton *ZP*, 4 – oxygen *OX*, 5 – biogens *BG*; (*b*) the concentrations in terms of the control on the amount of oxygen: 1 – *BR*, 2 – *LF*, 3 – *ZP*, 4 – *OX*, 5 – *BG*

As it can be seen from comparison of scenarios in Fig. 5.4, *a, b*, as a result of environment protection actions, the minimum in concentrations of bioresources increased from 4.5 to 6 units. As a

313

result of management, the loss of biological resources' volume, due to a period of hypoxia, decreased by 3 times.

In a general case, a scenario of fish stocks in the sea is a complex function of many of variables, which is impossible to predict without use of a dynamic model of marine ecosystem. The dynamic model of marine ecosystem, which we considered in this section, is a tool to support decision making in management of integrated processes in coastal zone area. For practical use of such models, extensive and painstaking work on initialization of the model's parameters is required, as all coefficients must "reflect" specific conditions of the marine ecosystem. However, before proceeding to adaptation of a model to real processes in the environment, one must be sure in appropriateness of creation of the model at all. Such findings may only be obtained by simulation of development processes in various types of ecosystem models, which are created in terms of controlled numerical experiments.

5.4. Bioresources Consumption Management in a "Sea – Land" Ecological-   Economic System

In this chapter, we will take a look at the balance between consumption and reproduction of marine resources in the "sea - land" system. It means such an organization of fishing and processing of bioresources, at which the production subsystem of "land" allocates a part of its profits for environmental needs, which are supposed to support normal state of marine ecosystems. Therefore, the structure of management model of bioresources consumption, must contain continuous assessment of the ecosystem's health and continuous monitoring of allocation of necessary funds for environmental purposes.

Structure of Integrated Model of Environmental Management. Let us study an integrated model of nature management, built with the use of the *ABC-AGENT* technology. We will use an index of biodiversity of the natural environment Bd and the level of pollution Pl as indicators of ecological state of the system. The structure of the environmental management model is shown in Fig. 5.5.

In accordance with the above-considered technology, current demand for marine production D is compared in the block of model forecasts of development scenarios with existing reserves of finished products, and in the case when the product price P exceeds its cost E, the management agent gives an order to sell some portion of products S, and to produce an additional portion of it (if necessary) V. Making a comparison between sales prices and production costs, opens us a possibility to control the entire system, so that profitability is ensured thanks to maintaining ecological condition of the marine ecosystem at a proper level.

The size of investment assigned by the production system to acquire resources for production, serve to pursue this goal. In order to simplify the model, in the further discussion we will unite all kinds of resources consumed by the system into three

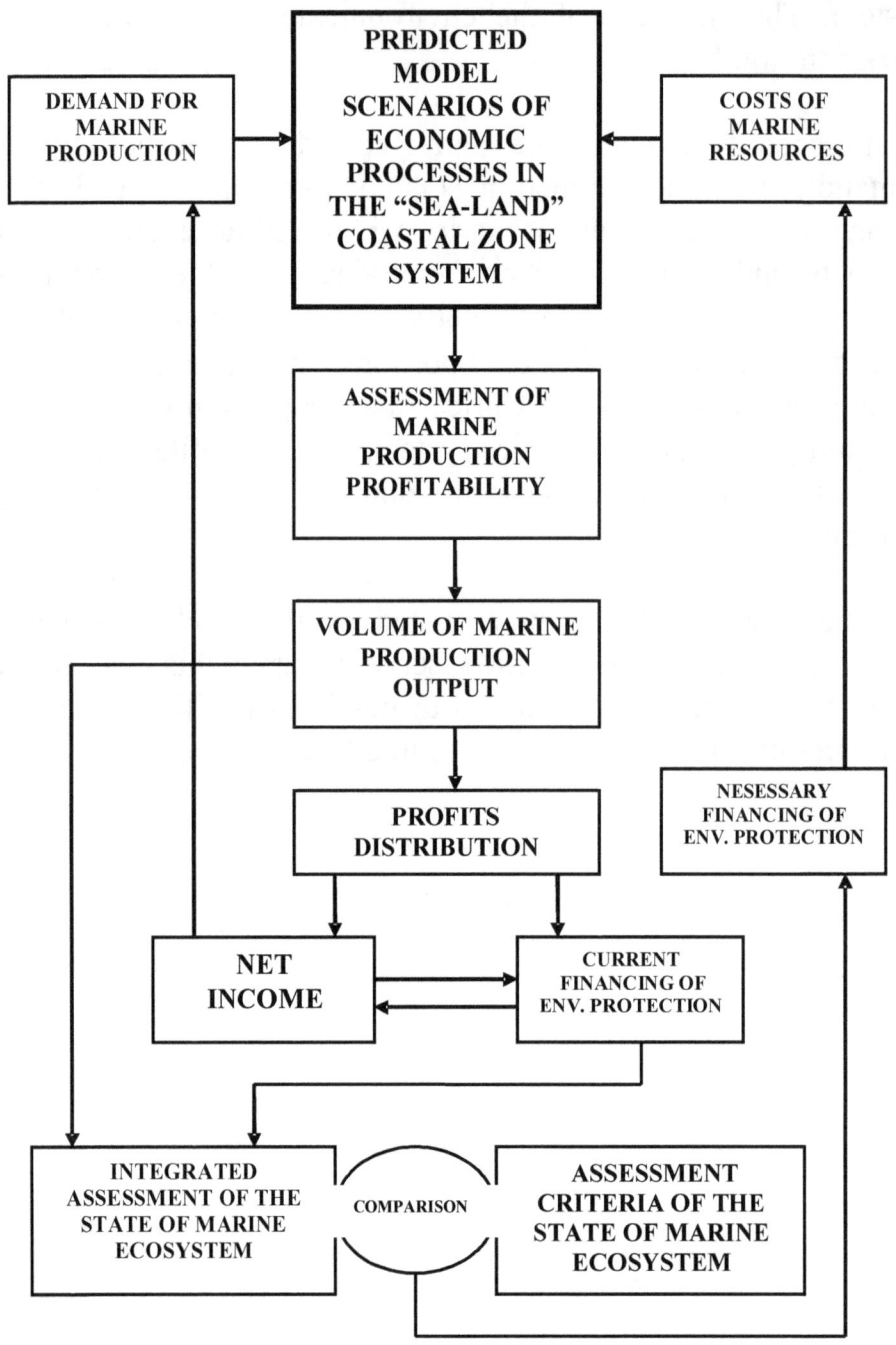

Fig. 5.5. Management structure of ecological-economic balance of environment   protection activities (env. – environment)

main categories: economic resources, biological resources and ecological resources. The economic resources of the *CZA* system determine production costs of seafood products. The biological resources include costs of the *CZA* monitoring, in order to assess current state of biological resources and to evaluate necessary environmental actions that support integrated level of biodiversity of the marine environment. The ecological resources relate to charges paid by the economic system for monitoring of water area contamination levels with production waste, as well as for environmental activity which reduces that level to some average long-term values.

Prediction of Economic Processes in "Sea-Land" Systems. The economic system invests its funds into acquisition of each of these types of resources; moreover, the size of investment is determined with the help of management agents that imitate reaction of the model to environment protection activities, while the cost of these actions is covered by a part of the profit obtained by the economic system. The management agents in the economic system's model structure compare current levels of funding of the production system – its working capital H_2 with the funds required to cover the costs for environmental purposes.

In the block of prediction of development scenarios, one can see an agent which determines amounts of loan funds that the economic system requires to cover the insufficient funds. It is assumed that the total amount of credit funds H_3, accumulated by the system by a certain date, must not exceed a certain limit of admissible value of H_3^*. This value can be used to control consumption of marine resources, because it limits economic activity of the system by setting the maximum allowable credit for acquisition of resources required for production of marine products.

These considerations make it possible to apply the model of *ABC-AGENT* technology (considered above in Chapter IV), to build a set of equations for economic processes accompanying consumption of the *CZA* marine bioresources and seafood products' manufacturing. Let us denote X_1 as a dimensionless variable representing demand D for seafood products of the *CZA* economic system. Then, the equation for the demand function can be written as

$$\frac{dX_1}{dt} = X_1 \left[1 - 2 \left(X_1 + a_{12} X_2 - a_{1F_1} F_D \right) \right],$$ (5.9)

in which X_2 is a dimensionless product price p, and F_D is an external influence defining the dynamics of demand. Similarly, let us represent an equation for the price X_2 and for X_3 – the dimensionless costs of product e

$$\frac{dX_2}{dt} = X_2 \left[1 - 2 \left(X_2 + a_{23} X_3 \right) \right],$$ (5.10)

$$\frac{dX_3}{dt} = X_3 \left[1 - 2 \left(X_3 - \sum_{i=1}^{n} y_i r_i - Q \right) \right],$$ (5.11)

where r_i are costs of resources, and Q is the product's added value.

Let us denote number of ready-for-sale products as H, by S we'll introduce the volume of sales per unit of time (e.g. per one day), and by V we'll denote production output for the same period of time. Then, the dynamic equation for finished products' volume takes the form

$$\frac{dH}{dt} = H \left[1 - 2 \left(H - V + S \right) \right],$$ (5.12)

$$S = IF \ (p < e; \ 0; \ R), \qquad R = IF \ (D < H; \ D; \ H),$$

$$V = IF \ (D < H; \ 0; \ M), \quad M = IF \ (D - H < M; \ D - H; \ M),$$

$$M = \min \ (m_1; \ m_2; \dots, m_n), \qquad m_i = H_{1i} / y_i \ \ (i = 1, 2, \dots, n),$$

where H_{1i} are the volumes of resources stocks available for production, y_i are the amounts of each kind of resources needed to produce a single product.

Similarly, we can represent dynamic equations and logical operators for inventory control of production resources and for working capital H_2 [181, 183]

$$\frac{dH_{1i}}{dt} = H_{1i}\left[1 - 2\left(H_{1i} - V_{1i} + S_{1i}\right)\right], \quad (i = 1, 2, \ldots, n),$$
(5.13)

$$V_{1i} = IF[D - H_{1i} < 0; 0; IF[y_i(D - H) < H_{1i}; 0; U_{1i}]],$$

$$U_{1i} = IF[y_i(D - H) - H_{1i} < \rho_i H_2 / r_i; y_i(D - H) - H_{1i};$$
$$IF[\rho_i(H_3^* - H_3) < 0; 0; U_{1i}^*]],$$

$$S_{1i} = IF[D - H < 0; 0; IF[y_i(D - H) < H_{1i}; y_i(D - H); H_{1i}]],$$

$$\frac{dH_2}{dt} = H_2\left[1 - 2\left(H_2 - pS + \sum_{i=1}^{n} S_2^i + S_3 + \chi H_2\right)\right],$$
(5.14)

$$S_2^i = IF[r_i y_i(D - H) - H_1^i < \rho_i H_2; r_i y_i(D - H) - H_1^i; \rho_i H_2],$$

$$S_3 = IF[\theta H_3 < H_2; \theta H_3; H_2],$$

$$\rho_i = \frac{r_i y_i}{r_1 y_1 + \ldots + r_n y_n}, \qquad (i = 1, 2, \ldots, n),$$

where χ represents a percentage of funds derived from turnover, i.e. the net profit of the system, and θ is a percentage of redemption of accumulated credit.

Let us denote volumes of resources purchased on credit as $V_{11}, \ldots, V_{1n}$. Then, the equation for accumulated credit H_3 takes the form [181]

$$\frac{dH_3}{dt} = H_3\left[1 - 2\left(H_3 - \sum_{i=1}^{n} r_i V_{1i} + S_3\right)\right],$$
(5.15)

319

$$V_{1i} = IF[(D-H)y_i < H_{1i}; 0; F_i]_,$$

$$F_i = IF[r_i(y_iD-H_{1i}) < \rho_iH_2; y_iD-H_{1i}; F_i^*], \ (i = 1, 2, \ldots, n),$$

Functions U_{1i}^* and F_i^* in equations (5.14) and (5.15), represent management actions that restrain amounts of resources which can be purchased by credit.

Profitability of production is determined by ratio of the income generated during a set time interval, to the total expenditure made during this time interval

$$\varphi = \ln \frac{[pS+1]}{[10 + \sum_{i=1}^{3} S_2^{\ i} + S_3 + \chi H_2]} \cdot$$

It is obvious that the value of production profitability will be positive when the income exceeds the expenditure, and it will be negative in the opposite case.

5.5. Scenarios of Marine Bioresources Consumption Considering Marine Biodiversity Conservation Expenditure

The bioproductivity of a coastal zone area strongly depends on pollution level of the marine environment. To control the state of the marine ecosystem, we will choose two of the integrated criteria: the index of marine biodiversity Bd, and the level of pollution with substances that are harmful to marine organisms Pl. To introduce a biodiversity index, we will use a weighted sum of concentrations Br_i of major marine organisms inhabiting a particular coastal zone area, averaged over the waters of this zone

$$Bd = \sum_{i=1}^{N} g_i Br_i . \tag{5.16}$$

We will assume that the average long-term variability of this index is known, and that it is being formed, for instance, by monthly averaged concentrations of major elements of a marine food chain Br_{im}, characteristic for the zone of the sea. Current index values for each year undergo a deviation from the average long-term variability. As the key factors shaping these deviations, external influences on the marine ecosystem are considered: the weather conditions (variations in air temperature, wind speed modulus, illumination, etc. around the respective mean annual values), and anthropogenic impacts (removal of pollutants into the sea from land, marine accidents happening to vessels, etc.).

Concentration of contaminants Pl plays a significant role in shaping values of biodiversity index. In order to introduce into the model the process of biodiversity index' reconstruction to a set level by means of environment conservation activities, we must used management agent $A(Bd, Bd_m)$ in equation for the index Bd. Thus, the dynamic equation of biodiversity index has the form

$$\frac{dBd}{dt} = Bd \left\{ 1 - 2 \left[Bd + a_{Bd/Pl} Pl - A\left(Bd, Bd_m\right) \right] \right\};$$ (5.17)

$$A\left(Bd, Bd_m\right) = IF \left[Bd > Bd_m; 0; a_{Br/y_2} y_2 \left(1 - e^{-\alpha_{Bd}\tau}\right) \right];$$

Parameter $a_{Br/y_2} y_2$ determines the degree of influence of the means assigned by the *CZA* economic system for environment protection actions, on biological diversity index. Parameter α_{Bd} specifies growth rate of the index.

The average for the considered coastal zone area level of pollution integrally submits to the weighted sum of concentrations of major pollutant types Pl_i, which are characteristic for this zone

$$Pl = \sum_{i=1}^{M} h_i Pl_i .$$ (5.18)

It forms an average long-term annual course of the parameter Pl_m, which is appropriate to utilize as an integrated criterion for evaluation of the ecosystem's pollution (contamination). Current concentrations of marine pollution Pl are compared in the model with an average long-term norm Pl_m. Consequently, an agent designated as $A(Pl, Pl_m)$, must simulate the necessary volume of the environmental protection activity.

We will assume that the level of contamination is proportional to the economic activity in the *CZA*, and in particular, to the volume of produced seafood $a_{Pl/S}S$. Then, dynamic equation for the level of pollution Pl should take into account the natural self-cleaning of the marine environment $a_{Pl}Pl$, as well as the effect of environmental protection activities simulated by the agent $A(Pl, Pl_m)$. Taking the above assumptions, the dynamics of pollution level is described by the equation

$$\frac{dPl}{dt} = Pl\left\{1 - 2\left[Pl - a_{Pl/S}S + a_{Pl}Pl + A(Pl, Pl_m)\right]\right\}, \qquad (5.19)$$

$$A(Pl, Pl_m) = IF\left[Pl < Pl_m; 0; a_{Pl_m}Pl\left(1 - e^{-\alpha_{Pl_m}\tau}\right)\right].$$

Simulation Experiments with Ecological – Economic Model "Sea-Land". When conducting simulation experiments, the ecological-economic model "sea-land" which we built, was considered as a tool to search for such nature management scenarios that ensure balance between economic benefit and a normal (for middle-term annual criteria Bd_m and Pl_m) state of the marine ecosystem. Calculations were performed on 500 dimensionless time steps (days). For better visualization, the scenarios of the predicted processes Bd and Bd were brought to a dimensionless scale of variability (0, 10).

Some parameters that characterize economic processes were also presented in a dimensionless form by normalization to the corresponding scale coefficients. Below, there are results of two series of computational experiments, in which the following parameters were established: $y_1 = 1,3$; $y_2 = 3$; $y_3 = 2$; $r_1 = 0,8$; $r_2 = 0,6$; $r_3 = 1,1$. Average cost of seafood was 7 arbitrary units, and its market price was 9 arbitrary units. Percentage of repayment θ was 10% of the accumulated credit value.

In the first series of experiments, we dealt with a situation where the economic system could obtain pure profit of 20% from its working capital, and it experienced no obligation to assign funds for environmental purposes. The only condition restricting its production activity was that it could not use loans to expand consumption of marine resources in excess of some limit value H_3^*. This limit value went down each time when the level of the marine environment pollution increased.

Scenarios of processes obtained in the first series of calculations are shown in Fig. 5.6. In Fig. 5.6a simulated demand for seafood and daily volumes of seafood production are presented.

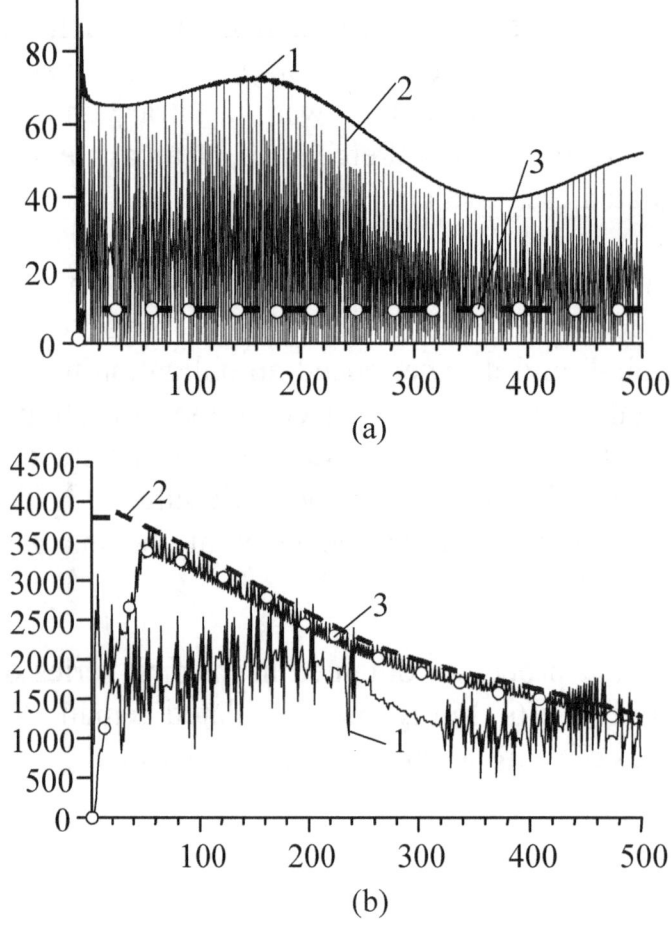

Fig. 5.6. Scenarios of development processes, obtained without taking into
account financing of environment protection actions: (*a*) 1 – demand D, 2 – offers of seafood products by the economic system S, 3 – cost of   seafood P; (*b*) 1 – current assets of working capital H_2, 2 – maximum allowable investments H_3^*, 3 – accumulated credit H_3

We can gather from these graphs that production and sales of seafood took place during the whole period of the experiment, but the complete satisfaction of demand for it was achieved only on the certain days of the experiment, when the economic system was equipped with all needed kinds of resources. Scenarios of working capital and accumulated credits are presented in Fig. 5.6, *b* from which it follows that consumption of biological resources and

324

production of seafood were limited due to the credit policy in relation to the economic system.

In Fig. 5.7, *a* scenarios of integrated characteristics of the marine ecosystem's state are presented.

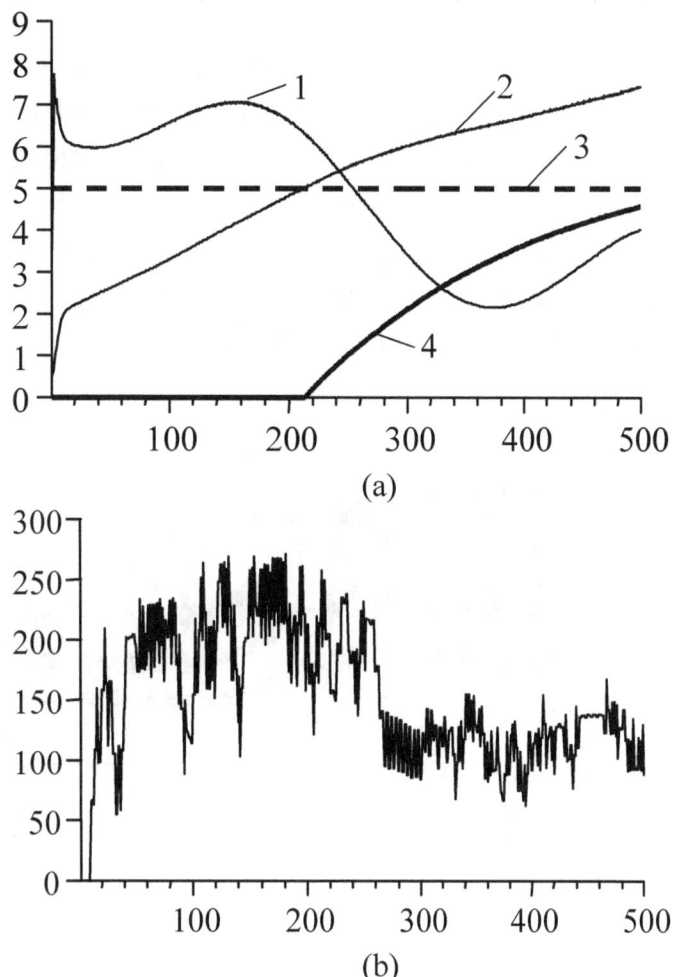

(a)

(b)

Fig. 5.7. Scenarios of development processes, obtained without taking into

account financing of environment protection actions: (*a*) 1 – index of

biodiversity *Bd*, 2 – level of pollution *Pl*, 3 – long-term averaged pollution level Pl_{m}, 4 – function of the agent of environment protection   actions on pollution $A(Pl,\ Pl_{\mathrm{m}})$; (*b*) Averaged for 10 days profitability of seafood production φ;

The graph of calculated biodiversity index in its form repeats the graph of demand for Bioresources (see Fig. 5.6, *a*), and the level of pollution scenario is experiencing the steady growth due to the lack of environment protection actions. The dotted line represents the average multi-year level of pollution, which the calculated script *Pl* had significantly exceeded in the second half-time of computing. This led to a decrease in biodiversity index and provoked a reaction of the agent of environment protection actions $A(Pl, Pl_m)$, which had imitated the necessary nature protection actions against of pollution (see curve 4 in Fig. 5.7, *a* and Fig. 5.8, *a*).

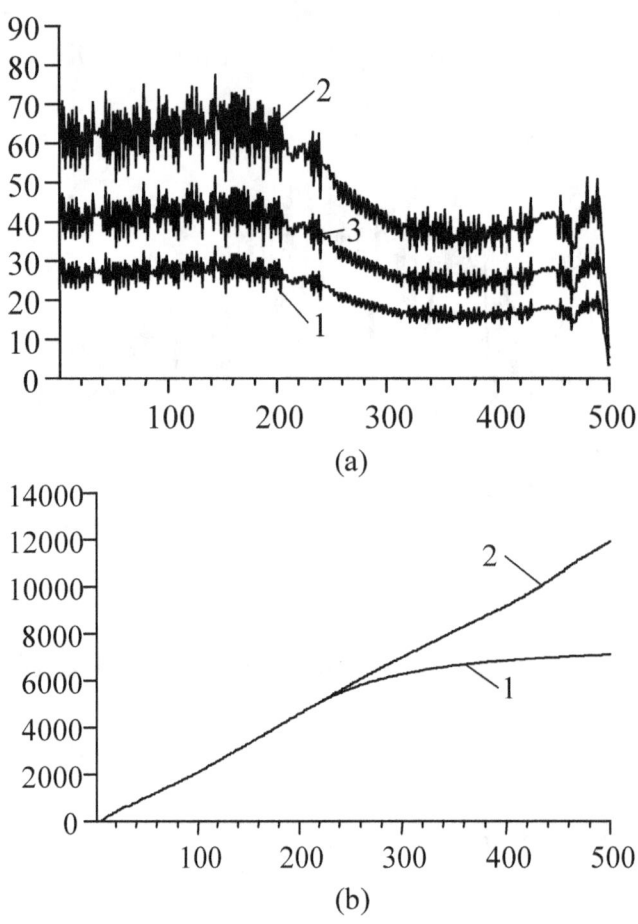

(a)

(b)

326

Fig.5.8. Averaged for 10 days, amounts of resources purchased for credit, and al location of funds withdrawing from working capital, obtained without

taking into account financing of environment protection actions by the economic system: (*a*): 1 – purchased economic resources V_{11}, 2 – purchased bioresources V_{12} , 3 – purchased ecological resources V_{13} ;

(*b*) 1 – net profit with account of the costs of environment protection

actions, 2 –net profit without account of the costs of environment

protection actions

However, the conditions set by the experiment were not implemented, and the economic system continued to obtain net profit and to invest working capital only into the resources which it needed for production. This is evident from the graphs of its cost-effectiveness (see Fig. 5.7, *b*), and from the graph of purchased resources volume (see Fig. 5.8, *a*).

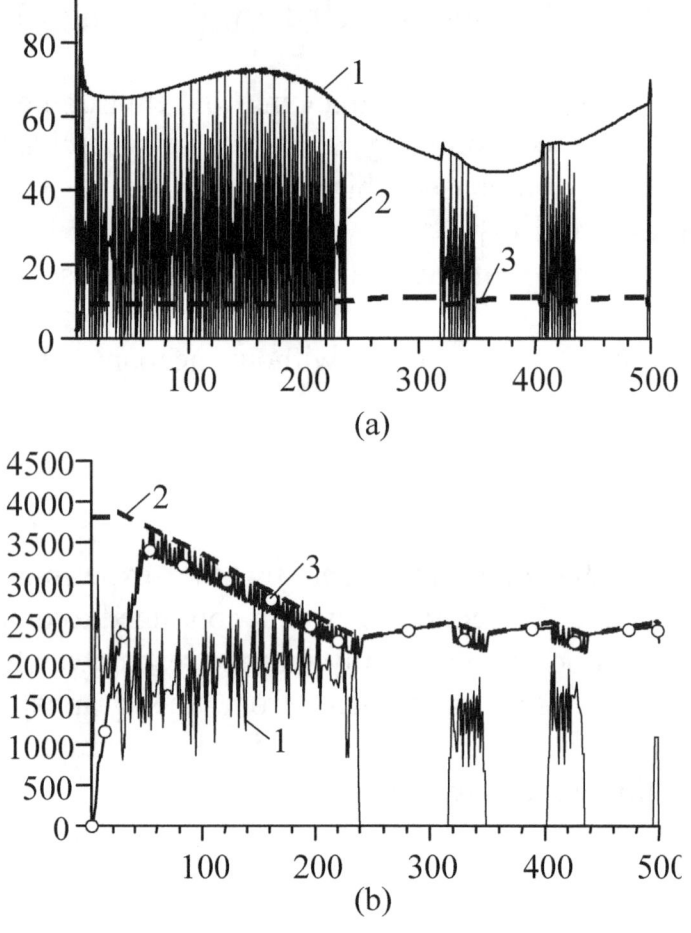

Fig. 5.9. Scenarios of development processes, including factor of financing

environment protection actions by the economic system:

(*a*) 1 – demand D, 2 – offers of seafood products by the economic

system S, 3 – cost of sea food P; (*b*) 1 – current assets of working

capital H_2 , 2 – maximum allowable investments H_3^*, 3 – accumulated   credit H_3

Fig. 5.8, *b* presents graphs of actual net profits earned by the economic system (curve 2), and the part of the profit that was to be preserved after financing of the environment protection actions (curve 1). The main conclusion to be drawn from this series of experiments is the fact that restriction of investment for consumption of marine resources (e.g., through licensing of their volumes) is not always a sufficient way to maintain ecological state of the marine environment.

In the second series of experiments, a more radical way of managing the nature environment was used, when the *CZA* economic system was supposed to spend a part of its profit on environment protection actions. It was supposed to do this right away, as soon as the level of marine pollution exceeds the average long-term value Pl_m = 3. Under these conditions, the economic activity of resources consumption was limited, and the demand for seafood was met only in certain periods of time (see the scenarios in Fig. 5.9, *a* and *b*).

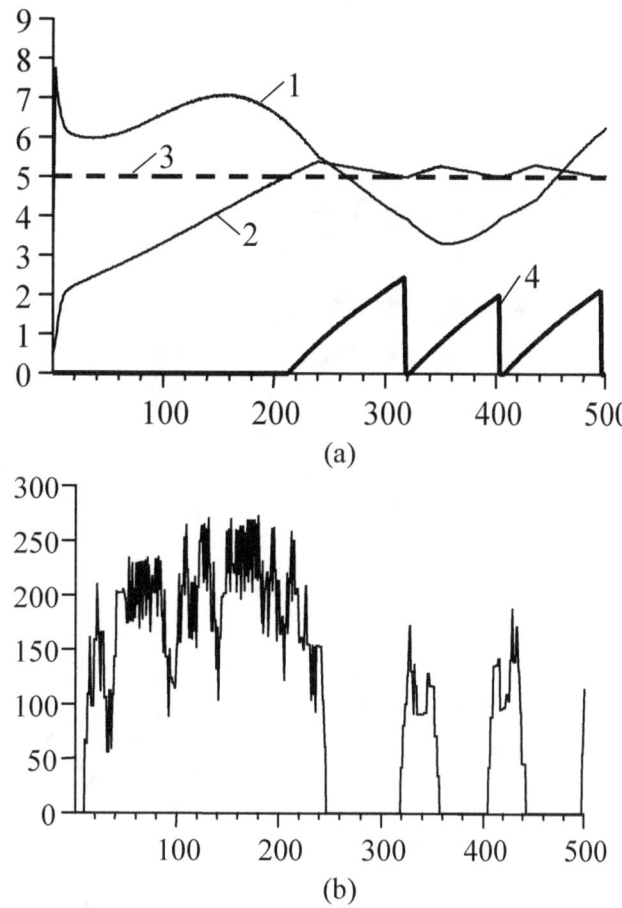

(a)

(b)

Fig. 5.10. Scenarios of development processes, including factor of financing

329

environment protection actions by the economic system: (*a*) 1 – index of biodiversity *Bd*, 2 – level of pollution *Pl*, 3 – long-term averaged

pollution level Pl_m, 4 – function of the agent of environment protection   actions on pollution $A(Pl, Pl_m)$; (*b*) Averaged for 10 days profitability of seafood production φ;

A part of the profit allocated to combat pollution of the marine environment, was proportional to the magnitude of the excess of this level's average long-term value. As it can be seen from Fig. 5.10, *a* and *f*, the agent of environmental activities $A(Pl, Pl_m)$ directed a part of the profit to cover the costs of pollution control operations; it happened each time when the level of pollution raised higher than the value $Pl_m = 3$.

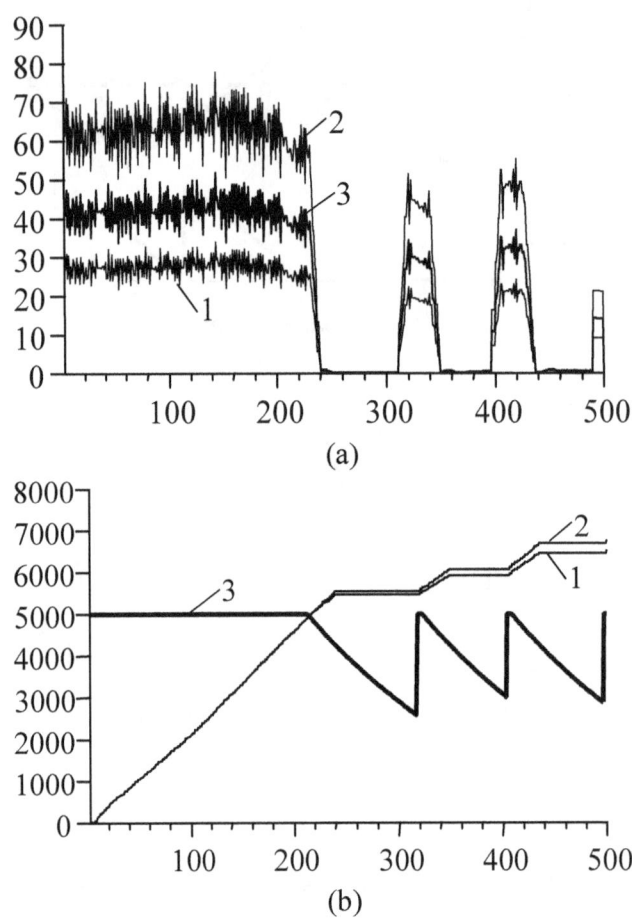

(a)

(b)

Fig. 5.11. Averaged for 10 days amounts of purchased on credit resources and

allocation of funds withdrawing from working capital, obtained without taking into account the financing of environment protection actions by the economic system: (*a*) 1 – purchased economic resources V_{11},

2 – purchased bioresources V_{12}, 3 – purchased ecological resources V_{13}; (*b*) 1 – net profit with account of the costs of environment protection

actions, 2 – net profit without account of the costs of environment

protection actions, 3 – function of the agent that controls financing of environment protection actions

The parameters characterizing the environment protection actions were selected on the basis of pollution level decrease rate equal to 1% per day. Therefore, to keep the scenario level of pollution near the long-term average values (see Fig. 5.10, *a*), the economic system periodically had to suffer additional expenses, which significantly increased the seafood production cost (see Fig. 5.11, *b*). In these cases, its seafood production was stopped because the production became unprofitable (see Fig. 5.9, *a, b*; Fig. 5.10, *b* and Fig. 5.11, *a*).

Computational experiments demonstrated that utilizing the information management technology of ecological and economic processes in coastal zone area allows to obtain visual representations of development processes' scenarios by varying a large number of model parameters, which play an important role in administrative decision-making about the use of the *CZA's* resources. These modeling experiments allow to perform evaluation of necessary environment protection actions that are required to support normal state of marine ecosystems, and ensure the continuous cost-effective production of seafood by local economic systems.

5.6. Assessment of Spatial Distributions of Marine Bioresources by

Remote Sensing Observations

Sustainable development of a coastal zone requires existence of information management technologies aimed at performing control over the state of marine resources and their consumption. Systems approach to this problem lies in the rational use of all available information for prediction of possible scenarios of the processes in the coastal marine ecosystem through the assimilation of observational data in their dynamic models [178]. One of the promising ways to solve this task is to use causal relationships between processes in marine ecosystems. They are well-known from research, and can be corrected by the information accumulated during continuous observations.

The principle of adaptive balance of causes, used in the *ABC*-method, provides continuous adaptation of time-scenarios of processes in marine ecosystems, which are linked causally, and hence have some correlation between them. Both, observed and unobserved (calculated on the model) spatial distributions of ecosystem's parameters, are connected with each other by spatial correlation functions, as well. Therefore, the *ABC*-model should be equally applicable to temporary developing processes observed in some field locations in marine ecosystems, and to spatial realizations of these fields. For example, having a map of surface temperature of the sea, one can use it in a spatial version of the *ABC*-model, to take into account the influence of the temperature on distribution of concentrations of oxygen, zooplankton and other variables of the ecosystem.

Since satellite observations are a major source of information about the state of the sea surface, these considerations open a possibility to estimate unobserved (or difficult to observe) parameters' distributions of the ecosystem along the span of the satellite trajectory by adapting their model values to remote sensing data. This section deals with practical aspects of implementation of the spatial version of a simplified *ABC*-model of the marine ecosystem. Mapping distributions of oxygen, zooplankton and bioresources concentrations can be performed by using remote sensing data fields of phytoplankton (chlorophyll) and sea surface temperature.

The systems principle of adaptive balance of causes [181] postulates the desire of all natural systems to reach a state of dynamic equilibrium with each other and with external influences applied to them. The existence of equilibrium can be explained with balance of positive and negative feedback connections, which are formed inside the system, due to causal relationships between the processes.

Grounding on this principle, the method of adaptive balance of causes (*ABC*-method), discussed in Chapter Two, was developed. In Chapter Three, we studied the reaction of marine ecosystem models to temporal scenarios of external impacts, which were mainly these: the annual course of the upper sea layer temperature; its illumination with day light; wind speed, etc. In these cases, the modular equation of the *ABC*-method linked the change rates of integrated ecosystem parameters with the values of these parameters and with external influences applied to them.

However, taking into account the above concept of the *ABC*-method, nothing prevents the use of this method to obtain profiles of spatial distributions of marine ecosystem parameters. The only difference in the *ABC*-model designed to calculate spatial distributions, is this: instead of dependence on the variables of time t, it depends on the spatial coordinate which we will identify as l. This model reflects adaptive balance (consistency) of spatial scenarios between themselves and with the given profiles of external influences.

Let us explain this idea on an example of a simplified model of marine ecosystem. In this case study we will set the task to develop estimates of spatial bioresource concentration' distributions on the remote sensing data of chlorophyll concentration and the of sea surface temperature. Let $BR(l)$ be the bioresource concentration along the trajectory l on the surface of the sea. The *ABC*-model of the marine ecosystem that is used to assess this parameter, must contain the key processes that ensure the existence of bioresources in the surface layer.

As the state parameters of the ecosystem, we will use averaged concentrations of phytoplankton PP and zooplankton ZP, oxygen OX and bioresources BR, on the upper layer of the sea. We mean by this the net concentration of all living organisms, located above zooplankton in the marine food chain. In order to simplify the consideration, we will not take into account the presence of nutrients and carbon dioxide.

Let us use field observations of chlorophyll concentration in the right side of the equation for phytoplankton, suggesting that they represent the final result of influences on the phytoplankton concentration of all the processes which are not included into the model. In addition to this, we will take into account the influence of sea surface temperature on the state of the ecosystem, as these data are usually available for analysis. Observations of temperature and chlorophyll concentration can be taken, for example, from maps of the fields, constructed on remote sensing data of the sea surface [143].

For a more realistic description of the processes formation conditions the management agents should be included in a structure of the ecosystem model. We will use resource limitation agents for concentrations of zooplankton on phytoplankton and oxygen $A_{ZP}(PP,OX)$, as well as the agents for bioresources' concentration on zooplankton and oxygen $A_{BR}(ZP,OX)$. In addition, we will introduce agents that provide the dependence of oxygen, zooplankton and bioresources concentrations on the sea temperature.

A conceptual model of an ecosystem can be represented by the scheme shown in Fig. 5.12.

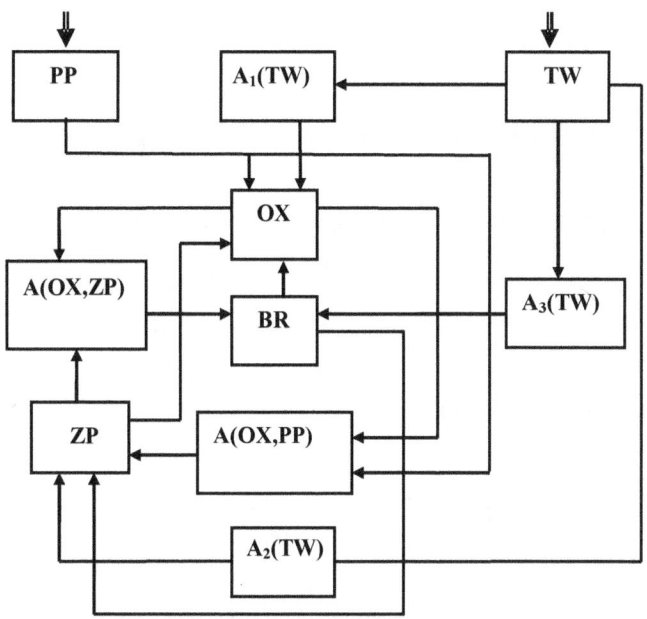

Fig. 5.12. Adaptive model of ecosystem for assessment of spatial distribution of   the ecosystem's parameters by remote sensing data of temperature and   chlorophyll   concentrations   in   the   upper layer of the sea

The following set of equations of the spatial *ABC*-model in the direction *l* at the sea surface corresponds to this conceptual model

$$\frac{dPP}{dl} = PP\{1 - 2[PP - a_{PP/CH}CH(l)]\} , \qquad (5.20)$$

$$\frac{dOX}{dl} = OX\{1 - 2[OX + a_{OX/BR}BR + a_{OX/ZP}ZP - a_{OX/PP}PP + a_{OX/TW}TW]\} ,$$

$$\frac{dZP}{dl} = ZP\{1 - 2[ZP + a_{ZP/BR}BR - A_{ZP}(PP, OX)$$
$$-a_{ZP/TW}\exp[-\alpha_{ZP}(TW - TW_{ZP}*)^2]]\} \qquad ,$$

$$\frac{dBR}{dl} = BR\{1 - 2[BR - A_{BR}(ZP, OX)$$
$$-a_{BR/TW}\exp[-\alpha_{BR}(TW - TW_{BR}*)^2]]\} \qquad ,$$

$$A_{ZP}(PP,OX) = a_{ZP/PP}PP(l)A_{ZP/PP}(l) + a_{ZP/OX}OX(l)A_{ZP/OX}(l)$$
,

$$a_{ZP/PP}PP(l) = IF[M_{ZP}(l) = PP(l); a_{ZP/PP}PP(l); 0]$$
,

$$A_{ZP/PP}(l) = IF[PP(l) < PP_c; 0; 1)]$$
,

$$a_{ZP/OX}OX(l) = IF[M_{ZP}(l) = OX(l); a_{ZP/OX}OX(l); 0]$$
,

$$A_{ZP/OX}(l) = IF[OX(l) < OX_c; 0; 1)]$$
,

$$M_{ZP} = \arg\min\{PP(l); OX(l)\}$$
,

$$A_{BR}(ZP,OX) = a_{BR/ZP}ZP(l)A_{BR/ZP}(l) + a_{BR/OX}OX(l)A_{BR/OX}(l)$$
,

$$a_{BR/ZP}ZP(l) = IF[M_{BR}(l) = ZP(l); a_{BR/ZP}ZP(l); 0]$$
,

$$A_{BR/ZP}(l) = IF[ZP(l) < ZP_c; 0; 1]$$
,

$$a_{BR/OX}OX(l) = IF[M_{BR}(l) = OX(l); a_{BR/OX}OX(l); 0]$$
,

$$A_{BR/OX}(l) = IF[OX(l) < OX_c; 0; 1]$$
,

$$M_{BR} = \arg\min\{ZP(l); OX(l)\}$$
.

The lower borders of minimum allowable concentrations of phytoplankton PP_c, zooplankton ZP_c and oxygen OX_c were set in the model. If the ecosystem's parameters at a given point of spatial scenario fell below the limit values, the management agents would direct them to zero. The most favorable upper-layer temperatures for development of zooplankton and bioresources were marked as $TW_{ZP}*$ and $TW_{BR}*$, accordingly.

Computational Experiments with the Ecosystem's Model. In the first computational experiment, testing of the ecosystem's model was performed under conditions where spatial profile measurements of surface temperature and chlorophyll concentrations were simple harmonic functions, as shown in Fig. 5.13, *a*.

All model variables were reduced to a common dimensionless scale of variability (0, 10), using a linear transformation and normalization to the maximum values of the real dimensional variables (see (2.22)). The influence coefficients in the dynamic equations of the model were selected within the range of values [0,1; 0,5]. The model parameters were set as follows: $a_{ZP/BR} = a_{BR/TW} = 2$, $\alpha_{ZP} = \alpha_{BR} = 0,001$, $TW_{ZP}^{*} = TW_{BR}^{*} = 1,5$. The lower allowable concentrations were not installed: $PP_{c} = ZP_{c} = OX_{c} = 0$.

The spatial scenarios, obtained by the model in this experiment, are shown in Fig. 5.13, *b*. First of all, our attention was drawn to complexity and variability of these scenarios, compared to those simple harmonic functions of the external influences that had been submitted to the input of computational algorithm of the model. The dominant influence on formation of the oxygen concentration scenario was made by the spatial distribution of temperature. The second factor of influence was production of oxygen by phytoplankton, which greatly complicated this scenario.

In formation of the spatial distribution of zooplankton, the leading role belonged to the agent of limitation $A_{ZP}(PP, OX)$, which connected current values of concentration $ZP(l)$ (when driving in a field along the simulated trajectory) with a minimum of two values: $PP(l)$ and $OX(l)$.

Similarly, formation of the bioresource concentration' spatial script took place. The agent of limitation $A_{BR}(ZP, OX)$ selected the smallest of the two values $ZP(l)$ and $OX(l)$, for example, as shown in Fig. 13, *b*, - the scenarios for a portion of the trajectory between steps 190 and 220. At this space interval, zooplankton concentration was lower than oxygen concentration. As a result, the bioresource's scenario was subordinated to the scenario of zooplankton.

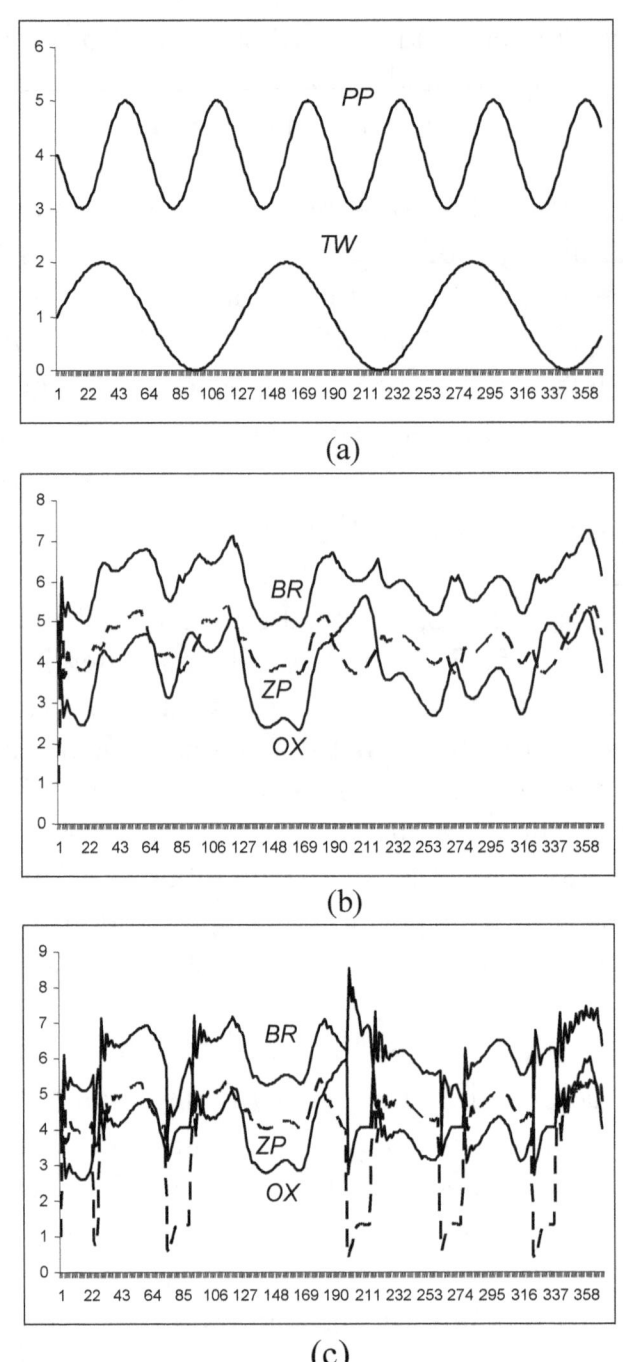

Fig. 5.13. Experimental calculations results for spatial distributions of
unobservable parameters of the ecosystem, with *ABC*-model

In the second experiment, additional condition was supplied: the concentration of phytoplankton was set not to fall below the minimum permissible value $PP_c = 3,3$. As it can be seen from Fig. 5.13, *a*, under this condition, the agent $A_{ZP/PP}(l)$ in the equation for zooplankton concentration had to sharply limit the value of this concentration, which led to further significant complication of all the calculated spatial scenarios. This is shown in Fig. 5.13, *c*, at which the zones of low concentration of phytoplankton corresponds with the zones of substantial fall in concentration of zooplankton and bioresources, accompanied by a rise of concentrations of oxygen.

Assessment of Spatial Distributions of the Marine Ecosystem's Parameters by Using Remote Sensing Data of Sea Surface. The *ABC*-model of the marine ecosystem (5.20) discussed above was used during computational experiments using data from remote sensing of the north-western region of the Black Sea derived from satellite measurements for winter conditions (February). These observations constituted the field measurements of chlorophyll and sea surface temperature, given to the nodes of a square grid covering the area.

To perform an objective assessment of influence coefficients in the equations of the ecosystem's model by formulas (2.37), it is necessary to take a series of observations of the ecosystem's parameters along a trajectory of the satellite. To accomplish this task, time series of observations in a separate point of the field can be used (though with a certain error). For example, we can take Taylor's hypothesis of existence of the relationship between spatial and temporal correlation functions of random fields [114]. In reality, such data require conducting of special field experiments in the sea, coordinated with satellite observations.

Due to the absence of such observations, the influence coefficients of the model (10) were chosen as a compromise between the model's sensitivity to satellite measurements of *CH* and *TW* fields, and the stability of the model's computational algorithm. An indirect confirmation of validity of this choice can be found out from literature which provides data on dependence between zooplankton and bioresources' concentrations in the north-west shelf of the Black Sea and nutrition and oxygen concentrations during winter season [93]. The coefficients used in calculations are given in Table 5.1.

In Fig. 5.14, *a* and *b* the fields of sea surface temperature and chlorophyll concentration are shown constructed according to remote sensing of the sea surface. The data were used in computational experiments to assess spatial distributions of the ecosystem's parameters by the *ABC*-method.

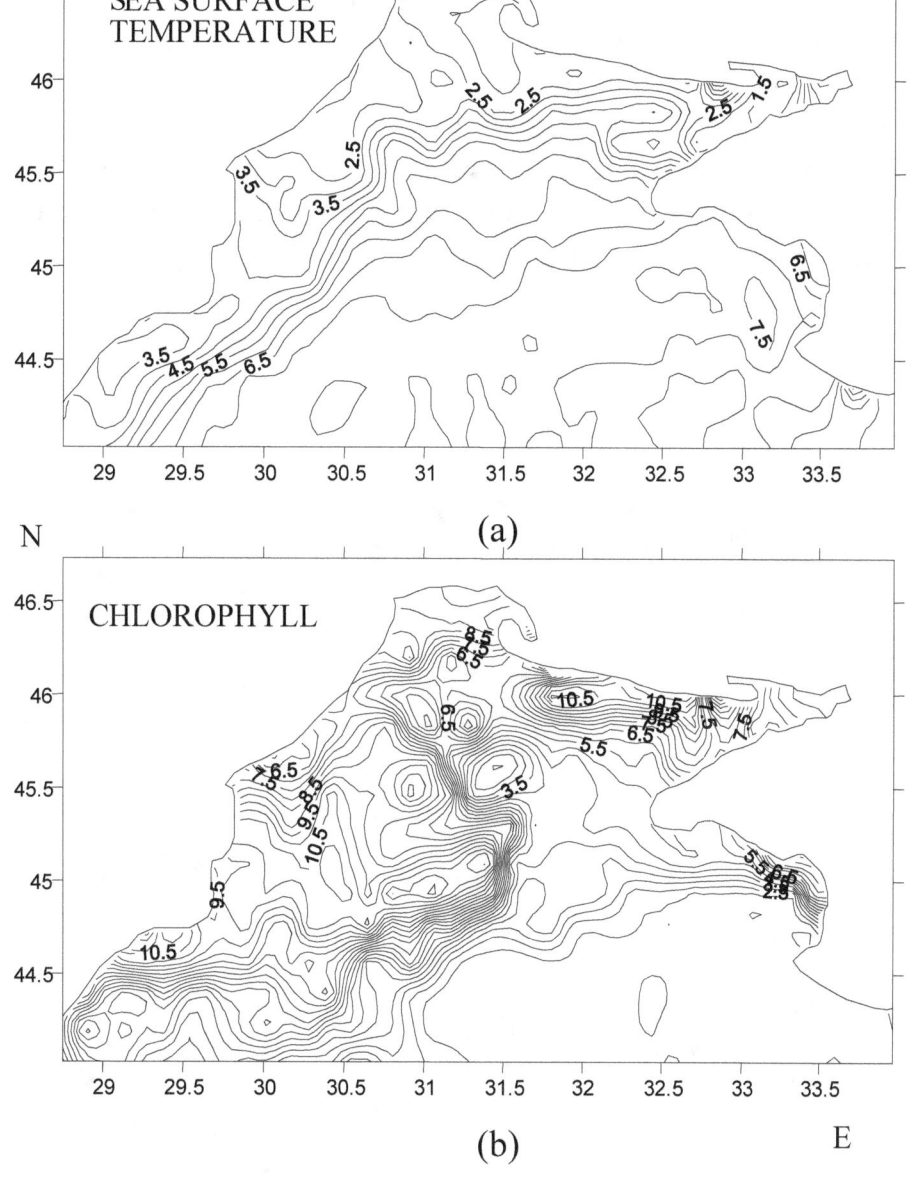

(a)

(b)

E

Fig. 5.14. Charts of sea surface temperature field (*a*) and chlorophyll

concentration (*b*) based on remote sensing data of the sea surface, which were used in calculations of the ecosystem's parameters

For the ease of graphic representation of simulation results, all the modeled processes, as well as all observational data, were reduced to a common dimensionless scale of variability [0, 10]. In the first series of computational experiments, profiles of spatial distributions of the ecosystem's parameters along certain directions were constructed. Zonal profiles of these parameters were obtained, and calculated for the latitude of $45^0\,47^{/}$, as shown in Fig. 5.15, *a* and *b*. To simplify calculations, border zones of homeostasis for the living ecosystem's objects were not installed. However, agents of limiting concentrations of zooplankton $A_{ZP}(PP,OX)$ and bioresources $A_{ZP}(PP,OX)$ were switched on.

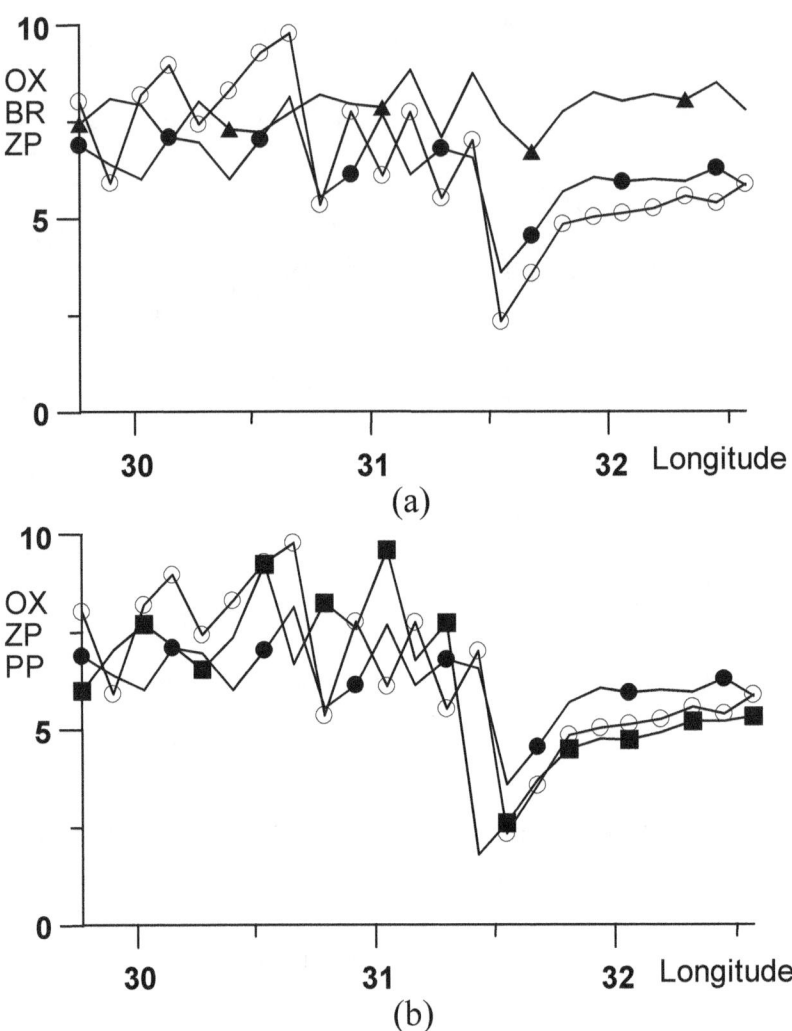

Fig. 5.15. Spatial distributions of the ecosystem's parameters at latitude $45^0\,47^{/}$, calculated by the model (10) with the use of

observational data presented   in Fig. 5.14: (a) ○ – *OX*, ● – *ZP*, ▲ – *BR*; (b) ○ – *OX*, ● – *ZP*, ■ – *PP*

Comparing the spatial variability of concentrations of oxygen, phytoplankton and zooplankton, as well as bioresource on the latitude of 45^0 $47^/$, with the initial fields of sea temperature and concentration of chlorophyll, shown in Fig. 5.14, we

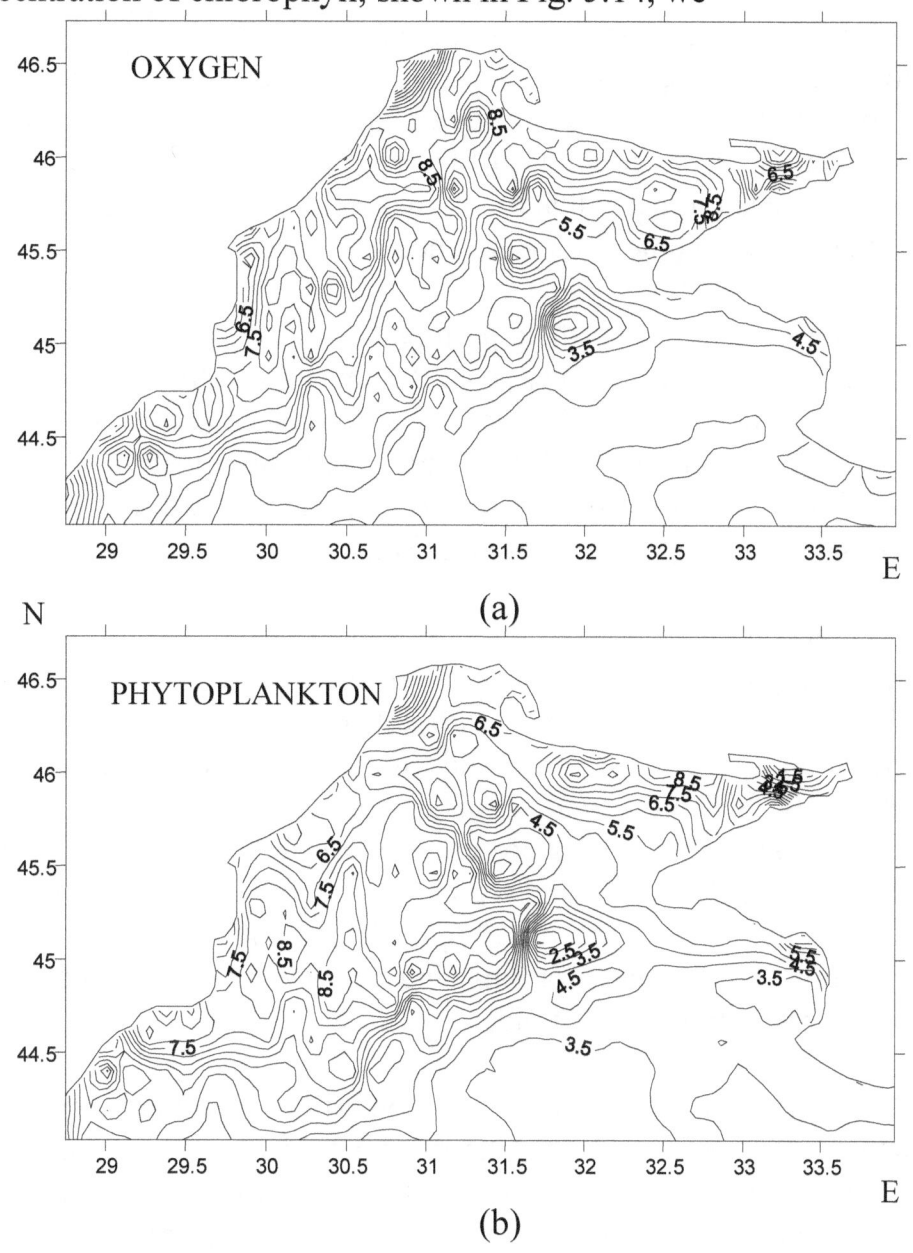

(a)

(b)

Fig. 5.16. Maps of oxygen concentrations (a) and phytoplankton concentrations (b), constructed with use of the Adaptive Balance of Causes ecosystem's model (5.20)

can see that they are in qualitative agreement with the distributions of source fields. The results of the actions of agents of limitation can be seen in profiles,

shown in Fig. 5.15. For example, the graph of zooplankton concentration in Fig. 5.15, *b* clearly follows down to a minimum of two of its resource values: phytoplankton, or oxygen. Lack of significant variations in values of the parameters in reference to their mean estimates indicates that the model influence coefficients used in calculations (see Table 5.1.) were chosen correctly.

In the second series of experiments, an adaptive model of the ecosystem (5.20) with influence coefficients from the Table 5.1 was consistently used at all latitudinal transects in fields of surface temperature and chlorophyll concentration, shown in Fig. 5.14. This allowed us to construct maps of spatial distribution of the ecosystem's parameters presented in Fig. 5.16 and 5.17.

In accordance with the model (5.20), distribution of phytoplankton concentrations is determined solely by the data of satellite observations of chlorophyll concentrations. Therefore, maps of these fields (see Fig. 5.14, *b* and Fig. 5.16, *b*) are almost identical. A different situation was typical for distributions of oxygen concentrations, since they were under the influence of observational data on both, chlorophyll and temperature. Therefore, concentrations of oxygen map shown in Fig. 5.16, *a* differ noticeably from the map of phytoplankton concentrations and reflects the influence of the isolines of the temperature field.

The map of zooplankton concentrations, formed under the influence of the agent, limited these concentrations on phytoplankton and oxygen, as we can see in Fig. 5.17, *a*. It further differs from the above charts, as in the background of its low-amplitude variability of random nature, it contains a number of local zones of extreme values. In particular, these zones are located in the vicinity of 46^0 N and

30,5^0 east longitude, as well as in the area of 45^0 N and 45^0 E, and can also be found on the map of bioresources concentration, shown in Fig. 5.17, *b*. Thus, the proposed algorithm for obtaining estimates of spatial fields allowed us to transform the information contained in observational data of surface temperature and chlorophyll concentrations into the information about other, unobservable parameters of the ecosystem.

Table 5.1

a / *MM/NN*	*P*	*X*	*Z* / *P*	*B* / *R*	*H*	*T* / *W*
P *P*		,6	−0,4		,5	−0,3
O *X*	,6		−0,2	−0,2		−0,2
Z *P*		,4	1	−0,4		0,1
B *R*		,4	0 ,4	1		0,1

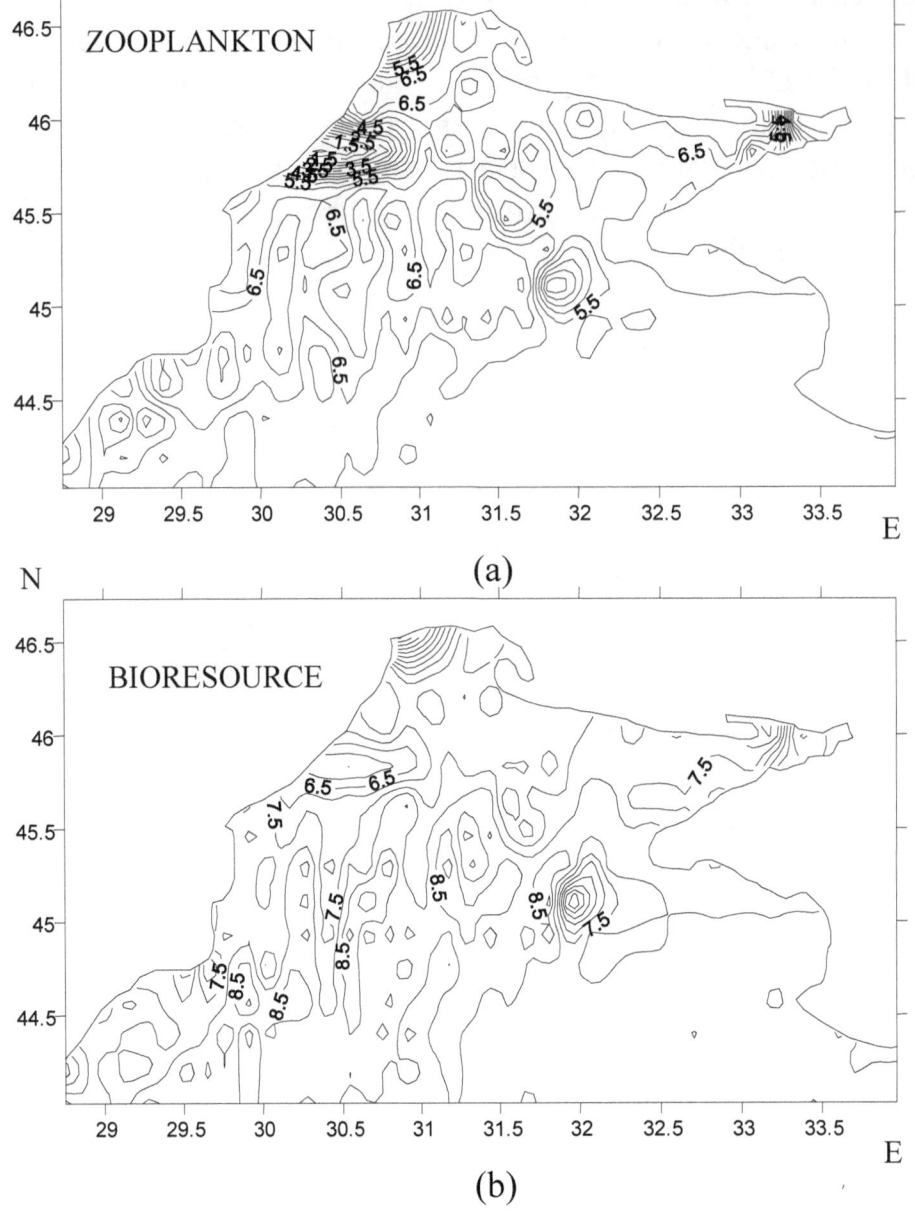

Fig. 5.17. Maps of zooplankton concentrations (a) and bioresource concentrations (b), constructed with use of the Adaptive Balance of Causes ecosystem's model (5.20)

The results of computational experiments can be summarized as follows. In accordance with the systems concept of adaptive balance of causes, variability of the fields of ecosystem's parameters is a

346

result of the ecosystem's aspiration to adapt to varying external influences. Therefore, spatial-temporal fields of the marine ecosystem's parameters are in dynamic equilibrium with each other and with external influences, applied to the ecosystem. The nature of spatial and temporal variability of the fields depends strongly on the habitat of living organisms, i.e. resources supply for their existence and development. This explains complex temporal scenarios of the ecosystem's processes and structures of spatial distributions of their parameters.

As it was mentioned above, the method of adaptive balance of causes takes into account properties of mutual adaptation of processes in complex natural systems, since it uses the equations of logistic type, which have fast convergence to stable solutions. A feature of this method is an ability to determine coefficients of logistic equation according to observations, using the expertise or analytical (probabilistic) approach. Therefore, the *ABC*-method allows introduction of variable coefficients into the model, the coefficients should be adapted to these observations. In summary, we will formulate two principal conclusions:

1. Profiles of spatial distributions of the marine ecosystem's parameters, obtained as solutions of adaptive model of the ecosystem, are shaped with complex nonlinear interactions that are due to cause-effect relationships and resource constraints in the development process.

2. *ABC*-models of marine ecosystems may be used for estimation of unobserved parameters of marine ecosystems by assimilation of remote sensing data along the projections of satellite trajectories on the sea surface.

Chapter VI. RECREATIONAL RESOURCES MANADGEMENT IN A COASTAL ZONE AREA

6.1. Factors of Recreational Attractiveness of a Coastal Zone Area

Recreational resources of coastal zone areas, which are actively used as resorts by local residents and numerous visiting tourists, strongly depend on natural environment, economic infrastructure, services, and general condition of socio-economic system of the whole *CZA* region. There is a considerable number of scientific publications devoted to management of *CZA* recreational resources. Various methods of economic and ecological-economic assessment of recreational resources have been developed [115]. It was noted that recreational attractiveness of local *CZA* sites is largely determined by quality and condition of the beaches, as well as their equipment and protection from waves and storm surges.

Sustainable development of recreational areas today has to deal with the same range of problems which are typical for all coastal areas. One of the common problems for the entire coastal zone is the riding of sea level, which has been observed during the latest decade, (up to 1 cm per year [22]). Besides this, wind and waves have always been constant sources of coastal erosion, washing away or changing stocks of sand, destroying breakwaters and other structures intended to reinforce the coastal line. Among additional problems, we can mention commercial removal of sand from beach zones for use in construction industry, as well as improper construction of protective structures, which change the dynamics of wave movements and cause removal of sand, which leads to destruction of beaches.

The sand stripes of beaches themselves present natural protection for coastal line from destructive action of waves, since the sand extinguishes the power of the surf. Therefore, conservation of sand beaches can provide not purely recreational, but also positive economic effect. In tropical areas, beach zones are ecosystems used by various inhabitants of the marine environment in their life cycles. Sand crabs, mollusks, tortoises and many other marine animals have

348

adapted to constant change of tides in beach areas, and can not exist outside of such areas.

Recreational resources play an important role in the development of seasonal tourism around the globe. Tourism is one of the fastest growing areas of economies of many coastal countries located in tropical and subtropical areas of the Earth. Along with the inflow of tourists, the infrastructure for recreation develops as well: it is followed by construction of hotels and campsites, development of roads, improvement of services. Simultaneously with these actions, potential danger of overpopulating coastal zone areas is growing. The issue of overpopulation comes up with double power in the midst of the tourist season annually, which inevitably leads to increased pollution of beaches and adjacent sea area. Therefore, the need in a comprehensive, system approach to assessment of ecological-economic balance of consumption and reproduction of recreational resources becomes apparent.

Grounding on the systems research methodology, we will consider recreational resources as an integral characteristic of the natural-economic system of *CZA*. Availability of recreational resources in a *CZA* points out at possibility to produce "recreational services": medical treatment, recreation, development of positive emotional and psychological effect on people. The main structural components of this system are:

- The natural environment: beaches, coastal parks, coastal landscapes, flora and fauna, air, weather and climate, etc., as well as its ecological state: pollution level of the near-coast marine environment, of coastal land areas and the air above them (contaminated with harmful chemical substances and bacteria), the quality of drinking water, cleanness of the territory, and others;

- The economic infrastructure of recreational services in the area: buildings, providing facilities, beach equipment, energy supply, transportation, information, communications, etc.;

– Organization and management of recreational services: treatment, nutrition, recreation, entertainments, sports, tourism, safety, etc.;

– Social ecological-economic system of the region including the coastal zone area: transportation accessibility, standards of living, presence or absence of social and ethnic tension, crime situation, risks of natural or technological disasters, historical background, culture, sightseeing attractiveness of the region, traditional demand for its recreational services, prices of the services, etc.

These four groups of factors determine general recreational attractiveness of local "sea – land" systems, which affects the demand for recreational services, and thus creates preconditions for successful economic activity. However, the implementation of recreational services is related to the cost of consumption and reproduction of resources, which also involves environment protection activities.

Many studies have highlighted the need in developing systems approach [82] to the problem of recreational resources management. In this connection, we need to talk about optimization of complex natural-economic systems, simultaneously by a number of quality criteria. Therefore, it is necessary to create a model of recreational resources management, which would take into account a large number of interrelated development processes and which would be able to predict scenarios of these processes under the influence of various conditions and factors.

To quantify expert judgments about recreational attractiveness of a local site of a coastal zone, we introduced an index z; to characterize each type of recreational resources that define its value, we introduced corresponding indexes: y_1, y_2, y_3, y_4, as shown in Fig. 6.1.

Then, the index of recreational attractiveness can be represented as a weighted sum of indices of recreational resources

$$z = c_1 y_1 + c_2 y_2 + c_3 y_3 + c_4 y_4 \qquad (6.1)$$

In this expression, the weights c_i have the meaning of priorities, provided (integrally) to each of the types of recreational resources, and taking into account their role in shaping the consumer's demand for recreational services in the region. Each of the types of the *CZA* recreational resources is connected with each other. Natural recreational properties traditionally attract investment in economic infrastructure of the *CZA*, which opens up opportunities for the development of recreational services. In its turn, the state of infrastructure (and services) affects the ecological state of the *CZA* environment. The system of mutual influences changes with time, which ultimately determines the dynamics of the *CZA*'s recreational attractiveness index. This circumstance makes us to consider the variables y_1, y_2, y_3, y_4 as a system of interconnected processes, and to develop a dynamic model for prediction of possible changes in the index of recreational attractiveness, depending on internal influence factors.

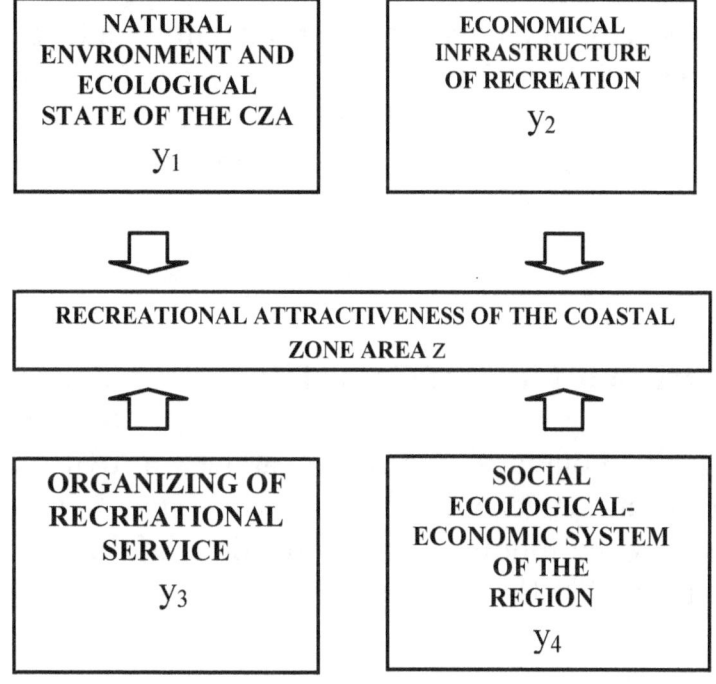

Fig. 6.1. Components of the natural-economic complex of the coastal zone area, forming its recreational attractiveness

We must also take into account external influences coming from the environment system of the region (a large area, a separate country), to which the *CZA* system "sea – land" belongs. From a consumer's (tourists') point of view, natural and environmental resources of the region are influenced by seasonal climate and weather factors: annual temperature variation of the sea and air, the number of sunny days, humidity, wind regime of the site, and the state of the sea water. These external influences should be taken into account in our model by means of introducing an influence function F_1.

Investment into the development of recreational basis of the region influence the economic infrastructure of recreation y_2 and the services y_3. To account for such effects, it is necessary to introduce functions F_2 and F_3. Changes in the socio-political situation of the country, natural and technological disasters, epidemic diseases, and other influences are able to substantially reduce recreational attractiveness of the region. To adapt to these external influences, we will introduce a factor F_4, affecting the state of social ecological-economic system of the region, i.e. affecting index y_4.

6.2. Dynamic Model of Coastal Zone Recreational Resources

Using the notations introduced above, a dynamic model of recreational attractiveness of the "sea - land" system can be represented by the following set of equations of the adaptive balance of causes method [181]

$$\frac{dy_1}{dt} = y_1 \left[1 - 2 \left(y_1 - b_{12} y_2 - b_{13} y_3 - b_{14} y_4 - F_1 \right) \right],$$

$$\frac{dy_2}{dt} = y_2 \left[1 - 2\left(y_2 - b_{21}y_1 - b_{23}y_3 - b_{24}y_4 - F_2\right)\right],$$

(6.2)

$$\frac{dy_3}{dt} = y_3 \left[1 - 2\left(y_3 - b_{31}y_1 - b_{32}y_2 - b_{34}y_4 - F_3\right)\right],$$

$$\frac{dy_4}{dt} = y_4 \left[1 - 2\left(y_4 - b_{41}y_1 - b_{42}y_2 - b_{43}y_3 - F_4\right)\right].$$

If we set model coefficients b_{ij} $(i,j = 1, 2, 3, 4)$ and external influences F_i, then solutions of the equations (6.2) will produce prognostic scenarios of recreational resources' variability. Assessment of values of the coefficients has important influence on the quality of prediction. As it was noted above, these coefficients are determined by experts, who are usually guided with rich experience in providing such assessments. In order to improve the quality of such peer review, one can use the method of analytical hierarchy process by Saaty [150], which was discussed above in section 3.3.

For example, let us consider an assessment of coefficients in the first equation of the model (6.2). In this equation, coefficients b_{ij} $(i,j = 1, 2, 3, 4)$ take into account the degree of influence of each factor y_2, y_3, y_4 on the value of y_1. The qualitative judgments of experts about the extent of these degrees, is represented with numbers ranged along a scale of $1 - 10$, as follows: 1 – very weak, 3 – weak, 5 – average, 7 – strong, 9 – very strong.

Let us suppose that, in the experts' opinion, the influence of economic infrastructure of recreation y_2 on the natural environment and ecological state of the "sea - land» system y_1 is estimated as average (i.e. 5 points), and the influence of service for recreation is estimated as strong (7 points). The ratio of these influences 5 / 7 is the result of a pair-wise comparison and the number should be included in the matrix A of such comparisons as the element a_{12}.

Let us further suppose that the influence of social ecological-economic system in the region y_4 on the natural environment and ecological state of the CZA y_1 is assessed as weak (3 points). Then, the elements of the matrix of paired comparisons for the first equation of the model (6.2) take the values shown in Table 6.1.

Table 6.1. Expert assessments of paired comparisons for the first equation of the model (6.2)

Y_1	Y_2	Y_3	Y_4
Y_2	1	5/7	5/3
Y_3	7/5	1	7/3
Y_4	3/5	3/7	1

The corresponding matrix of paired comparisons for the first equation takes the form A_1

$$A_1 = \begin{pmatrix} 1 & 0.7143 & 1.6667 \\ 1.4000 & 1 & 2.3333 \\ 0.6000 & 0.4286 & 1 \end{pmatrix} \qquad (6.3)$$

Evaluation of the elements of the eigenvector of the matrix by the method of successive approximation gives the following results, which determine the values of the influence coefficients in the first equation of the model (6.2)

$$b_{12} = 0{,}3333; \quad 0{,}3333; \quad 0{,}3333; \qquad (6.4)$$

$$b_{13} = 0{,}4666; \quad 0{,}4666; \quad 0{,}4666;$$

$$b_{14} = 0{,}2000; \quad 0{,}2000; \quad 0{,}2000.$$

Rapid convergence of the estimates in this case demonstrates full consistency of the expert estimates of degrees of the processes' impact on each other.

Similarly, we can determine coefficients in all other equations of the model (6.2). Thus, the renewed information on the resource properties of the "sea - land" system, grounding on expert evaluation, will be included into the model. To identify remaining coefficients, we will assume that the influence of natural environment y_1 on economic infrastructure of recreation y_2 was estimated by experts as equal to 4 points, the influence of recreational services y_3 was evaluated as 7.5 points, while the influence of social ecological-economic system of the region y_4 was equaled to 3,5 points.

To estimate influence of natural environment y_1 on services y_3, we will take the index of 2.7 points, for the influence of economic infrastructure of recreation, we'll take 6.3 points, while the influence of social ecological-economic system of the region y_4 will be equal to 5,2 points. Finally, we will evaluate the impact of the environment y_1 on the state of social ecological-economic system of the region y_4 as 6.2 points, the impact of infrastructure of recreation y_2 as 7.8 points, and service for recreation y_3 as 4,8 points.

Under these conditions, the matrix of paired comparisons for 2, 3 and 4 of the equations of the model (6.2) should be taken as follows:

$$A_2 = \begin{pmatrix} 1 & 0.5333 & 1.1428 \\ 1.8750 & 1 & 2.1428 \\ 0.8750 & 0.4666 & 1 \end{pmatrix};$$

$$A_3 = \begin{pmatrix} 1 & 0.4286 & 0.5192 \\ 2.3333 & 1 & 1.2115 \\ 1.9259 & 0.8254 & 1 \end{pmatrix};$$

$$A_4 = \begin{pmatrix} 1 & 0.7949 & 1.2917 \\ 1.2581 & 1 & 1.6250 \\ 0.7742 & 0.6154 & 1 \end{pmatrix}.$$

The influence coefficients for these equations, calculated by the method of analysis of hierarchies, are presented in Table 6.2.

We will now turn our attention to integrated assessment of recreational attractiveness of the *CZA* (6.1), which determines the demand for recreational services in this part of the coast. In addition to the values of variables y_1, y_2, y_3, y_4 sensitive to the properties of this territory, the index of its attractiveness z in the "sea - land» system depends on the preferences that potential participants of recreational process (tourists) have in selecting a suitable recreational facilities for them . These preferences are taken into account when setting the weights c_i in the formula (6.1).

Table 6.2. Influence coefficients, calculated by the method of analysis of hierarchies

b_{ij}	1	2	3	4
1	1	0,33	0,47	0,2
2	0,27	1	0,5	0,28
3	0,19	0,44	1	0,37
4	0,33	0,41	0,26	1

We will assume that the demand for recreational services in the *CZA* is determined by the preferences of the tourists visiting the area. We will also assume that the natural environment and ecological conditions of the "sea – land" system were estimated by them as 7,5 points, the economic infrastructure of recreation as 5,5 points, the services as 6 points, and the state of social ecological-economic system of the *CZA* region as 2 points. Then, with the method of analysis of hierarchies, we can easily obtain the following estimates of the consumers' preferences in equation (6.1)

$$z = 0{,}36\,y_1 + 0{,}26\,y_2 + 0{,}29\,y_3 + 0{,}09y_4$$
(6.5)

After the coefficients are identified, the model (6.1) – (6.2) becomes an instrument of administrative-managed simulation. It provides an opportunity to analyze various scenarios of the index of recreational attractiveness of the "sea - land" system, which reflect the reaction of the model to predicted changes in external conditions.

6.3. Prediction of Demand for Recreational Resources with Variable External Conditions

As it was discussed above, dynamics of the recreational attractiveness index for the "sea - land" system depends on the predicted changes in external conditions F_i ($i = 1, 2, 3, 4$). To provide some examples, we will study two possible scenarios of the developing events. In the first scenario, we will assume that market conditions of the demand for recreational resources are changing, and are not favorable for the index of CZA recreational attractiveness. In the second scenario, we will simulate efforts of the administration of the "sea – land" system, aimed at increasing of the CZA tourists' demand for resources, developing in these conditions.

Let us choose a two-year period and consider discrete sequences of the simulated processes' values with a time interval of one day between samplings. We will present the equations of the model (6.2) in finite differences, and replace the time intervals between samplings with dimensionless steps of calculations. For the ease of simulation of recreational resources, we will use dimensionless (reduced) values of A that are associated with corresponding actual dimensional quantities A' by the dependence (2.22) $A = 5\,A'\,[\mathrm{M}\,(A')]^{-1}$, where $[\mathrm{M}\,(A')]$ is the mean value of the interval of variability of the corresponding dimensional quantities A'. This transformation limits the variability of scenarios' values by the interval (0, 10).

357

As an external influence of recreational resources on the natural components, we will use a sine function of time F_c, simulating long-term averaged annual temperature regime of the "sea – land" system. To conduct the first computational experiment, we will distort this function to simulate deterioration of weather conditions in the region during the summer of a certain year. This simulation will be achieved if we replace the function of the external influence of F_c by the function F_1, which sets the lower-than-normal temperatures for the sea and the air, a reduced number of sunny days, etc. Both of these functions are shown in Fig. 6.2.

Furthermore, we will assume that the state of social ecological-economic system of the region, which includes this coastal zone area, was also deteriorated in this period of time due to some growth of social (or ethnic) tension. This situation imitates the function of external influence F_4 (see Fig. 6.2, a). We will also assume that under these conditions, the administration of the "sea – land" system made efforts to increase levels of economic well-being and tourist services quality, but the resulting levels of recreational services were turned to be random fluctuations. The functions F_2 and F_3 are shown in Fig. 6.2, a.

Scenarios of the components of the attractiveness index, calculated by the model (6.1) – (6.2) with coefficients of influences that are listed in Table 1, are shown in Fig. 6.2, b. Comparing functions of external influences with scenarios of corresponding indices of recreational resources, we must pay attention to differences between them, which are the result of close interdependence between these processes. The most noticeable effect of external conditions can be seen on the example of influence on the economic infrastructure of recreation y_2 resulting from increase of social tension in the area, which occurred in the second half of the simulation period. This influence can also be seen in the resulting index of recreational attractiveness z.

In the second experiment, additional efforts of the administration of the "sea – land" system were imitated. The measures to enhance the attractiveness of recreational resorts in the winter time were taken by the administrative bodies of the area. For this purpose, a linear growth of economic resources (the curve F_2 in Fig. 6.3, *a*) was introduced in the model. Also, a saturated increase in socio-economic situation in the region (the curve F_4 in Fig. 6.3, *a*) was introduced.

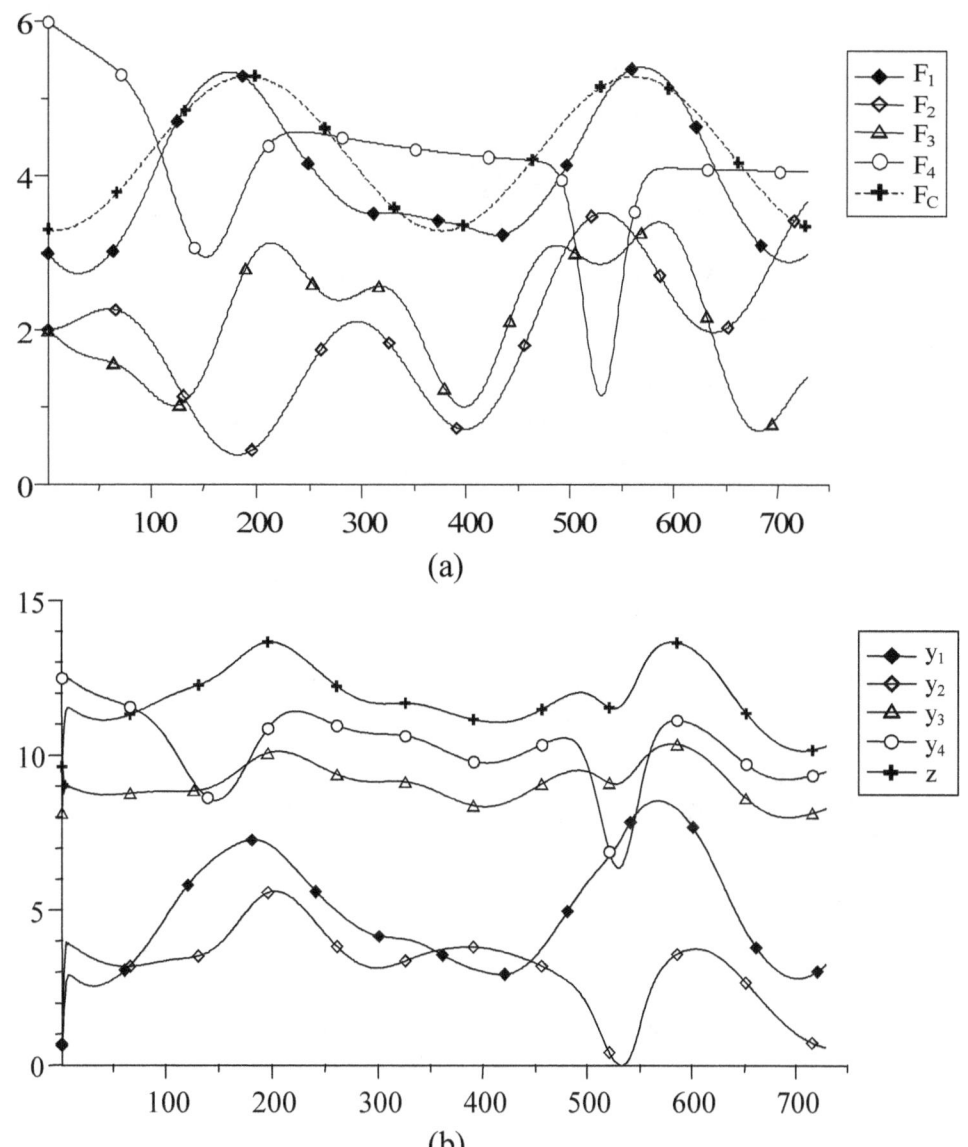

(a)

(b)

Fig. 6.2. (*a*) Simulated external influences on the "sea – land" system;

(*b*) Variability scenarios of indices of four types of recreational

359

resources, forming a recreational attractiveness of the coastal territory

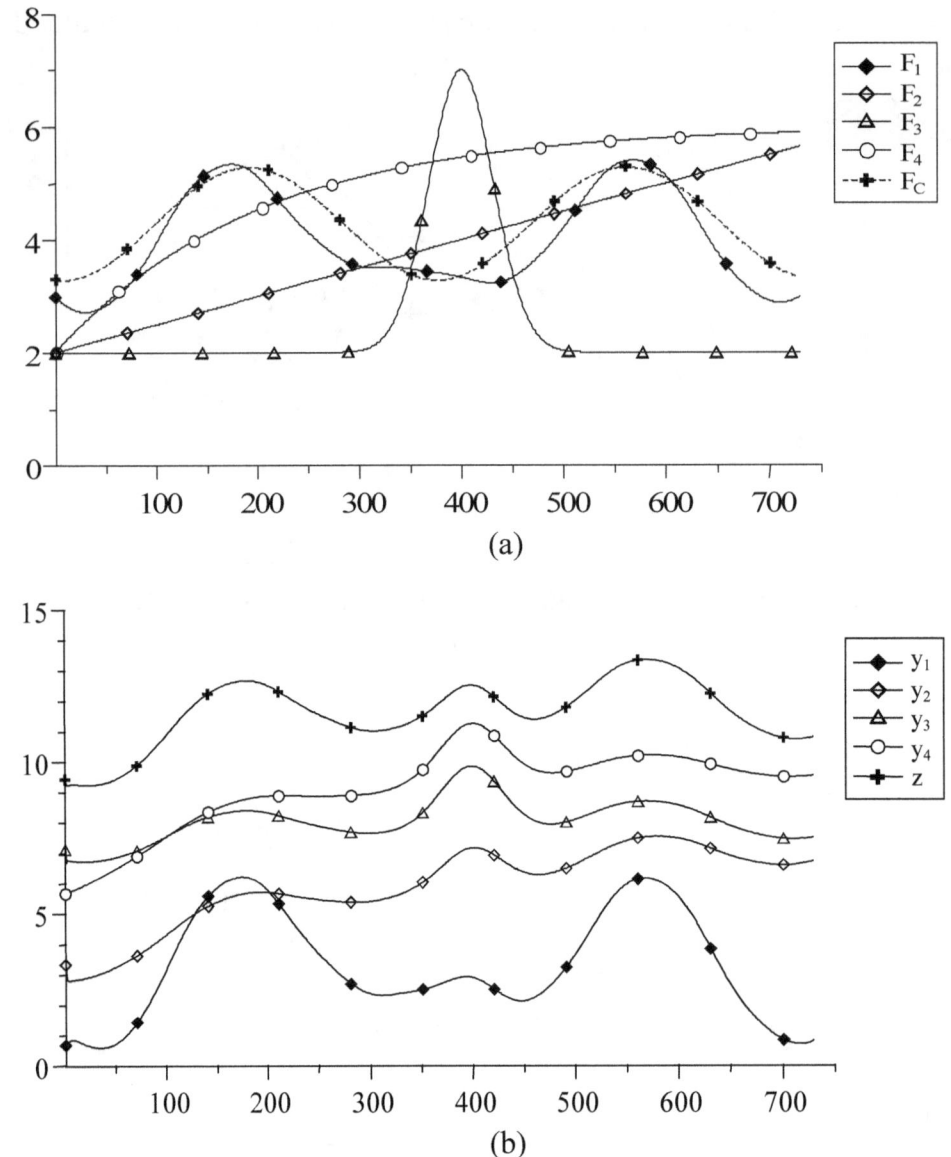

Fig. 6.3. (*a*) Simulated external influences on the "sea – land" system, including increased services F_3; (*b*) scenarios of variability indices of four types of recreational resources, forming recreational attractiveness in these
conditions

In addition to this, we simulated a sharp increase in amount of services during the winter time (curve F_3 in Fig. 6.3, a). These actions of the administration of the "sea – land" system had a positive impact on the scenarios of all components of recreational resources, as shown in Fig. 6.3, b. The resulting scenario of recreational attractiveness acquired an additional maximum, which falls on the winter season.

The experiments have demonstrated the ability to manage recreational resources' scenarios of the "sea - land" system, by choosing and introducing external influences into the model. But in reality, these control actions can only be meaningful in cases when profitability of the entire system of recreational services has been ensured. Therefore, an economic model of recreational resources consumption which would allow us to select cost-effective management scenarios, is required.

The model of recreational attractiveness of the "sea - land" system which we constructed above, provides an important "input action" for the economic model, because the index of recreational attractiveness z determines the demand for recreational services. Below, we will show that the simulation technology *ABC-AGENT,* discussed in Chapter Four, can be applied for evaluation of economic efficiency of utilizing *CZA*'s recreational resources.

6.4. Economic Model of Recreational Resources Consumption

Having the information about the four types of recreational resources y_1, y_2, y_3, y_4 of a certain "sea – land" system, which represents a particular local site of coastal zone, we can build estimates regarding the amount of recreational services which this area has capacity to provide. To do this, we will first of all define the meaning of the concept of "recreational services". We will assume that the role of recreational services is connected with "consumption" of sets of recreational resources $(y_1^0, y_2^0, y_3^0, y_4^0)$ by tourists. Then, the ratio y_1 / y_1^0 will show how many tourists can simultaneously enjoy the natural and ecological resources of the "sea – land" system. The ratio y_2 / y_2^0 will provide an assessment of number of tourists, who have the ability to simultaneously use the recreational infrastructure of the area. Since each of the tourists will possibly be using all types of the available recreational resources, the total number N of tourists will experience limitations in the types of resources which in this "sea - land" system is in the smallest volume. Therefore, it will be the following condition

$$N = \min\{y_1 / y_1^0; y_2 / y_2^0; y_3 / y_3^0; y_4 / y_4^0\} \qquad (6.6)$$

In economic sense, N characterizes the degree of use of the *CZA* recreational potential, as it represents the number of ready-to-be-provided recreational services. If the existing resource capabilities of the complex "sea - land" system are limited by any type of recreational resources, it is possible to increase the amount of recreational services of this type $(y_1^0, y_2^0, y_3^0, y_4^0)$ by providing additional investments and thus, by increasing the volumes of the respective types of resources. However, this way would cause some increase in costs of the recreational services, and would cause a temporary decline in economic viability of recreation.

Profitability of recreation as a form of socio-economic services is an important indicator of effectiveness of the "sea – land" system's resources use. It depends on many factors that determine position of the system in the market of recreational services. Therefore, to carry out management of recreational resources, it is necessary to forecast profitability scenarios and to make decisions about investments, which may help to increase recreational capacity of the "sea – land" system. To construct an economic model of recreational services, we will use a typical model of economic systems management – the informational technology *ABC-AGENT* which was discussed above, in Chapter Four.

The possibility to use it can be explained by the following circumstances. Each economic system is aimed at acquiring resources for production and realilzation of the product or service in the market, and at obtaining profits. Its structure should involve a stock of finished products, some supplies of production resources and some financial resources – the working capital of its production. Management of such system means tracking and maintaining balance between resources and finished products, and making decisions about supply of new resources or funds, if this is effective in terms of obtaining profit in future. The main components of managerial decision-making carried out by control agents is outlined in the general structure of *ABC-AGENT* informational technology, which is illustrated by Fig. 4.9.

Production of recreational services, which is based on consumption of recreational resources, represents continuous, dynamic contact between two markets: a resources market and a sales market where producers can sell their products. Therefore, management of the *CZA* recreational resources requires constant monitoring of dynamics of prices for these resources and for changes that occur in the market of recreational services. We will assume that a value H determines the existing quantity (amount) of ready-to-implement recreational services of the "sea - land" system in the sense that it's how many tourists can be provided with the necessary set of resources in required quantities (amounts) $(y_1^0, y_2^0, y_3^0, y_4^0)$ at a certain moment of time. This value characterizes the current stock of finished product (vouchers): for example, the number of transfers to motels and recreation hotels of the "sea - land" system, which can be provided to tourists instantly.

Let us denote current demand for recreational services as D. If the demanded quantity is less than existing number of vouchers H, then all potential tourists will receive the anticipated recreational services. Otherwise, the number of services will be limited to their amount N, but the insufficient number of vouchers can be provided by assigning additional costs to supply the requested amount of recreational resources. This circumstance requires inclusion of an operator-agent in the dynamic equation of recreational services, which can compare current values of supply and demand, and take decision about arranging additional services at the moment when the demand for them exceeds the offer.

Another logical control operator must monitor profitability of recreational services, taking into account their changing cost. If the index I, which characterizes the profit obtained from provided recreational services, falls below zero (income earned is less than the cost), then provision of services should be suspended. The equation of dynamics of the provided recreational services, constructed by the method of adaptive balance of causes [181], takes the form

$$\frac{dH}{dt} = H\left[1 - 2\left(H - V + S\right)\right] \qquad (6.7)$$

In this equation, the rate of return I is controlled by the "agent of sale of products" S

$$S = \text{IF}\ (N < 0;\ 0;\ R\ ),$$

$$R = \text{IF}\ (D < H;\ D;\ H),$$

and production of additional recreational services is controlled by the "agent of production output" V, which takes into account the resource potential of the "sea – land " system

$$V = \text{IF}\ (D < H;\ 0;\ M),$$

$$M = \text{IF}\ (D - H < N;\ D - H;\ N).$$

In order to meet current demand for services, the economic "sea - land" system buys insufficient resources with its existing working capital H_2. The dynamics of the working capital depends on the amount of current profit I, loans H_3 and the costs of acquisition of resources S_2

$$\frac{dH_2}{dt} = H_2\left[1 - 2\left(H_2 - I - H_3 + S_2\right)\right]. \qquad (6.8)$$

We will assume that the working capital H_2 is spent for acquisition of additional recreational resources in the same proportion as the costs $(y_1^0, y_2^0, y_3^0, y_4^0)$ are to the cost of recreational services. Let us introduce notations for prices per unit of each type of resources: p_1, p_2, p_3, p_4. Then, each type of resources will take a share in general production cost of the services, which can be represented by the ratio

$$\rho_i = \frac{p_i y_i^0}{\sum\limits_{i=1}^{4} p_i y_i^0}, \quad (i = 1, 2, 3, 4).$$

To purchase i – a type of recreational resource, the economic "sea - land" system has $\square_i H_2$ part of their working capital. If this amount is not enough to fully meet the demand for recreational services, the system can acquire resources per credit, by using bank (or other) loans. However, then a new variable H_3 arises in the economic model of the "sea – land" system, which is a debt for using resources per credit, or the credit which has accumulated by the current moment of time, and requires to be gradually compensated from the available working capital.

In this case, in the round parentheses of equation (6.8), there is an additional term S_3, which identifies expenses for repayment of the accumulated credit. If we denote percentage of loan repayment as $\square$, then the logical management agent S_3 takes the form

$$S_3 = \text{IF} \ (H_3 \square < H_2; \ H_3 \ \square; \ H_2).$$

In this expression, we included an option of priority debt repayment, which may be weakened.

The total amount of current expenditure for acquisition of resources is

$$V_3 = p_1 V_{11} + p_2 V_{12} + p_3 V_{13} + p_4 V_{14},$$

where $V_{11}, V_{12}, V_{13}, V_{14}$ were quantities of resources purchased per credit are denoted. The current value of accumulated credit will be expressed by the ABC-equation

$$\frac{dH_3}{dt} = H_3\left[1-2\left(H_3-V_3+S_3\right)\right]. \qquad (6.9)$$

Now, let us consider the dynamics of inventories of recreational resources available to the "sea – land" system. These reserves are used when recreational services are provided, and are compensated continuously by purchasing additional amounts of resources. We will introduce this value as H_{1i}, which denotes the number of units of i – a type of resource that the "sea - land" system has available at a certain moment in time. In the dynamic regime, these quantities reflect the following equations of the ABC-model:

$$\frac{dH_{1i}}{dt} = H_{1i}\left[1-2\left(H_{1i}-V_{1i}+S_{1i}\right)\right], \quad (i=1, 2, 3, 4)$$
(6.10)

In these equations, expenditure of each type of resources is proportional to total volume of provided recreational services $S_{1i} = V_i\, y_i^0$. In cases, when the available amounts of recreational resources are sufficient to meet the customers' demand, purchasing of additional resources does not occur. Otherwise, a number of resource F_i has to be purchased, and this is a way to contribute a part of the working capital (or credited funds) into improvement of natural environment, into economic infrastructure and services of the resort, which helps increase recreational attractiveness of the "sea - land" system

$$V_{1i} = \text{IF } ((D - H)\, y_i^0 < H_{1i}\,;\, 0\,;\, F_i\,).$$

Function F_i manages the acquisition of resources. It takes into account financial resources which the economic system has in its possession. If the amount of available capital funds assigned for the acquisition of i–type of resource equal to $\square_i H_2$, and the cost of the missing amount of i–type of resource is equal to $p_i\,(y_i^0 D - H_{1i})$, then the logical condition for this function takes the form

$$F_i = \text{IF } (p_i\,(y_i^0 D - H_{1i}) < \square_i H_2;\, y_i^0 D - H_{1i}\,;\, R_i\,).$$

366

Acquisition of resources is limited by function R_i to the extent that the value of the credit, accumulated by the economic "sea - land" system is approaching its maximum permissible value $H_3{}^*$. It is established by an organization which finances the "sea – land" system, or by an administrative authority which regulates consumption of resources in the region.

As we have noted above in Chapter Four, the scheme of integrated management of coastal zone resources consumption can be grounded on limitation of production by regulating the magnitude of $H_3{}^*$. The function sets the maximum allowable amount of payments for the use of resources in the "sea - land" system, which defines the administrative bodies that monitor the state of resources of the system. This means that the natural and economic resources belonging to the society (the state), are gave to a private economic system in debt in the form of accumulated credits, compensating through the gradual repayment of their value. Setting limits for the accumulated credit $H_3{}^*$, the administration of the region (or its subordinate administration of the local *CZA* site) limits consumption of resources and thereby reduces human pressure on the environment. We will use this scheme to assess cost-effectiveness of environmental management of a "sea - land" system maintaining its ecological condition.

6.5. Model of Balance of Recreational Services and Environmental Activity

The aim of recreational resources management is to find some environmentally justified volume of consumption of these resources, i.e. to define such volume of providing recreational services, where the levels of environment pollution do not exceed some prescribed limits. Conceptual model of ecological-economic system of recreational resources consumption is shown in Fig. 6.4.

To define an integral feature of ecological state of natural environment, we will use the recreational facility index *PL*, which is a weighted sum of concentrations of the most harmful contaminants in the air, in the marine environment and on the territory occupied by the recreational system. The task of management is to ensure that the economic system of recreation assigns a part of its profit for environment protection actions in such quantities which are sufficient to compensate harmful effect caused to the environment by this system.

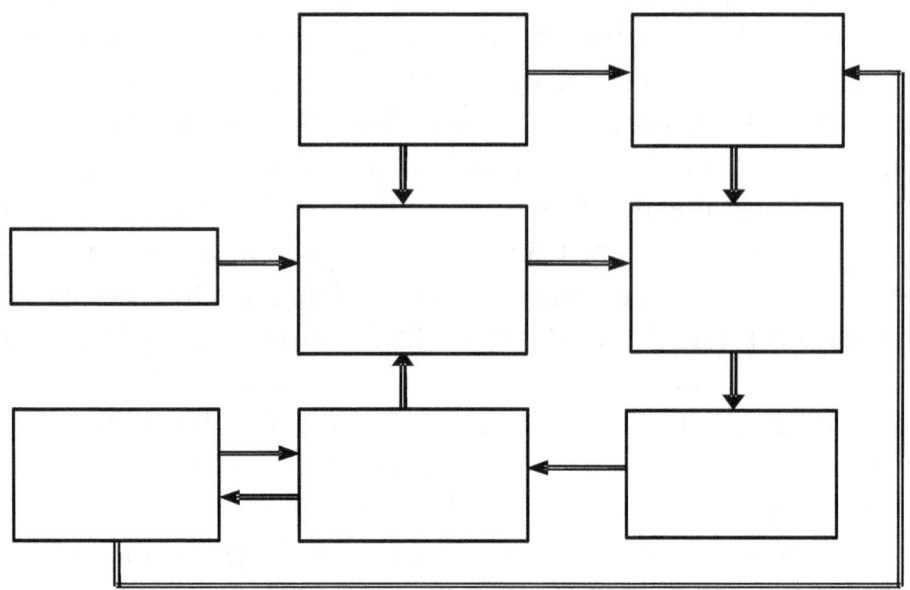

Fig. 6.4. Conceptual model of balance of recreational resources consumption and   environmental protection actions

Recreational potential of the territory *Q* determines its recreational attractiveness *RA* and simultaneously generates the price of recreational services *P*, by which we will mean a voucher for spending a certain time in a recreation hotel by a customer. The demand for recreational services is denoted as *D*. With the growing volume of services *V* increases pollution *PL*. To deal with pollution, environment protection actions *EP* are necessary. They are funded by a part of the economic system's profit. As recreational attractiveness depends on seasonal weather factors, this factor is reflected in the model by external influence function *SF*. Pollution monitoring is performed by the agent $A(PL_c, I_{PL})$. The *ABC*-model corresponding to Figure 6.4, reflects dynamic balance between recreational resources consumption and environment protection actions

$$\frac{dRA}{dt} = RA[1 - 2(RA + a_{RA/PL}PL - a_{RA/SF}SF)],$$

$$\frac{dPL}{dt} = PL[1 - 2(PL + a_{PL/EM}EM - a_{PL/S}S)],$$

$$\frac{dP}{dt} = P[1 - 2(P - a_{P/Q}Q - a_{P/EM}EM)],$$ (6.11)

$$\frac{dEP}{dt} = EP\{1 - 2[EP - A(PL_c, I_{PL})]\},$$

$$\frac{dD}{dt} = D[1 - 2(D - a_{D/RA}RA + a_{D/P}P)],$$

$$A(PL_c, I_{PL}) = IF[PL < PL_c; 0; a_{PL}I_{PL}(1 - e^{-\alpha_{PL}\tau})],$$

$$S = F(D, H_2, H_3^*, \varphi).$$

To build a model of economic processes of recreation, we will apply informational technology *ABC-AGENT*, which was discussed above, in Chapter Four. Its basic equations have the following form:

$$\frac{dX_3}{dt} = X_3\left[1 - 2\left(X_3 - \sum_{i=1}^{3} y_i r_i - Q\right)\right],$$

$$\frac{dH}{dt} = H\left[1 - 2(H - V + S)\right],$$

$$S = IF\ (p < e;\ 0;\ R\ ),\ R = IF\ (D < H;\ D;\ H),$$

$$V = IF\ (D < H;\ 0;\ M),\ M = IF\ (D - H < M;\ D - H;\ M),$$

$$M = \min\ (\ m_1;\ m_2;\ m_3),\quad m_i = H_{1i}/y_i\ ,$$

$$\frac{dH_{1i}}{dt} = H_{1i}\left[1 - 2\left(H_{1i} - V_{1i} + S_{1i}\right)\right], \tag{6.12}$$

$$V_{1i} = IF\{D - H_{1i} < 0; 0; \quad IF[y_i(D-H) < H_{1i}; 0; U_{1i}]\},$$

$$U_{1i} = IF\{y_i(D-H) - H_{1i} < \rho_i H_2 / r_i; y_i(D-H) - H_{1i};$$
$$IF[\rho_i(H_3^* - H_3) < 0; 0; U_{1i}^*]\},$$

$$S_{1i} = IF\{D - H < 0; 0; IF[y_i(D-H) < H_{1i}; y_i(D-H); H_{1i}]\},$$

$$\frac{dH_2}{dt} = H_2\left[1 - 2\left(H_2 - pS + \sum_{i=1}^{3} S_2^{\;i} + S_3 + \chi H_2\right)\right],$$

$$S_2^{\;i} = IF[r_i y_i(D-H) - H_1^{\;i} < \rho_i H_2; r_i y_i(D-H) - H_1^1; \rho_i H_2],$$

$$S_3 = IF[\theta H_3 < H_2; \theta H_3; H_2],$$

$$\rho_i = \frac{r_i y_i}{r_1 y_1 + r_2 y_2 + r_3 y_3}, \; (i = 1,2,3),$$

$$\frac{dH_3}{dt} = H_3\left[1 - 2\left(H_3 - \sum_{i=1}^{3} r_i V_{1i} + S_3\right)\right],$$

$$V_{1i} = IF[(D-H)y_i < H_{1i}; 0; F_i],$$

$$F_i = IF[r_i(y_i D - H_{1i}) < \rho_i H_2; y_i D - H_{1i}; F_i^*],$$

$$\varphi = \ln\frac{[pS+1]}{[10 + \sum_{i=1}^{3} S_2^{\;i} + S_3 + \chi H_2]},$$

where X_3 identifies costs of recreational services, V is the volume of these services, S is the number of units of purchased vouchers for recreational services, H_{1i} are volumes consumed by industrial, natural and environmental resources – the numbers of each kind of resources which are needed to provide a single service, V_{11}, V_{12}, V_{13} are volumes of resources purchased for credit, H_3^*, F^*, U_{1i}^* are control functions, φ is profitability of recreational services, $\square$ is the net profit.

6.6. Ecological-Economic Balance of Recreational Resources Consumption

The model of balance between recreational services and environment protection activity considered in the previous section was built with the goal to analyse possible scenarios of recreational resources consumption in terms of control over the level of environmental pollution. It was assumed that, with the help of the model, an administrative body of recreational management which maintains environment of the local *CZA* area, will be able to choose a financial policy for environment protection actions which will ensure economic viability of the resort and resumption of recreational resources.

To determine the capacity of the model, computational experiments were carried out on 500 dimensionless time steps (days), during which the model's reaction on external changes was studied. Initially, we accepted a strategy of environmental actions, where the economic system of recreation could increase volumes of recreational services to a certain constant level which depended only on the value of accumulated credits. Before this level was achieved, the system had been able to acquire loans for consumption of recreational resources, and the rate of repayment of accumulated credit had been 5% at each step.

The level of pollution was considered to be proportional to the volume of the recreational services, and was considered to have no direct influence on their intensity. With growing volume of recreational services, the amount of funds, invested by the economic system of recreation into usage of resources, increased. Therefore, an assumption was made that any restriction of investment into acquisition of resources, (i.e. accumulated credit), would make the

371

economic system of recreation to reduce their economic activity, and hence, the level of environment pollution will decrease. In addition to redemption of credit, at each step of calculations, 20% of the working capital of the economic system was confiscated as net profit.

The scenarios of processes in ecological-economic model of recreational resources, which were obtained in a series of experiments when the environmental management made credit limitations grow proportionally to the growth of the *CZA's* environment pollution, are shown in Fig. 6.5, *a, b* and Fig. 6.6, *a, b*. In order to limit investments, a maximum admissible amount of accumulated credit H_3^* was set-up, which was amounted to 1 200 conventional units. The influence of weather conditions was given in the form of seasonal variability of the function of recreational attractiveness *RA*, as it is shown in Fig. 6.6, *a*.

As it can be seen from Fig. 6.5, *a*, the demand for recreational services was primarily formed due to seasonal factors *SF* and the ratio of sales cost (price) of the service ($P = 10,2$ conditional units) and its costs ($E = 1,4$ conditional units) was chosen in such a way that the economic system made a sustainable profit during the entire period of calculations. Dynamics of working capital is shown in Fig. 6.5, *b*. The same figure shows maximum allowable credit accumulated by the economic system: $H_3^* = 1200$ conditional units.

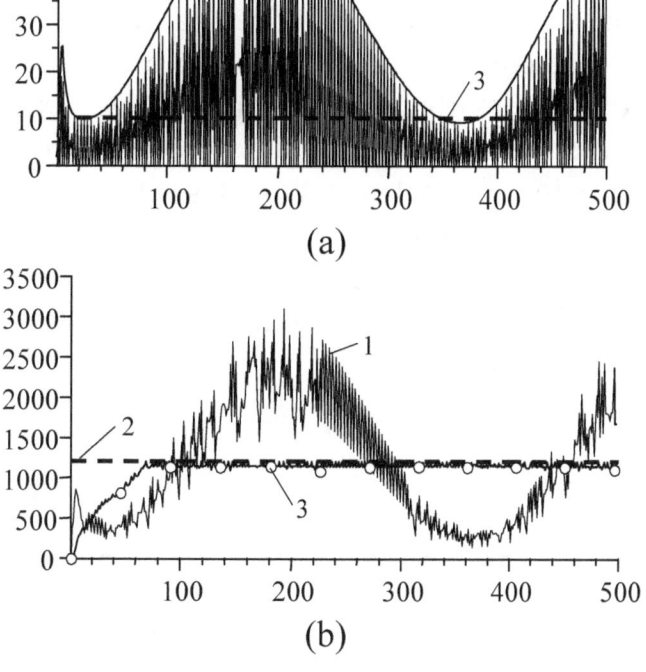

Fig. 6.5. (*a*) 1 – demand D, 2 – implementation of recreational services S,

3 – price of recreational service p; (*b*) 1 – working capital H_2,

2 – maximum permissible investment H_3^*, 3 – accumulated credit H_3

The system was receiving credits to acquire resources, starting with the first step of calculations, and up to step 70, the accumulated credit reached its allowable limit.

Scenarios of recreational attractiveness and the level of pollution (contamination) (see Fig. 6.6, *a*) follows the scenario of demand. Moreover, the intensity of resources consumption was so high that in the period of the highest demand, the level of pollution of the *CZA* environment exceeded the maximum allowable value. Averaged for 10 days, profitability of resources consumption φ (shown in Fig. 6.6, *b*), remained relatively high.

It follows from the analysis of the scenarios that, during the whole time of the experiment, profitability of recreational services was totally dependent on demand for these services. The services' variability traced seasonal factors of recreational attractiveness. The used scheme of credit limitation (see Fig. 6.5, *b*) turned out to be inefficient from the environmental point of view, as introduction of a "ceiling" for the value of accumulated credit could not reduce the intensity of the recreational services. Therefore, a conclusion was made that stronger economic levers are needed to ensure environmental protection in the cases where the level of pollution of the *CZA* natural environment requires reduction of commercial activity of the economic system.

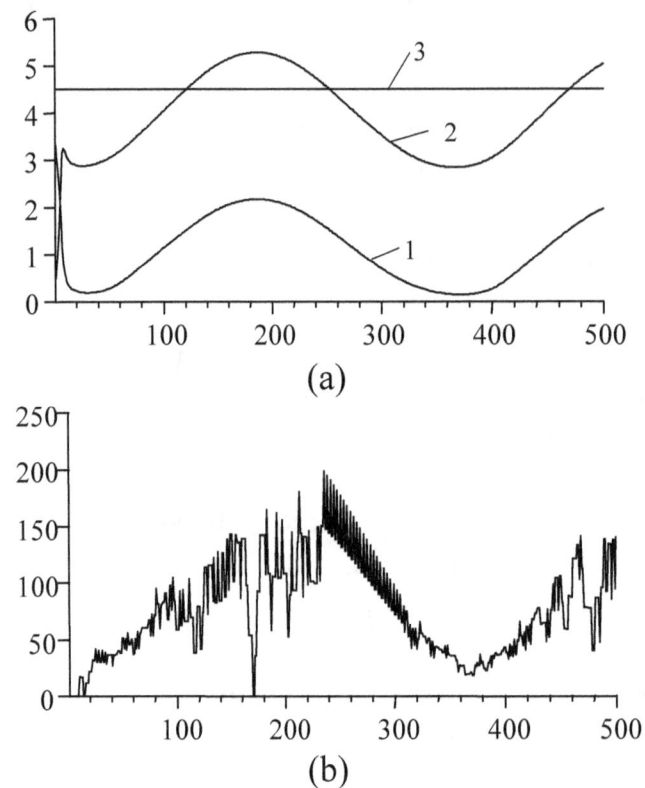

(a)

(b)

Fig. 6.6. (*a*) 1 – recreational attractiveness *RA*, 2 – level of pollution *PL*,

3 – maximum allowable level of pollution PL_c; (*b*) averaged over 10 days costs-effectiveness of consumption of recreational resources φ

In the second simulation experiment, to provide more rigid tactics of control over the level of pollution (contamination) control, some sanctions were imposed on the economic system of recreation. The results of this experiment are shown in Fig. 6.7, *a, b*. The funding provided by the economic system for pollution control and protection I_{PL}, was functionally linked to the level of pollution PL and the magnitude of the working capital H_2, as follows:

$$I_{PL} = 0,05(PL)H_2.$$ (6.13)

Since working capital typically increases along with the volumes of recreational services, the formula (6.13) represents an environmental tax that takes into account both, the intensity of recreational resources consumption, and the actual level of the natural environment pollution. While the environmental degradation is growing, the agent $A(PL_c, I_{PL})$ in the equation for the volume of environment protection actions in the model (6.11), was increasing the cost of environmental resources. Some part of the net profit was spent on environment protection actions.

As another way to control the ecological state of the environment, some steps towards management of the size of maximum permissible loan H_3^* of the economic system of recreation were taken. For this purpose, another management agent was introduced. It was to control allocation of credits (licenses) for consumption of recreational resources. The function of this agent was to lower the "ceiling" of maximum allowable credits for acquiring recreational resources in terms of deterioration of the environment, where the level of pollution is above the critical PL_c. These operations were performed according to the following formula

$$H_3^* = IF[PL < PL_c; H_3^*; H_3^* e^{-\alpha_{PL}\tau}].$$ (6.14)

As it can be seen from Fig. 6.6, *a*, the demand for recreational services grew considerably in the summer time due to seasonal factors. However, in contrast to the results of the first experiment (see Fig. 6.5, *a*), in the second experiment the possibility of the

375

economic system to use the demand for making profit, was significantly limited. With the increasing number of recreational services, the level of pollution has also increased. As the level of contamination came close to the limiting values (see Fig. 6.7, b), the management agents significantly restricted the values of obtainable credits and raised the costs of recreational services. As a result, the service was suspended until the time (step 260), when the level of environmental pollution was returned to acceptable standards.

The dynamics of working capital of recreation is presented in Fig. 6.8, b. The same figure shows the maximum allowable amount of credit accumulated by the economic system: $H_3^* = 1200$.

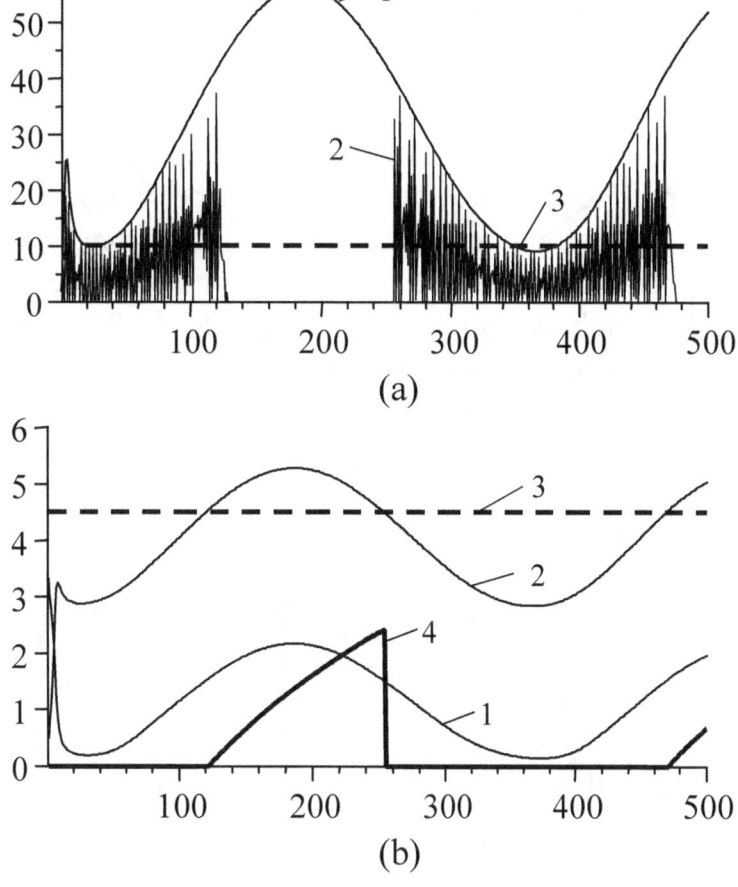

Fig. 6.7. (*a*) 1 – demand *D*, 2 – implementation of recreational services *S*,

3 – price of recreational service *p*; (*b*) 1 – recreational attractiveness *RA*, 2 – level of pollution *PL*, 3 – maximum allowable level of pollution PL_c, 4 – necessary amount of environmental protection actions *EP*

The economic system of recreation had used loans to acquire resources, starting with the first step and up to step 70 of calculations, the accumulated credit had reached its limit value (see curve 3 in Fig. 6.8, *a*). However, the recreational services output continued to grow till step 120, when the level of pollution reached a critical value (see curve 2 in Fig. 6.8, *a*). At that moment in time, management agent $A(PL_c, I_{PL})$ began to increase environmental taxes, which was intended to provide the necessary level of environment protection actions *EP* (see curve 4 in Fig. 6.7, *b*).

Simultaneously, the agent (6.14) began to sharply lower the maximum allowable level of accumulated credit H_3^*. In such conditions, production of recreational services became unprofitable and had to be terminated. This is obvious from examining the scenario of recreational resources consumption φ (averaged over 10 time steps), which is shown in Figure 6.8, *b*. The production of services was restarted when the level of pollution decreased and became lower than the critical allowable amount (see curve 2 in Fig. 6.7, *b*).

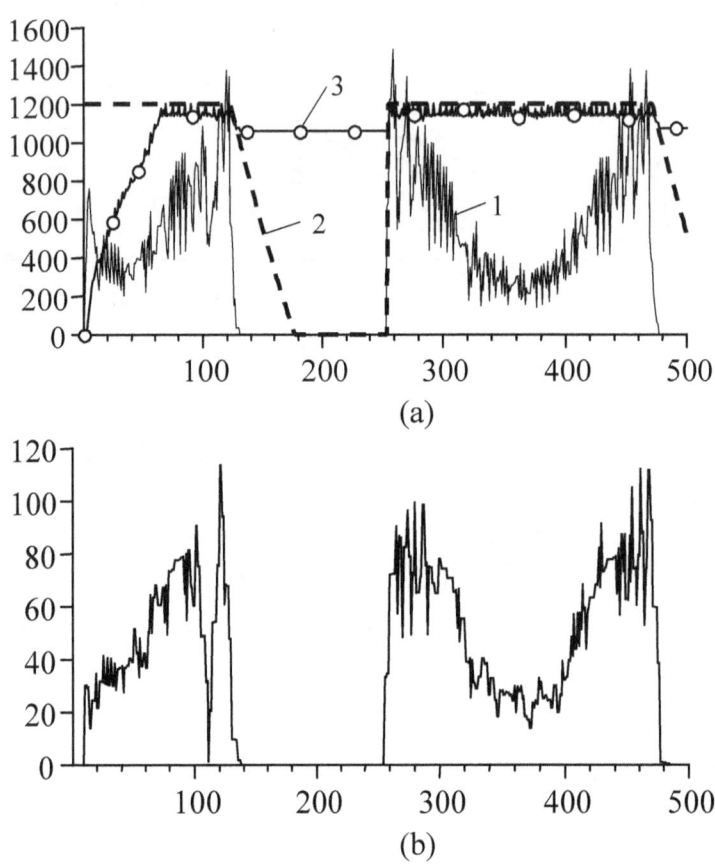

Fig. 6.8. (*a*) 1 – working capital H_2, 2 – maximum permissible investment H_3^*,

3 – accumulated credit H_3; (*b*) averaged over 10 days costs-effectiveness   of consumption of recreational resources φ

A comparison between the results of the experiments shows that the suggested model could be used as a tool for choice of management scenarios for recreational system, in which the economic part of the system will maximize its incomes, and sufficient funds will be allocated for environment protection actions. This could maintain the level of environment pollution below critical values. Computational experiments performed with the model, allowed us to build predictive scenarios of resources consumption to meet the demand for recreational services. The analysis of the scenario confirms the possibility of using dynamic models of "sea - land" systems for creation of informational technologies for providing decision-making support and management of recreational processes. In their turn, such technologies can serve as a basis for dynamic inventory of recreational resources, which are specialized geo-informational systems of the "sea – land" type, suggesting options of effective dialogue between consumers of resources and administrations of environmental organizations.

In terms of economic use of recreational resources, such models make it possible to determine which of the control actions can provide profitable consumption of all kinds of the *CZA* resources, what is especially important when planning investing into infrastructure and development of new resorts. From the view point of administrations of the environment protection agencies, such models can assist in performing control and maintaining the volumes of recreational services. They can help in taking decisions about limiting amounts of recreational services, to bring the whole system into accordance with environmental standards.

6.7. Transformation of Demand for Energy Production in Recreational Areas in the Light of Environmental State Management

One of the most significant problems of *CZA's* sustainable development relates to supplying these areas with energy, which inevitably affects the ecology of the surrounding environment. As recreational attractiveness of CZAs is growing, the energy consumption tends to increase. In most cases the energy is obtained by one of the traditional ways: by burning fossil fuels in boilers, and using thermal power. We know how bad is the pollution caused with air emissions of greenhouse gases that accompany traditional methods of energy supply. We also know that alternative energy sources – solar panels, wind generators, heat pumps and others are considerably inferior to traditional sources of energy in cost, and hence, in profitability of production. Therefore, the traditional sources of energy prevail in CZAs almost everywhere.

The problem of traditional energy supply for *CZA* potentially limits the use of its recreational potential. Using alternative energy sources seems to be more beneficial for the local *CZA* sites with high recreational potential, despite its higher cost. To make decisions related to development of recreational establishments, having an informational technology becomes crucial, as it can help to carry out comparative assessment of economic benefits of traditional and/or alternative methods of energy consumption in view of the main parameters of ecological state of the environment.

In this section, we will consider an adaptive transformation model of demand for energy supply of recreational services in conditions of management of balance between consumption of traditional and alternative energy. As the main incentive for the transition from traditional to alternative energy suppliers, we will take the following objective of development management: reduction of hazardous caused by traditional energy production, through application of economic sanctions over any violations of the established requirements. As a further incentive, we will consider a factor of growing demand for development of environmental awareness among people who visit *CZAs* for leisure and travel, as they must be aware of the risk caused to their health in association with traditional energy utilization .

A model of *CZA* energy supply should contain a flexible system of environmental fines, which would support measures that cause increase of prices for traditional energy suppliers, in cases when energy production deteriorates natural environment. Keeping this in mind, we will use the abovementionned systems concept of ecological and economic balance between *CZA* development processes, which determines basic blocks of the simulation model of management. The most important of the blocks will be the economic blocks of traditional and alternative energy suppliers, a block of integrated assessment of ecological conditions in the region, a block describing characteristics of demand for energy suppliers, and block for controlling and setting penalty sanctions for environmental pollution. The general scheme of communication between the blocks is shown in Fig. 1

Aggregated demand for energy is distributed between traditional and alternative energy suppliers under the influence of price control mechanisms and ecological state of surrounding environment. Constantly updated information about the level of pollution, provided by a system of territorial monitoring, is the basis for management of traditional energy costs, which can be done by means of introduction of differentiated environmental taxes. The tax rate for «traditional» energy should significantly increase and take the form of environmental penalty when its cumulative impact on the environment grows higher than the maximum allowable level of pollution.

Comparing the costs of «traditional» and «alternative» energy should result in turning the aggregated demand for energy towards increasing use of alternative energy sources. Information about the benefits of ecologically-clean energy, as well as providing the cost-effective transition to this type of energy supply, is capable to convince the public to shift local investment policy towards extensive use of alternative energy suppliers.

Transformation of aggregated demand leads to redistribution of energy production and to achieving dynamic balance between traditional and alternative energy supply in the region. The idea of achieving and maintaining dynamic balance within a system is the essence of systems modeling concept. Therefore, to build a simulation model of transformation of the regional power economy, we used systems method of adaptive balance of causes (*ABC*-method) [181].

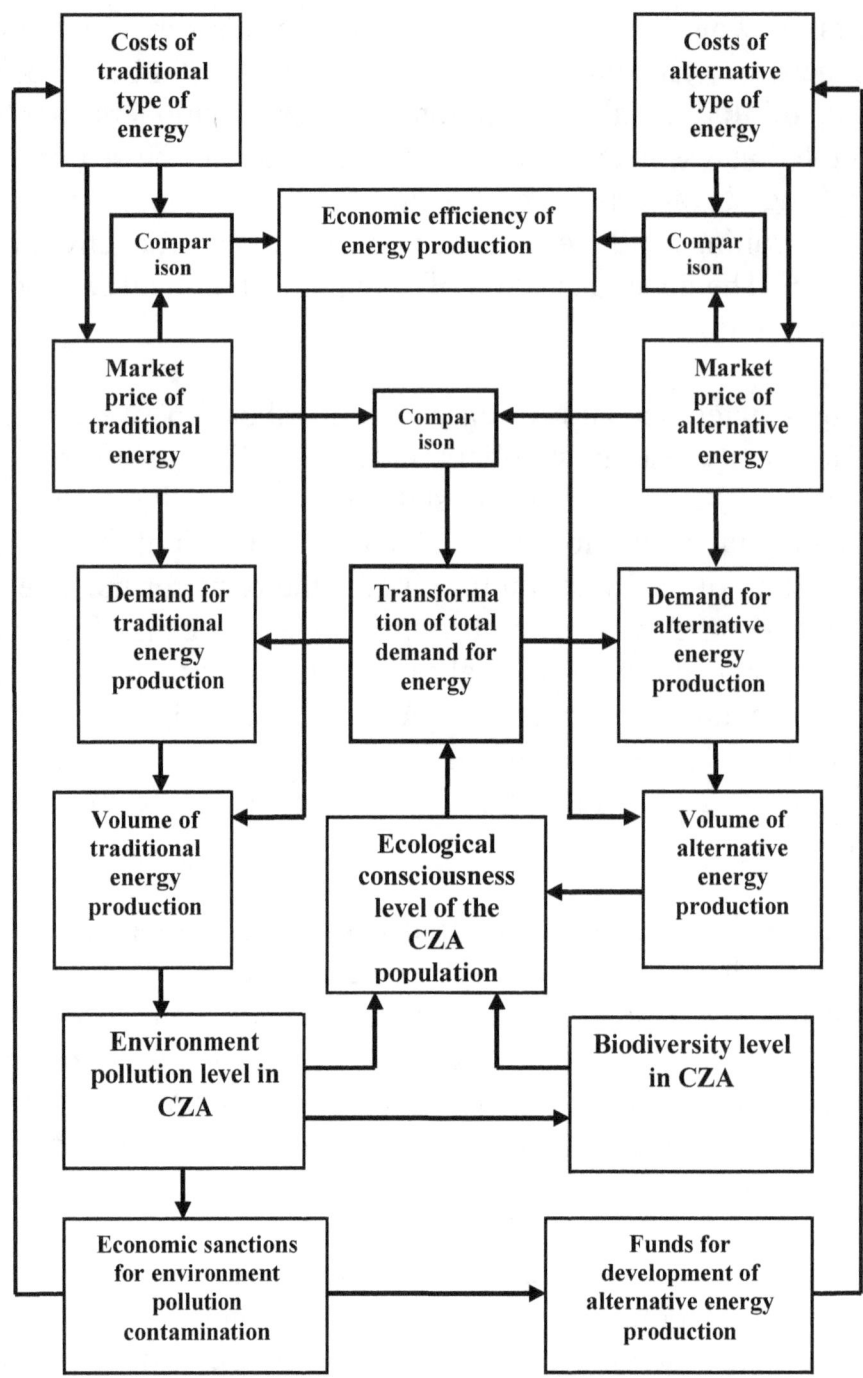

Fig. 6.9. Conceptual model of aggregated demand transformation of energy
supply in recreational area

Simulation Model of Transformation of Demand for Energy Supply in Recreational Areas. In accordance with the conceptual model shown in Fig. 6.9, we developed a structure of dynamic equations for economic blocs of traditional and alternative energy supply. Grounding on the conclusions made in Chapter Four, we will use standard equation, typical for operations taking place in any economic system. The key processes that characterize dynamics of economic systems are: demand for a product's output, its price, cost, etc. They can be represented in a form of dynamic equations of informational technology *ABC-AGENT*, as we discussed in section 4.3. The variables in the equations of the model of traditional and alternative energy suppliers will be different in superscripts: we'll use "*T*" for the processes related to traditional energy suppliers, and respectively, "*A*" – for the processes related to alternative energy suppliers.

For convenience of modeling, we will use the dimensionless (reduced) value of A, which are connected with the corresponding actual dimensional quantities A' by dependence (2.22)

$$A = 5\,A'\,[\text{M}\,(A')]^{-1},$$

where [M (A')] is the average interval of variability of the corresponding dimensional value A'. This limits variability of the model scenarios for ecological and economic processes at intervals of dimensionless quantities [0, 10].

In order to simplify our task, we will assume that production of traditional and alternative energy meets total demand for energy completely in the *CZA*. The ecological state of the region is a very complex function containing many influence factors, as well as their geographical distribution and temporal variability. The average pollution across the *CZA* is represented by an integral, weighted sum of concentrations of major pollutants Pl_i contained in the air, land surface or surface waters, which enter the environment in course of

383

production of traditional energy sources. We will assume that there is a system for monitoring these types of pollution, which provides estimates of typical (for this *CZA*) temporal variations of integral parameter *Pl*

$$Pl = \sum_{i=1}^{M} h_i Pl_i .$$

Furthermore, we will assume that, according to the archived records of territorial monitoring taken over the recent few years, we know scenarios of mean values of this parameter, plotted with a month interval of the moving averaging. Then, the dynamics of contamination level can be represented by equation

$$\frac{dPl}{dt} = Pl[1 - 2(Pl - a_{Pl/V^T}V^T + a_{Pl}Pl)] , \qquad (6.15)$$

in which the term $a_{Pl/V^T}V^T$ connects the level of pollution with production volumes of «traditional» energy, and the term $a_{Pl}Pl$ represents self-purification processes in natural environment (reduction of contaminants' concentrations), for example, happening due to turbulent mixing and replacement of air.

For biodiversity index, we will use the weighted sum of concentrations of major flora and fauna species of the region Br_i, which was averaged over its territory

$$Bd = \sum_{i=1}^{N} g_i Br_i . \qquad (6.16)$$

The same way as in the case of integrated pollution index *Pl*, it is assumed that we know the volatility of index *Bd*, which is formed, for example, by average monthly concentrations of Br_{im}. The current index value for each year undergoes deviations from the average long-term variability. Variations of air temperature, rainfall, wind speed modulus, illumination, etc. (compared to their multiyear averages), can be considered as major factors shaping these deviations. However, the emission caused by power plants and other energy facilities, also plays a significant role in shaping the index value of biodiversity.

In order to simplify the discussion, we will not take into account the weather factors in our model of ecological condition. We will focus mainly on the impact of the pollution level Bd. Therefore, the dynamic equation of biodiversity index will be used in the following form

$$\frac{dBd}{dt} = Bd[1 - 2(Bd + a_{Bd/Pl}Pl)]; \qquad (6.17)$$

The total demand for energy provision is divided between traditional and alternative energy suppliers, not only under the influence of price control mechanisms, but also due to environmental awareness of the population which depends on the information about pollution levels. Information about the benefits of clean energy, as well as the return on the expenditures invested into the transition to this type of power supply, can convince public to approve of the transition to alternative energy supply.

This index reflects the concerns, which population of a *CZA* experiences while watching the increasing pollution (contamination) of air, water, food products and other properties of the natural environment, which are crucially important for healthy well-being of local population. Contamination and pollution takes place largely due to burning carbonated materials at heating power stations, in energy power stations, in residential buildings and at industrial facilities.

Therefore, we will associate the value indicating transformation of the demand for energy supply in the region with the value of index *Ec* of population's environmental awareness level, as the population must be concerned with pollution and with negative trends in biodiversity dynamic of the region. This index reflects level of population's concern over deterioration of air quality, water, food, etc.

Let us introduce the equation for the index Ec in the following form

$$\frac{dEc}{dt} = Ec[1 - 2(Ec - a_{Ec/Pl}Pl + a_{Ec/Bd}Bd)].$$ (6.18)

As it follows from the above equation, with increasing pollution levels Pl and decreasing biodiversity of the environment Bd, level of environmental consciousness of CZA population Ec increases.

In order to control the volume of harmful substances' emissions into the environment, a part of profit of traditional energy types' producers should be directed at environmental purposes. A differentiated environmental tax can become a solution to this task. It can place production costs and, consequently, the price of production of traditional energy, into dependence of the environment pollution level.

Direct control of the environmental pollution provides a basis for determining the rate of environmental tax $T^T(Pl)$ that must be included in the equation of the cost of traditional energy, which we denoted as X_3

$$T^T(Pl) = a_{X_3/Pl}Pl.$$

During the monitoring experiments, the environmental tax rate should be increased or decreased, depending on the actual level of pollution, which should prepare the system for regional environmental monitoring. The growth Pl is a sign of deteriorating environmental situation in the CZA; to address that need, additional environmental actions must be performed. The expenditure should be paid from the profit of the economic system of energy supply. This type of expenditure represents differentiated environmental tax.

When the level of pollution exceeds the allowed maximum, the value which we denote as Pl^*, for more effective management of the environment, additional financial penalties should be applied to objects of energy production and supply. These penalties must also be included in the equation for production costs of traditional energy, in the form of control agent $A^T(Pl, Pl^*)$. The function of this agent can be seen in the following relations

$$A^T(Pl, Pl^*) = IF\{Pl < Pl^*; C_1; C_2\}, \qquad (6.19)$$

$$C_1 = a_{X_3^T/Pl_1} Pl_1 \exp(-\alpha_{C_1} \tau_{C_1}), \qquad (6.20)$$

$$C_2 = a_{X_3^T/Pl_2} Pl_2 [1 - \exp(-\alpha_{C_2} \tau_{C_2})]. \qquad (6.21)$$

In relations (6.20) and (6.21), Pl_1 designates pollution level in those periods in time when it falls below its maximum permissible value Pl^*, and Pl_2 identifies the cases when this level exceeds the permissible maximum. Time function C_2 begins to increase the amounts of environmental penalties, as soon as the concentration of pollutants exceeds the value of Pl^*, and it continues to operate until the time when the Pl_1 falls down to the level Pl^*. After this point in time, a function of time C_1 introduces some time delay (inertia) which corresponds to relief in environmental penalties, to ensure a more reliable management of Pl level.

Thus, by using a differentiated environmental tax and environmental fines, the cost of traditional energy sources can be increased substantially, when the state of environment is worsening compared with the average long-term statistics, or when pollution levels exceed maximum permissible values. The growing cost, in its turn, should reduce the demand for traditional forms of energy, and thus the demand for alternative types of energy will be increased.

Economic Model of Energy Supply for Recreational Areas. Key processes that characterize dynamics of economic systems of traditional and alternative energy production are: production and sales of finished products, reception and repayment of loans, acquisition and use of resources for production processes, and others – all these processes can be represented in form of dynamic equations of informational technology *ABC-AGENT*, discussed above in Chapter Four. Economic models of traditional and alternative energy have similar equations; in most cases their variables differ only by the superscript *"T"*, meaning «traditional», and *"A"*, meaning «alternative» energy.

Let us introduce the following notations for the model variables: X_1 – demand for energy; X_2 – its price; X_3 – cost of energy production; and a few types of consumed resources: H_{11} – investments, depreciation and repair, H_{12} – maintenance, fuel, solar radiation, H_{13} – environmental charges (or environmental taxes). Now, we'll introduce indications of prices for these resources: r_1 – investments, depreciation and repair; r_2 – maintenance, fuel, solar radiation, r_3 – environmental charges (environmental taxes). We will identify the aggregated demand for the energy as a sum $X_1 = X_1^T + X_1^A$, where the upper indices denote each of the corresponding types of energy.

Given the dependence of demand on prices, on the index of diversity and on contaminants concentration, we will obtain the following equations of the *ABC*-economic model of traditional energy for the dimensionless variable

$X_1^T = 5\ D[M(D)]^{-1}$, which represents the demand for the traditional form of energy

$$\frac{dX_1^T}{dT} = X_1^T \{1 - 2[X_1^T + a_{12}X_2^T + a_{1Ec}Ec - a_{1F_1}(F_D - X_1^A)]\},$$

(6.22)

$$\frac{dX_2^T}{dt} = X_2^T \left[1 - 2\left(X_2^T - a_{23}^T X_3^T - Q^T\right)\right],$$ (6.23)

$$\frac{dX_3^T}{dt} = X_3^T \{1 - 2[X_3^T - A^T(Pl, Pl^*) - E^T - T^T(Pl)]\},$$ (6.24)

$$E^T = E_0^T + \sum_i y_i r_i^T,$$

where $y_i r_i^T$ are the shares of each type of the resources utilized by traditional energy suppliers within the energy production cost; Q^T is the added value, $a_{1F_1}(F_D - X_1^A)$ is an influence function for the alternative energy supplies; F_D is a prognostic aggregated demand for energy; Ec is the level of environmental awareness of the *CZA* population; $A^T(Pl, Pl^*)$ is a control agent for production costs management by environmental criteria; E_0^T is the share of initial investment in the production cost of the industry; $T^T(Pl)$ is the environmental tax.

The economic model of alternative energy is similar in its function to the traditional model of supply. The minor differences are as follows: the function of ecological consciousness of the society increases the demand for alternative types of energy, instead of reducing it, as the traditional energy suppliers do. Therefore, the function of Ec influence enters the equation of the demand in the model of alternative energy with the opposite sign

$$\frac{dX_1^A}{dT} = X_1^A\{1 - 2[X_1^A + a_{12}X_2^A - a_{1Ec}Ec - a_{1F_1}(F_D - X_1^T)]\}$$

(6.25)

$$\frac{dX_2^A}{dt} = X_2^A\left[1 - 2\left(X_2^A - a_{23}^A X_3^A - Q^A\right)\right],$$ (6.26)

$$\frac{dX_3^A}{dt} = X_3^A\left[1 - 2\left(X_3^A - E^A\right)\right],$$ (6.27)

$$E^A = E_0^A + \sum_i y_i^A r_i^A$$

where $y_i^A r_i^A$ are costs of resources consumed by the alternative energy suppliers within its production cost structure; Q^A is its added value; $a_{1F_1}(F_D - X_1^T)$ is the influence function for traditional energy suppliers; E^A is the initial investments share within the cost of production.

Computational Experiments with Model of Demand Transformation for Energy Supply of Recreational Areas. To illustrate the use of our model, let us study the results of computational experiments carried out with the system of equations (6.22) – (6.27). Calculations for this model were performed for 730 steps (days) from initial arbitrary values of the simulated processes, since the technology *ABC-AGENT* provides rapid adaptation of processes to each other and to external influences. A significant role is given to the choice of values of influence coefficients in the equations, and to control operators' $A^T(Pl, Pl^*)$ parameters. Also, the role of projected total demand for energy is quite important. Some of the 5 coefficients used in the equations are provided in Table 6.3.

In the first series of the experiments, a growing demand for energy supply in recreational area was studied:

389

$$F_D = D[1 - \exp(-\alpha_{F_D} t)] ,$$

It is characteristic for periods of intensive construction of new recreational establishments. The graph of this process is shown in Fig. 6.10a. This figure also shows scenarios of environment pollution (contamination) levels, as well as biodiversity level, and environmental awareness of population, calculated by the model. Management of traditional energy suppliers by environmental criteria was not performed in this series of simulation scenarios. The above controls (environmental taxes, environmental fines, and influence of ecological consciousness of the population on the choice of energy sources for recreational areas) were not taken into account. In the absence of management actions, demand for energy in the recreational zone was satisfied mainly by traditional energy sources, because its cost

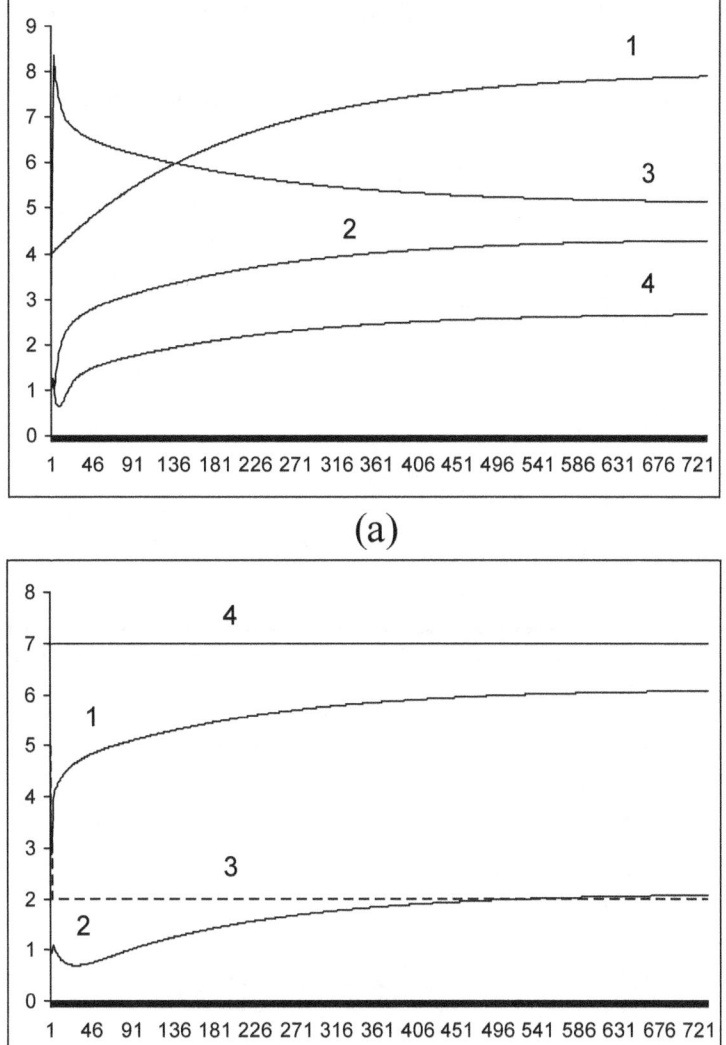

(a)

(b)

Fig.6.10. Scenarios for the model parameters in conditions of growing
    demand for energy supply, without traditional energy suppliers'
    management by environmental criteria (in the absence of taxes, penalties   and influence of public environmental awareness): (*a*) 1 – general
    demand for energy in *CZA*, 2 –   level of environment' pollution with
    traditional energy suppliers, 3 – environment biodiversity level,

4 – public environmental awareness in *CZA*; (*b*) 1 – demand for

traditional energy sources, 2 – demand for alternative energy sources,

3 – cost of traditional energy, 4 – cost of alternative energy.

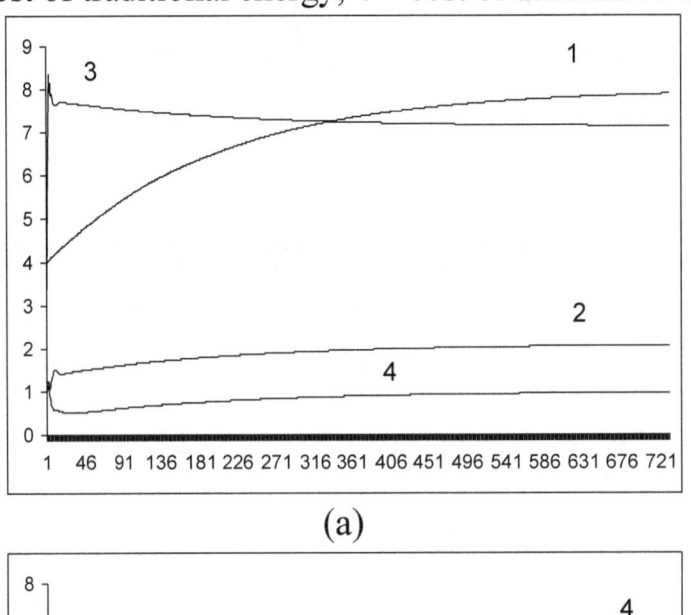

(a)

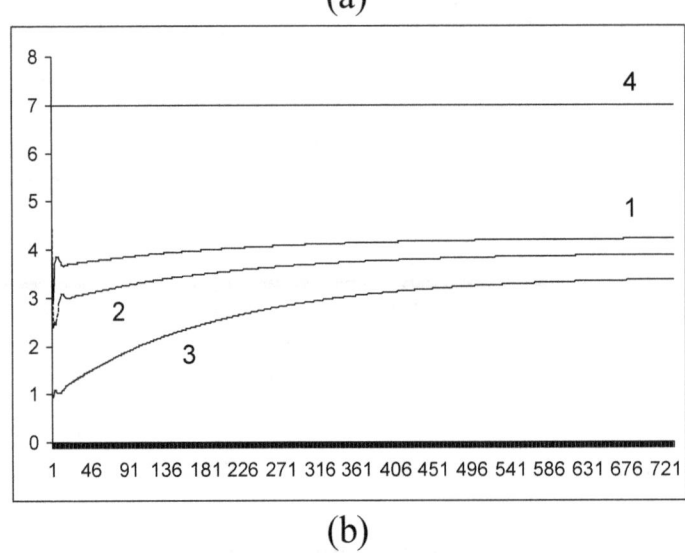

(b)

Fig.6.10. Applying tax for traditional energy production, which is

proportional to environment pollution level: (*a*) 1 – general demand for energy in *CZA*, 2 – level of environment' pollution with traditional

energy suppliers, 3 – environment biodiversity level, 4 – public environmental awareness in *CZA*; (*b*) 1 – demand for traditional energy sources, 2 – demand for alternative energy sources, 3 – cost of traditional energy, 4 – cost of alternative energy.

was significantly lower than the cost of alternative energy (see Figures 3 and 4 in Fig. 6.10, *b*). The constructed scenarios provided the basis for analysis of effectiveness of various management options.

Influence of environmental tax applied to traditional energy suppliers of *CZA*, is demonstrated by scenarios of ecological-economic processes, shown in Fig. 6.11.The cost of production of traditional energy was set proportional to the concentration of pollutants (contaminants). As a result, it grew up almost two times (curve 3 in Fig. 6.11, *b*). This led to redistribution of energy demand: the share of traditional energy (curve 1 in Fig. 6.11, *b*) decreased considerably compared to the previous calculation (see curve 1 in Fig. 6.10, *b*). The share of renewable energy in total energy demand increased by almost two times (curve 2 in Fig. 6.11, *b*). After applying the environmental tax, the level of environmental pollution (curve 2 in Fig. 6.11, *a*) decreased significantly.

The next computational experiment studied the role of environmental awareness of CZA' population in the transformation of demand from using traditional towards utilizing alternative sources of energy. The influence of pollution on production cost of traditional energy was not taken into account. Scenarios of the processes are shown in Fig. 6.12.

As it follows from the figures, the public awareness of the necessity to shift towards environmentally friendly methods of energy production, provides roughly the same effect as the cost (tax) limitation of production volumes of traditional energy. This time, the difference in costs of two kinds of energy (see Fig. 6.12, *b*) was similar to that of the first experiment, but the demand for alternative energy sources (curve 2 in Fig. 6.12, *b*) increased more than three times. This experiment confirmed, that the proposed model allows us to consider not only economic, but also social methods of control over the state of natural environment in the area of recreation.

The most difficult environmental situation arises when the level of environment pollution by traditional energy sources exceeds the maximum allowable level. In this case, management agent (6.19) comes into effect. It sets up penalties on production of traditional energy, thus increasing its production cost (see equation (6.24)). To check the function of the management method which applies penalties for environment pollution on traditional energy producers, we performed a second series of computational experiments.

In these experiments, we assumed that the overall demand for energy provision for recreational facilities undergoes random fluctuations around an average level. The time course of general demand is represented by curve 1 in Fig. 6.13. The dynamics of the other model variables is determined by fluctuations of the demand, as well as by selection of coefficients that link these variables between each other.

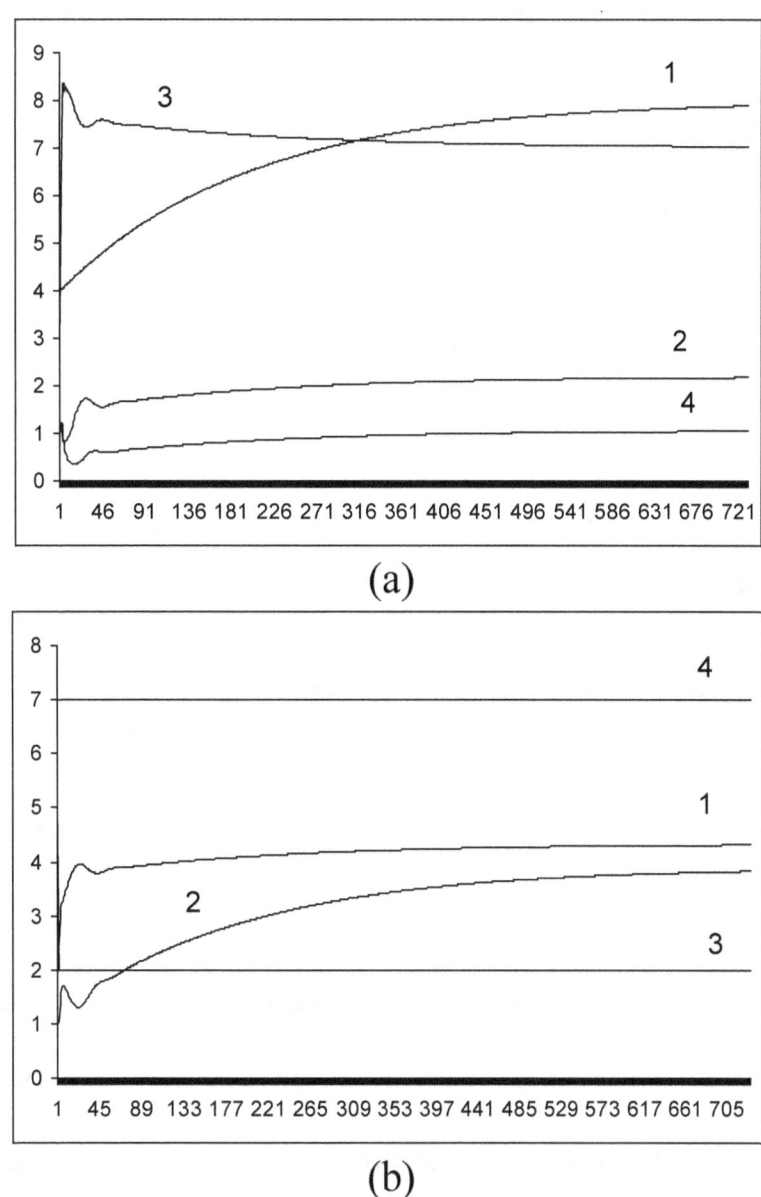

(a)

(b)

Fig. 6.12. Setting up dependence between demand for traditional and
alternative sources of energy and the level of CZA population's
environmental awareness: (*a*) 1 – general demand for energy in *CZA*,
2 – level of environment' pollution with traditional energy suppliers,

3 – environment biodiversity level, 4 – public environmental awareness   in *CZA*; (*b*) 1 – demand for traditional energy sources, 2 – demand for   alternative energy sources, 3 – cost of traditional energy, 4 – CZA

population's environmental awareness. (*b*) 1 – demand for traditional

energy sources, 2 – demand for alternative energy sources, 3 – cost of   traditional energy, 4 – cost of alternative energy.

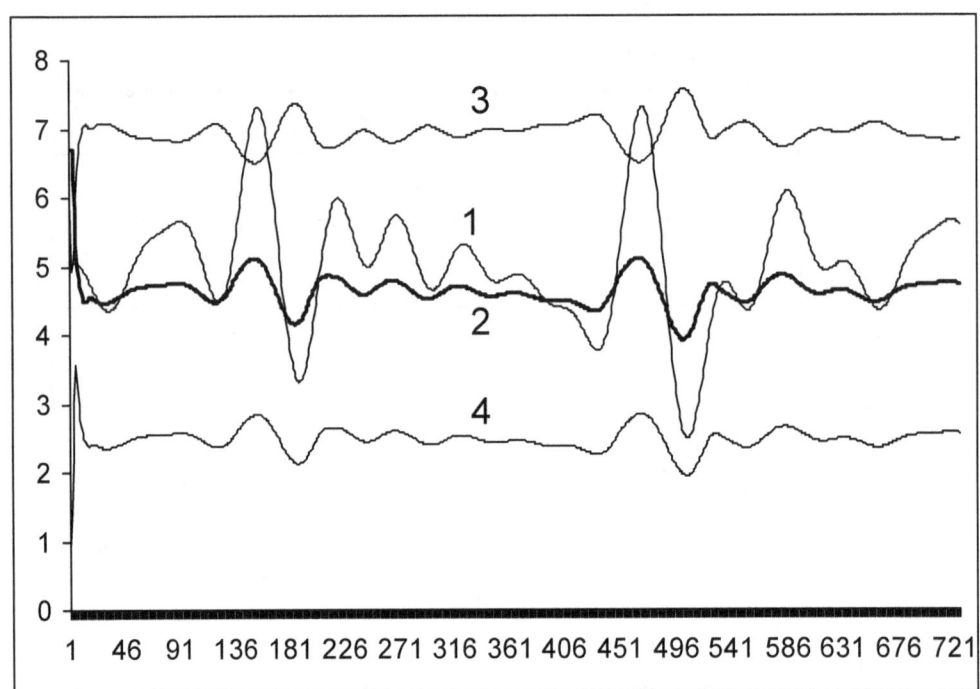

Fig.6.13. Reaction of systems model of energy supply for recreational zone, on   random fluctuations of the general demand for energy: 1 – general

demand, 2 – level of pollution caused by traditional energy suppliers,

3 – level of biodiversity of the environment, 4 – environmental

awareness of *CZA's* population

The model parameters utilised in this experiment, are provided below, in Table 6.3.

Table 6.3

PARAMETER	VALUE	PARAMETER	VALUE	PARAMETER	VALUE
a_{12}^T	1,2	Q^T	3,0	a_{Pl/X_1^T}	0,8
$a_{1F_D}^T$	0,7	Q^A	3,0	Pl_m	2,0
α_{F_D}	0,05	Pl^*	4,5	t_n	0,1*
$a_{1F_D}^A$	0,5	a_{12}^A	0,4	$\tau_{C_1}^n$	0,1*
$a_{X_3^T/X_1^T}$	0,4	$a_{1F_1}^A$	0,5	$\tau_{C_2}^n$	0,1*
$a_{X_3^T/Pl_1}$	2,9	a_{23}^T	0,5	α_{C_2}	0,05
α_{C_1}	1,0	a_{23}^A	0,4	$a_{1F_D}^A$	0,0
$a_{X_3^T/Pl_2}$	14,5	a_{12}^A	0,4	D	7,0

Fig.6.14 shows function of control operator $A^T(Pl, Pl^*)$, which set up environmental penalties (Curve 3) for exceeding maximum allowable pollution level $Pl_m = 4,5$. As it follows from the picture, the volumes and duration of fines must depend on the time periods when violation of maximum allowable pollution level is taking place. In other words, when the condition $Pl > Pl_m$ is implemented

397

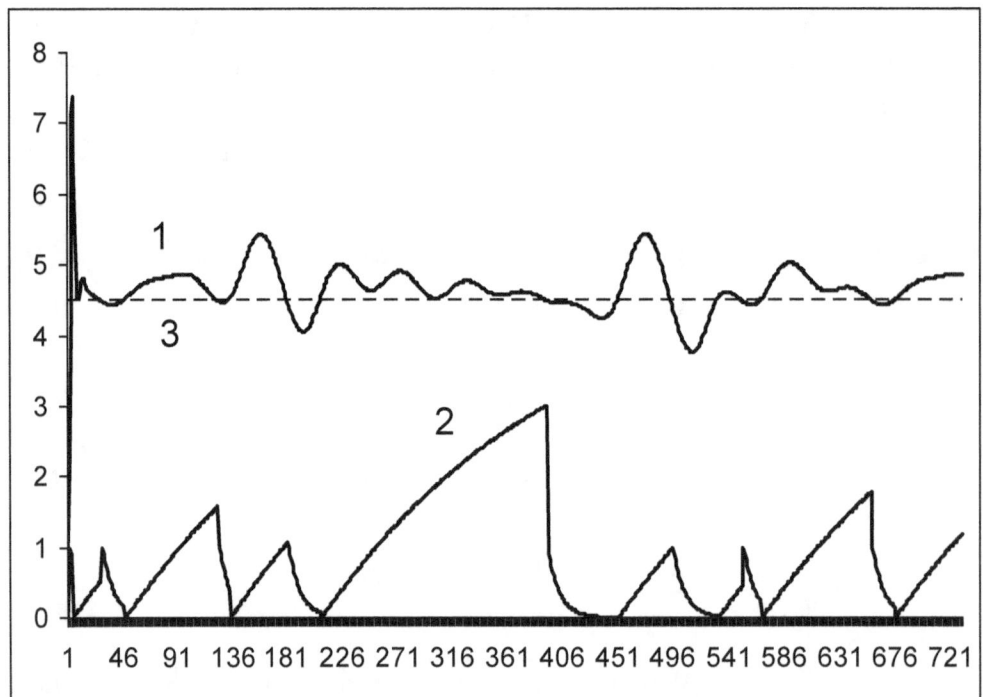

Fig. 6.14. The mechanism of applying environmental penalties for exceeding   maximum allowable pollution level: 1 – level of pollution in the absence of environmental control and penalties, 2 – function of penalty volumes   management, 3 –   maximum allowable pollution level

To assess the influence of environmental fines two versions of calculations were implemented. In the first version, the environmental fines were introduced but were not applied. The results of the experiment, where environmental penalties were not applied, are shown in Fig. 6.14. Management of pollution level was carried out only by means of collecting environmental tax, which proved to be insufficient to significantly reduce the pollution level.

The goal of environmental pollution management by means of applying such penalties is to ensure that the maximum permissible pollution level Pl_m is not exceeded for long and frequent periods of time. In the second version of the calculations, environmental penalties were introduced and collected along with environmental tax. The results are shown in Fig. 6.15. Applying environmental penalties in this experiment led to considerable reduction of time periods when the maximum permissible pollution level was exceeded (Curve 1 went down to position of Curve 4).

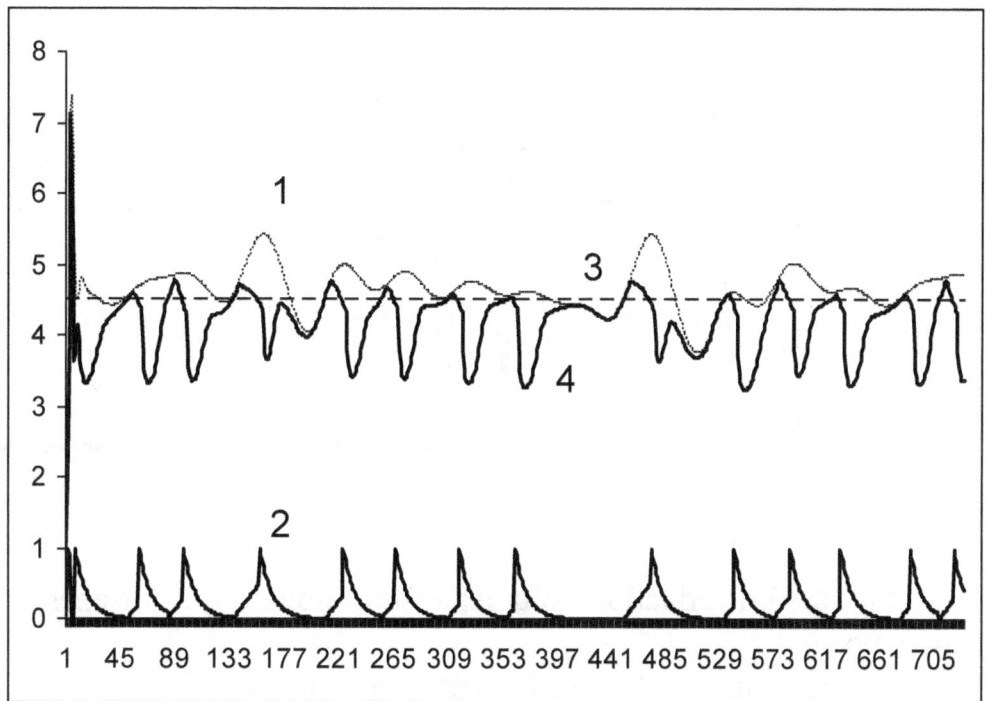

Fig. 6.15. Reduction of pollution level by means of applying fines and cost
limitations on production of traditional types of energy: 1 – pollution level without applying of fines in order to limit production volumes, 2 – amounts and duration of penalties aimed at reduction of pollution, 3 – maximum permissible pollution level, 4 – pollution level in conditions of applying of fines in order to limit production volumes

Fig. 6.16 shows scenarios of production volumes for traditional types of energy developing in two directions: with (Curve 2) and without (Curve 1) applying of environmental penalties. Comparison of the two graphs shows that environmental fines led to reduction of production of traditional types of energy approximately to 40%. These measures were carried out in order to reduce the level of pollution (contamination) of natural environment, as it is illustrated by Fig. 6.15.

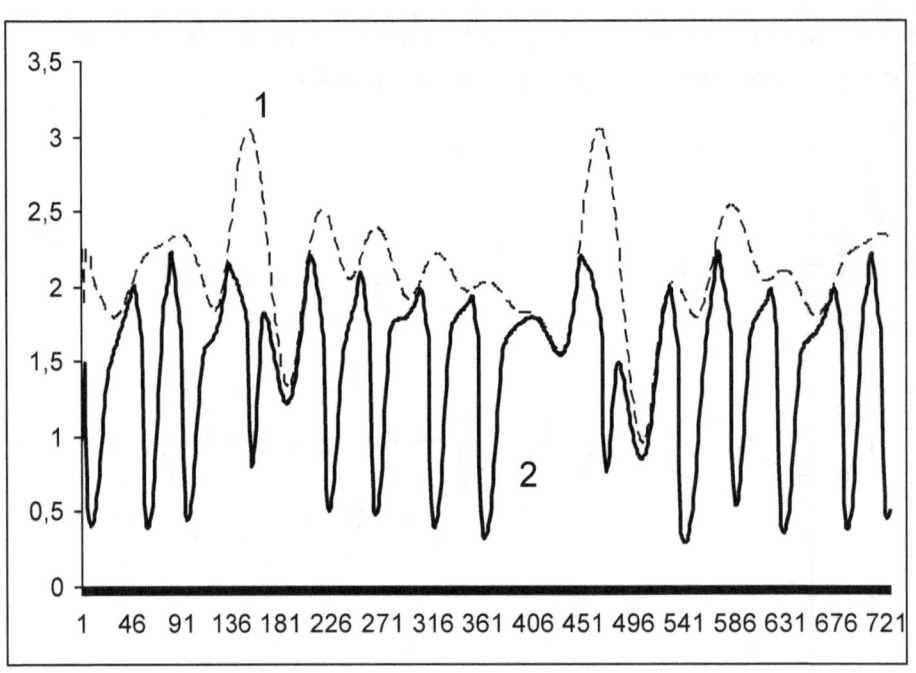

Fig. 6.16. Production volumes of traditional types of energy: 1 – with no
environmental fines, 2 – with application of fines

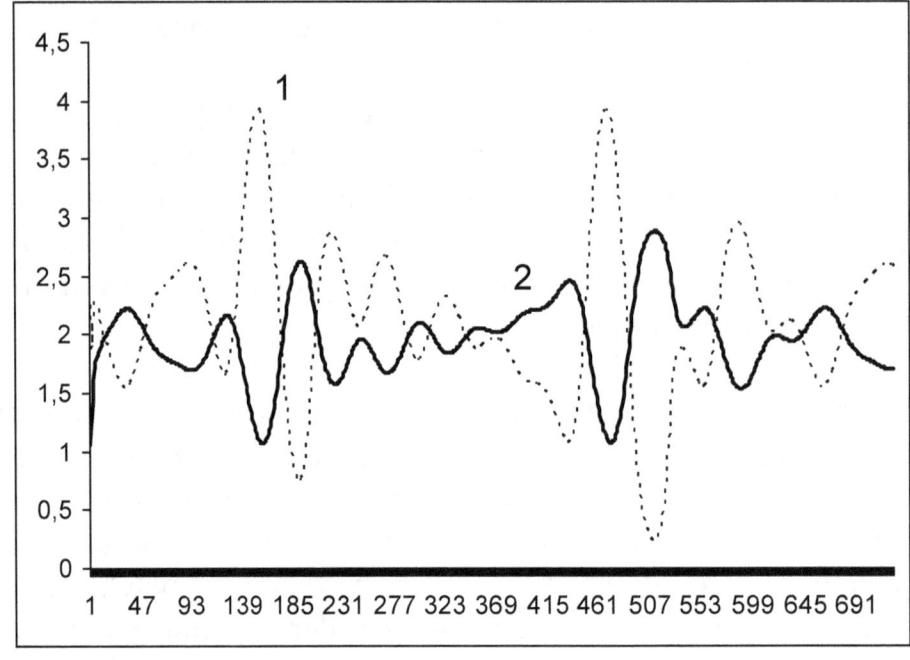

Fig. 6.17. Demand for production of two types of energy with no penalties applied on traditional energy producers for environmental pollution:

1 – traditional energy sources, 2 – alternative energy sources

In accordance with the above concept suggesting transformation of the energy-based economy of recreational areas, considerable reduction of volumes of traditional energy should lead to redistribution of the total energy demand in favor of greater use of alternative energy types. Fig. 6.17 shows scenarios of demand for these two types of energy. The scenarios were built prior to applying fines on traditional types of energy producers for environmental pollution. Then, application of the fines caused growth of prices for traditional types of energy and to corresponding decrease in demand for such energy. Therefore, after the fines were applied, the demand scenarios experienced changes, as it is shown in Fig. 6.17. The demand for traditional energy types reduced to nearly 25%. The demand for alternative energy suppliers grew approximately to the same amount.

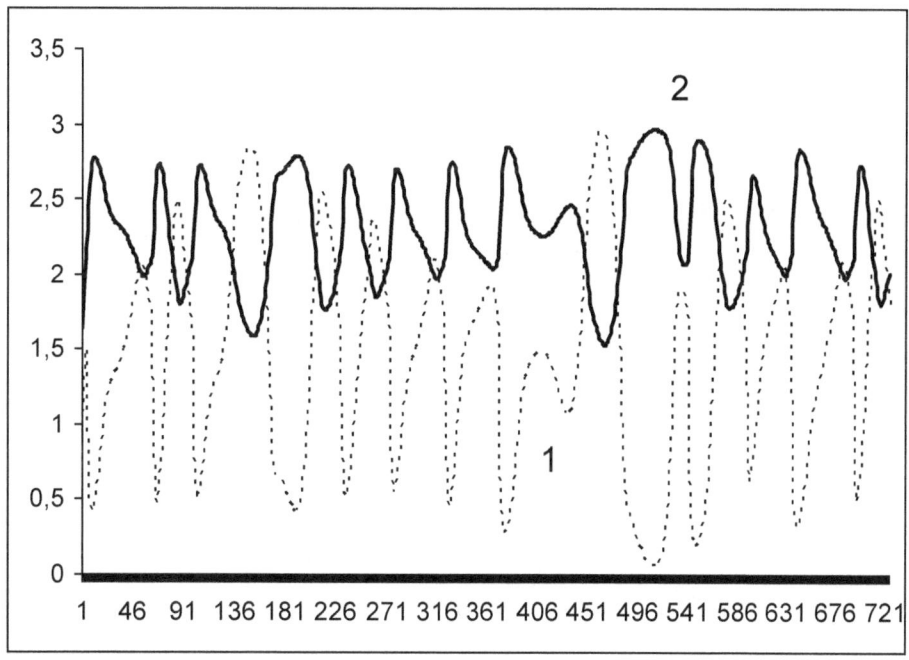

Fig. 6.18. Transformation of demand for two types of energy production,

401

accounting fines for pollution of environment: 1 – traditional energy sources, 2 – alternative energy sources

As the social ecological-economic system of *CZA* is an extremely complex and multifunctional system, the transformation of demand for energy supply of recreational services takes place under simultaneous influences of numerous factors. By changing modeling parameters, which are used to establish values of environmental fines $a_{X_3^T / Pl_1}$, $a_{X_3^T / Pl_2}$, α_{C_1}, α_{C_2} (see Table 6.3), we can considerably decrease the level of environment pollution by means of restriction of production volumes of traditional types of energy in recreational areas.

CHAPTER VII. COOPERATIVE MANAGEMENT OF INDUSTRIAL OPERATIONS AND ENVIRONMENTAL ACTIVITY WITHIN A "SEA – LAND" SYSTEM

7.1. Integrated Model of Industrial Operations in a Coastal Zone Area (*CZA*)

In previous chapters we considered consumption of two of the three main types of marine resources: biological and recreational resources. The third type of resources can be defined as a set of specific properties of marine environment which are currently used for making various industrial operations at sea. Among numerous industrial operations performed at sea, we can name utilizing of seas and oceans' aquatories, as well as using seabed for industrial purposes. Operations executed in marine waters are connected with the following main types of economic activities:

– Transportation of passengers and cargo,
– Mining of oil and gas on the shelf,
– Mining of minerals and construction materials,
– Obtaining alternative energy by using power of winds, tides and waves, and
– Disposal of industrial, agricultural and household waste.

All these types of industrial operations potentially bear a threat of pollution to the marine environment, and therefore, they conflict with the objectives of conservation and preservation of natural ecosystems of *CZAs*. The problem of finding a reasonable balance between economic benefits of industrial operations in marine waters and the environmental health of the latter is no less important than the consumption of maritime biological and recreational resources of the *CZA*.

The deterioration of ecological state of marine environment in the places, where industrial operations are carried out, should be regarded as the result of consumption of natural resources of marine environment by a production system. Therefore, each industrial operation in the sea must be accompanied by some compensation of the producer's economic profit, which should be directed at prevention of damage to the marine ecosystem. In other words, each economic system of marine production must acquire a certain amount of marine environmental resources to carry out its offshore industrial operations.

The above analysis of ecological-economic balance of economic uses of biological and recreational resources of the *CZA* provides a basis for constructing a general formal model of industrial operations in the coastal zone of the sea, which is intended to predict profitability of operations, taking into account environmental constraints and counteractions to emergencies. For this purpose, we will use the informational technology *ABC-AGENT* and the method of adaptive balance of causes, to build an integrated model of industrial operations in the coastal zone of the sea.

Let us introduce the following notations for the model variables:

D – demand for implementation of industrial operations in the coastal zone,

P – cost of performance of an industrial operation,

E – production cost of all operations,

H – anticipated volume of all industrial operations, provided by all types of resources,

S – current volume of industrial operations,

V – planned volume of operations, taking into account funds to purchase resources,

C – index of biodiversity, describing the state of the ecosystem,

PL – level of pollution of the marine environment during operations,

r_i – price per unit of resource ($i = 1, 2, 3$),

y_i – number of units of the i-type of resource needed for the performance of the unit industrial operation,

ρ_i – proportion of i-type of resource in the performance of the unit operation,

H_{11} – integrated parameter of the economic providing of operations,

H_{12} – integrated parameter of the ecological protection safety of the marine environment,

H_{13} – integrated parameter of the counteractions to mitigate the consequences of emergency situations,

Φ – economic profitability of offshore operations.

As previously mentioned, to simplify the modeling, we will use dimensionless (reduced) values that are associated with corresponding actual dimensional quantities A' by the dependence $A = 5 A' [M (A')]^{-1}$, where $[M (A')]$ is the average of variability interval of a dimensional quantity A'. As noted above, such a transformation limits variability of scenarios with the interval of values [0, 10].

Using the known dependence of demand on price and quality of goods (services) we will obtain the following balance equation of the *ABC*-model for the dimensionless variable $X_1 = 5 D [M (D)]^{-1}$, representing the dynamics of demand for operations at sea,

$$\frac{dX_1}{dT} = X_1 \left[1 - 2 \left(X_1 + a_{12} X_2 \right) \right],\tag{7.1}$$

where X_2 is dimensionless price of operation P. The cost of operations is determined by technological cost of X_3 and value-added cost Q, a part of which should be used for environment protection purposes. We will assume that the cost of environment protection actions is equivalent to acquisition of an "ecological" type of resource. Therefore, balance equation for the value can be written in the form

$$\frac{dX_2}{dT} = X_2 \left[1 - 2 \left(X_2 - a_{24} X_3 - Q \right) \right],\tag{7.2}$$

and in the equation for the costs of operations, we will include the costs of both, technological and environment' protection types of resources

$$\frac{dX_3}{dT} = X_3 \left[1 - 2 \left(X_3 - \sum_{i=1}^{n} a_{3i} r_i y_i \right) \right],\tag{7.3}$$

where n is the number of types of resources acquired by the economic system performing marine operations.

The demand for implementation of industrial operations is compared with the existing capacity of the system to carry, transport and ship cargoes. Let H be the number of individual industrial operations in a unit of time. The implementation of these operations is ensured by the existing carrying capacity of the system. Then, the rate of implementation of industrial operations can be represented by the following equation of the *ABC*-model

$$\frac{dH}{dT} = H\left[1 - 2\left(H - V + S\right)\right],\tag{7.4}$$

where V is the number of operations prepared to be carried out at current moment of time, and S is the number of performed operations.

Operating profit N, obtained from performing a single industrial operation at a moment of time t, will be determined by the difference of its price and cost: $N = P - E$. Execution of the operation becomes unprofitable when $N < 0$. Therefore, the expected rate of return must be monitored by a special management agent; its function is to compare current demand for operations at sea with current carrying capacity of the industrial system. It should carry out control over piloting of vessels and transshipment of cargo, and assess economic benefit of these operations performed at the current market price. In such a way, the operations are subject to the following economic conditions

$$S = \text{IF } (N < 0;\ 0;\ R),$$

$$R = \text{IF } (D < H;\ D;\ H),$$

where D is current demand for operations.

Along with management of economic operations, a thorough control of ecological condition of the marine environment in the zone of execution of operations must be performed. It must be done under constant surveillance of management agents. The volumes V of operations should be limited in cases when the marine ecosystem is subject to real risk of irreversible degradation due to insufficient environment protection actions. A similar restriction should take place each time when insufficient funds are assigned to ensure elimination of risks of natural and industrial disasters. Thus, in a general case of offshore operations, it is necessary to have three kinds of resources for corresponding manufacturing, environmental, and preventive measures. If we denote the minimal available amount (to the industrial system) of these three types of resources as M, then the volume of operations provided with resources, will be determined by the following logical relationships

$$V = \text{IF } (D < H; \; 0; \; M),$$
$$M = \text{IF } (D - H < M; \; D - H; \; M),$$

$$M = \min (\; m_1; \; m_2; \; m_3),$$

where m_i are the volumes of each of the types of resources

$$m_i = H_i / y_i \quad (i = 1, 2, 3).$$

In order to increase the number of operations to the level determined by current demand, it is necessary to increase funding for each of these kinds of activities, using working capital available to the economic system. We will denote current level of working capital of the economic system as H_2. Its variability is determined by the size of current profits I, investments H_3 and expenses S_2

$$\frac{dH_2}{dT} = H_2 \left[1 - 2 \left(H_2 - I - H_3 + S_2 + S_3 \right) \right]. \tag{7.5}$$

We have agreed to assume, that the costs of environment protection activities in the sea, being financed from the working capital of the economic system, is equivalent to acquiring of a certain number of the "ecological" resource. We will also introduce the expenses which should be assigned into a fund of prevention and mitigation of consequences of emergencies. This value is equivalent to the economic system to acquisition of a certain amount of resources of "counteractions to emergency situations".

Therefore, the available working capital H_2 of the economic system may be allocated for acquisition of economic, ecological types of resources and insurance services. This can be done in the same proportion in which each of these kinds of resources is required, in order to conduct operations at sea.

Let us determine a share of i- type of resources in production of one operation unit by the following relation

$$\rho_i = \frac{r_i y_i}{r_1 y_1 + r_2 y_2 + r_3 y_3} \ , \qquad (i = 1, 2, 3).$$

Then, for acquisition of resource of the i-type, a production company has $\square_i H_2$ part of its working capital funds. If this amount is insufficient, the production system can take a loan and acquire the resources per credit, provided that its accumulated debts H_3 (credit accumulated to the current time) does not exceed a certain pre-established amount H_3^*. Setting the maximum permissible credit H_3^* means, that the administrative body establishes limits for operations at sea, when additional investments into nature-protection measures are not covered with profits anticipated from the economic activity.

The economic system must allocate a part of their working capital to pay off the accumulated credit debt. We will assume that this repayment takes place regularly and continuously, and we'll denote the amount of money allocated for this purpose as S_3. Taking into account the interest paid on the loan $\square$, the logical conditions for the S_3 take the form

$$S_3 = IF(\theta H_3 < H_2; \theta H_3; H_2) \ .$$

If we denote the volumes of resources purchased per credit as V_{11}, V_{12}, V_{13}, then the total value of current investments of the economic system into acquisition of all types of the resources it needs, will be

$$V_3 = r_1 V_{11} + r_2 V_{12} + r_3 V_{13}.$$

The current value of the accumulated credit will be expressed by a balance equation-

$$\frac{dH_3}{dt} = H_3 \left[1 - 2 \left(H_3 - V_3 + S_3 \right) \right]. \qquad (7.6)$$

Thus, the balance equations of the *ABC*-model (7.1) – (7.6) represent the processes of production, as shown in Fig. 7.1 – Fig. 7.2.

Model of Resources' Costs Formation. To carry out economic activity in coastal off-shore areas, the local administrative authority has to manage funds allocated for acquisition of resources. We will introduce notation H_{11} for the amount of technical facilities for cargo transportation and shipment operations, which the economic system has for performing industrial operations at sea. The value of H_{11} is an assessment of available economic resources of the system. It can be expressed by a balance equation

$$\frac{dH_{11}}{dt} = H_{11}\left[1 - 2\left(H_{11} - V_{11} + S_{11}\right)\right], \tag{7.7}$$

in which S_{11} stands for operating charges of economic resources, and V_{11} designates current investments into regional infrastructure that manages industrial operations at sea. This investment is necessary for performing operations at sea, and is regarded as acquisition of insufficient economic resources.

Similarly, we can represent balances of the other two types of resources that provide conditions for operations at sea. Let H_{12} be a reserve of ecological type of resources, which reflects existing value of biodiversity index. The biodiversity level, in its turn, is supported by current level of environment protection activities in the area of operations. Then, the balance equation for ecological resources takes the form

$$\frac{dH_{12}}{dt} = H_{12}\left[1 - 2\left(H_{12} - V_{12} + S_{12}\right)\right].$$

(7.8)

In this equation, S_{12} means current consumption of ecological resources by the production system, i.e. deterioration of ecological state of marine environment in the area of operations, which depends on the volume of cargo operations performed in the area. V_{12} indicates improvement in the ecological state of marine environment resulting from total environment protection activities in the area of operations.

The existing fund of financial resources intended to combat the effects of emergencies H_{13} should be considered as a stock of resources that provide resistance to natural disasters and man-made emergency situations. The fund is to be spent on emergency prevention measures or on mitigation of negative effects caused by emergencies. We denoted current expenditure of the fund as S_{13}. Its replenishment is denoted as V_{13}. Then, the balance equation of the fund can be presented in the form

$$\frac{dH_{13}}{dt} = H_{13}\left[1 - 2\left(H_{13} - V_{13} + S_{13}\right)\right] \tag{7.9}$$

We will assume that total expenditure of all types of resources is in proportion to the total volume of industrial operations at sea. In the case where the available resources are insufficient to meet the planned volume of operations, resource stocks must be replenished by investment of adequate amounts of financial resources. The task of investment management system is to determine how many resources of each kind must be purchased at the expense of further investments. Conditions for determining of the missing amounts of each resource type represent the following relationships, which take into account financial constraints on acquisition of resources, faced by production system

$$V_{1i} = \text{IF } [(D - H)y_i < H_{1i} ; 0 ; F_i].$$
$$F_i = \text{IF } [r_i (y_i D - H_{1i}) < \square_i H_{2i} ; y_i D - H_{1i} ; R_i].$$
(7.10)
$$(i = 1, 2, 3).$$

In these expressions, function r_1 represents production cost (technological) of a single industrial operation in the sea; function r_2 reflects the cost (averaged) of a single protective measure aimed at conservation of biodiversity of the marine ecosystem; function r_3 provides the cost (averaged) of a single event, aimed at preventing negative consequences of emergency situations at sea. Since the invested funds must be obtained and used prior to performing operations at sea, they should be considered as loans acquired (on a percentage basis) by a system that carries out management of industrial operations.

This formulation of the investment management problem should consider the fact that the interest on loans increases the overall costs of industrial operations and, consequently, it may reduce their profitability. In addition to this, in the case of low cost operations, the total value of credit H_3 gained in the process of acquiring volumes of each type of resources H_{1i}, can not exceed a certain predetermined value H_{3i}^*. Thus, in formulas (7.10), function R_{ij} restricts the acquisition of i-type of resource to the extent to which the value of loan H_{3i} approaches the value H_{3i}^*.

Let us regard the cost-effectiveness of industrial operations as a ratio

$$\varphi = pS\,[U + (U^0 - U)\exp(-\alpha\,\tau)]^{-1}, \quad U = qV.$$
(7.11)

In this formula, the following notations were used: p is market price of an operation at sea; S is the number of operations performed per unit interval of time (day); V is the number of operations that current economic infrastructure can provide in case of demand for them; α is a parameter that sets the rate of the company's loan repayment (if the loan was taken by the company to put necessary economic infrastructure into operation); U^0 is the amount of initial investments; U denotes current expenses related to performing operations at sea; τ is time.

It should be assumed that, the higher is the value of φ, the higher is economic efficiency of an industry. The cost of developing of necessary infrastructure U^0 will be covered over time through ongoing deductions from income, and operating costs with time will acquire a value U, which is determined by total volume of operations V and by the cost of their production. Thus, the cost-effective use of resources is a variable function of time (the scenario). The logarithm of expression (7.11) $\Phi = \ln \varphi$ can be used to assess profitability of operations at sea, taking into account protection of the marine environment and prevention of emergency situations.

In Chapters Five and Six, we considered a scheme of integrated management of biological and recreational resources consumption in coastal zones. It was based on setting up maximum allowable quantities of accumulated credit H_{3i}^*. The magnitude of H_{3i}^* is defined by an administrative body that regulates consumption of natural resources in the area. Obviously, this kind of control can also be applied in cases of performing operations in *CZA*, because with decrease of the value of H_{3i}^*, the society restricts consumption of natural resources in order to protect the environment.

Profitability of industrial operations will depend on the dynamics of prices for the three main types of the resources consumed by the economic system. With deterioration of ecological situation in the zone of operations and with increasing risk of negative consequences from natural or technological emergencies, the prices for resources would rise. Under these conditions, the total amount of debt on investments H_3 can provoke the production system to limit the volume of production, even in terms of high economic profitability of their production. Therefore, the model experiments which allow to assess and to predict changes in ecological-economic balance of economic activity in the sea, definitely have certain value.

7.2. Simulation Experiments with Integrated Model of Industrial Operations in a Coastal Zone Area

In experiments conducted with the model of industrial operations (7.1) – (7.11), we set the task to obtain a basic scenario of economic processes taking place under controlled external influences affecting the system. We used the following external influences: variable demand $D(t)$ for industrial operations and dynamics of prices for resources r_1 and r_3. With regard to the value of ecological resource r_2, it has been assumed that it was inversely proportional to the index of marine biodiversity *BR* in the zone of performing operations. The level of demand has a positive impact on the income and consequently, on profitability of operations, while the value of ecological resources r_2 characterized expenses which the production system had to allocate for maintainance of biological diversity index at a required level.

In addition to this, we simulated conditions when the economic system had to invest a part of its profit into development of all three types of infrastructures: economic, ecological and the one responsible for prevention of emergency situations. It was agreed that 40% of the profit had to be directed to carry out these goals. The system could take loans to cover all of its expenses. The interest on loans was set at the level 0,4. To provide environmental state monitoring of the marine ecosystem, function H_3^*, which is the maximum allowable credit, was put into dependence of biodiversity index value BR. The value of BR was determined by the level of pollution of marine environment, which was considered proportional to the volume of industrial operations at sea. The equation for BR index contained a term, which takes into account the volume of environment protection activities in the sea. It was considered proportional to the investment V_{12} which was to cover the cost of these actions.

The abovementioned relationships were represented by following equations

$$\frac{dBR}{dt} = BR\left[1 - 2\left(BR - a_{CV}V_{12} + a_{CPL}PL\right)\right], \tag{7.12}$$

$$\frac{dPL}{dt} = PL\left[1 - 2\left(PL - a_{PL/S}S\right)\right], \tag{7.13}$$

$$\frac{dH_3^*}{dt} = H_3^*\left[1 - 2\left(H_3^* - a_{H_3^*/BR}BR\right)\right], \tag{7.14}$$

$$\frac{dr_2}{dt} = r_2\left[1 - 2\left(r_2 + a_{r_2/BR}BR\right)\right]. \tag{7.15}$$

To perform computational experiments, the equations of ecological-economic model (7.12) – (7.15) were represented in finite differences. The sampling interval of processes in time (one day) was adopted as a unit of time. Calculations were conducted during 30 time steps. The dynamics of demand was determined by an increasing function of time, which is shown in Fig. 7.1a.

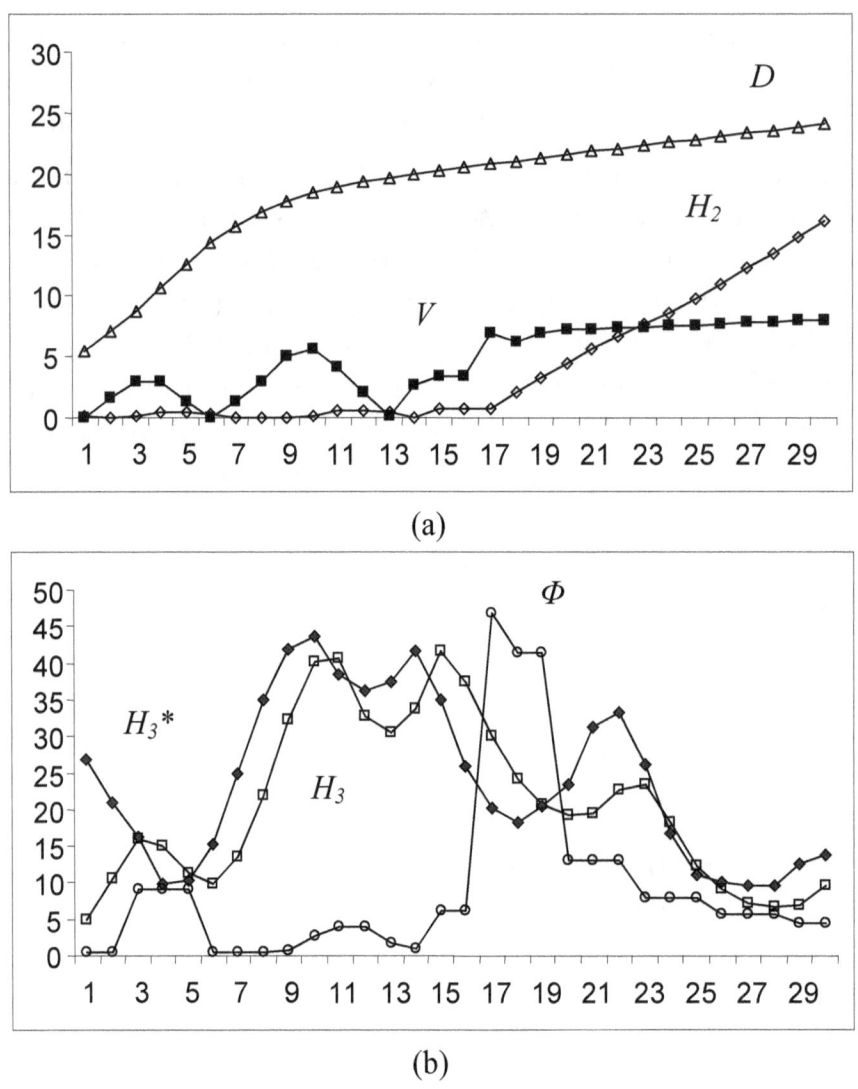

(a)

(b)

Fig. 7.1. Forecasted scenarios of economic processes at 30 steps forward, averaged over 3 time steps: (a) $5*10^{-2}$ H_2 – current working capital;

V – requests received by the economic system to perform operations;

D – demand for operations; (b) H_3^* – maximum allowable financing of   operations restricted by the state of the ecosystem; H_3 – current

investment into infrastructure of local industry, profitability of operations $2*10^{-1}$ $Ф$

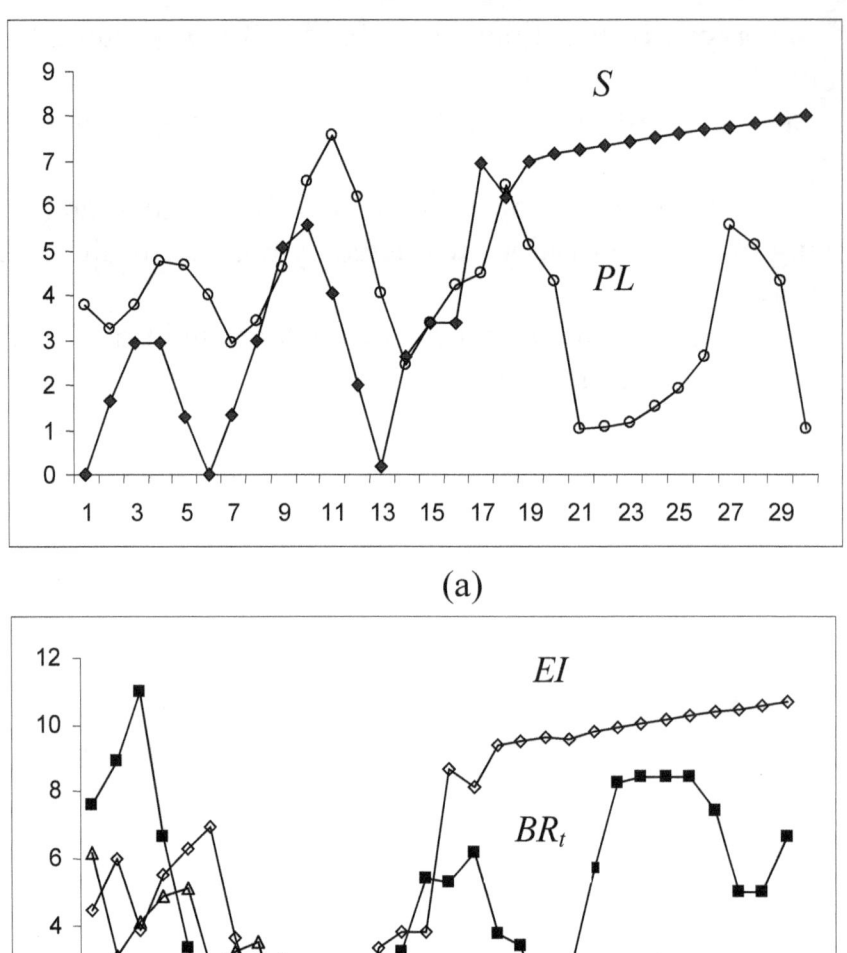

(a)

(b)

Fig. 7.2. Forecasted scenarios of ecological and economic processes at 30 steps   forward, averaged over 3 time steps: (*a*) $4\cdot10^{-1}S$ – amount of executed operations, *PL* – level of marine environment pollution;

(*b*) *EI* – investments in environment protection activities,

BR_t – index of biodiversity after the environment protection activities,

BR_0 – index of biodiversity before the environment protection
activities

The ratio between production cost of the operations and prices for resources was favorable for the economic system in conditions of the selected value of demand. Therefore, the amount of received orders for execution of operations V, starting from step 17, started to increase. Working capital of the system H_2 in this period did not increase because its revenue was almost entirely withdrawn for repayment of loans which had been taken for acquisition of resources.

Fig. 7.1, b shows current investments into the economic system's infrastructure H_3. The values of investments were limited in accordance with maximum possible values of loans for operations H_3^*, which in their turn, were dependent on the state of the ecosystem. Cost-effectiveness of production fell down dramatically in those periods when current accumulated investments into infrastructure of operations H_3 exceeded the ecologically-admissible norm H_3^*. This happened, for example, in the period from step 15 to step 20, when profitability of operations went down sharply (see fig. 7.1, b).

The decrease in profitability of production was caused by excessively large volume of cargo-carrying operations at sea during this period of time, which is illustrated by the curve S in Fig. 7.2, a. Increasing level of marine environment's pollution (curve PL) was the reason for this, as it provoked a sharp increase in investment into environment protection activities (see curve EI in Fig. 7.2, b). As a result, after 20 steps of calculations, pollution level PL reduced, and the index of biological diversity BR_t started to grow, as shown in Fig. 7.2, b. To draw a comparison, we provided a scenario of biological diversity index BR_0 in the same figure, This scenario would have occurred if the environment protection activities had not been performed in the sea.

417

Our computational experiments demonstrated possibility of using integrated models for simulation of three basic development processes in nature-economic systems of coastal zone areas: the level of economic viability of industrial operations at sea, the index of marine biodiversity in the area of industrial operations, and the level of marine pollution caused by both, industrial activities and emergency situations. The proposed ecological-economic model allowed us to evaluate profitability of performing industrial operations in conditions of continuous control over marine environment's pollution levels and general ecological state of the region. The model allows to take into account the effects caused by extreme weather conditions and human-induced factors. To counteract these factors, the economic system of marine operations has to finance the regional infrastructure.

To obtain practical recommendations about maintaining environmental and economic balance in *CZA* regions, the model's parameters must be adapted to real values indicating volumes of industrial operations. The biological diversity index should be represented as a weighted sum of concentrations of the major aquatic organisms that inhabit this region. Similarly, another index should be represented – the one that characterizes the level of pollution of marine environment. These objectives should become main objectives for further research, directed at creation of information management technology for operations in the sea.

7.3. Conceptual Model of Marine Industry Operations System in the Kerch Strait Area

One of the most common types of marine operations is transportation and transshipment of cargoes in maritime areas. In this section, we will develop a conceptual model for management of this type of operations, and we will use marine operations taking place in the Kerch Strait area as an example.

This area represents a complex ecological-economic system, in which intensive economic activity comes into contradiction with objectives of activities aimed at protection of the marine environment. To control industrial operations at the Kerch Strait area, a number of informational technologies can prove to be very supportive, once they are based on the integrated models of natural-economic systems, discussed above.

Despite substantial simplification of real processes occurring in the strait, such integrated models allow us to evaluate main factors of the region's economic development in conditions of preserving marine environment and its biodiversity. Integrated description of possible scenarios in the models of the Strait opens a possibility to take well-weighed decisions directed at reduction of risks of weather-related and human-related emergencies.

The main economic operations in the strait area are represented with transshipment of various cargoes coming from the shallow Azov Sea. In the southern part of the strait, there is a set of storage capacities, where oil is drained by small vessels, in order to transship it further by large tankers. In the northern part of the strait the transshipment of bulk cargoes if performed. According to Internet sources [17] in 2007, on the raids of the strait, the *TNK-BP*, *LUKoil* and *Rosneft* companies handled over seven million tons of oil. During the same time, it nearly three million tons of bulk cargo (sulfur, coal, grain, fertilizer) was handled in the same area.

Extreme weather conditions that occur from time to time, can significantly affect maritime operations in the strait. In the period from November to March, the whole strait area is affected with north-east winds, which are traditionally characterized with very high speed and duration. In the summer, winds from the south are quite frequent [23]. For example, during a violent storm of Nov. 11, 2007, four boats crashed in the Kerch Strait, which resulted in contamination of the marine environment with nearly 7 thousand tons of sulfur and up to 2 tons of petroleum products.

At the same time, the Kerch Strait presents a traditional route for fish migration between the Azov and the Black Sea. Many of the migrating species are rare, and are listed in so-called Red Data Book of the Black Sea region. A number of rare and endangered local species are also protected by International Union for Conservation of Nature. As a result of pollution of the sea with oil and hazardous chemicals, the fishery of the Azov and Black Sea basin was challenged with considerable damage. This is largely due to the fact that the management of cargo operations in the Strait is not related to ecological state management of marine ecosystem of the Strait. The funds allocated for prediction and prevention of emergency situations are scarce, no activities are preformed in order to prevent and eliminate devastating natural and industrial affect on the region.

In this section of our research, we intend to combine three major development processes of the natural-economic system of the Kerch Strait in one overall model: the level of economic profitability of industrial operations in the Strait, an index of biodiversity of the marine ecosystem in the Strait area, and the level of marine pollution originating from both, production activities and natural emergencies. The emphasis will be made on construction of scenarios of these processes, which will characterize conditions for preserving balance between ecological and economic processes taking place in the strait.

Conceptual Model of Ecological-Economic Balance of the Strait's Economy. The task of resource management in the Kerch Strait area is to provide conditions ensuring profitability of industrial operations. At the same time, along with acquiring the highest possible economic benefits, the local administrative bodies must carry out continuous control over the ecological state of marine environment in the region. It must be maintained at a sufficiently high level, and the risk of adverse effects of potential emergencies must be reduced to minimum. In this formulation of the problem, the profitability of operations depend on the costs of industrial operations conducted in the Strait area, on environment protection activities and on the measures taken to prevent emergency situations in the area. The interaction of three types of infrastructures, providing for ecological-economic balance of economic activity in the strait area, is shown in Fig. 7.3. Fig. 7.4

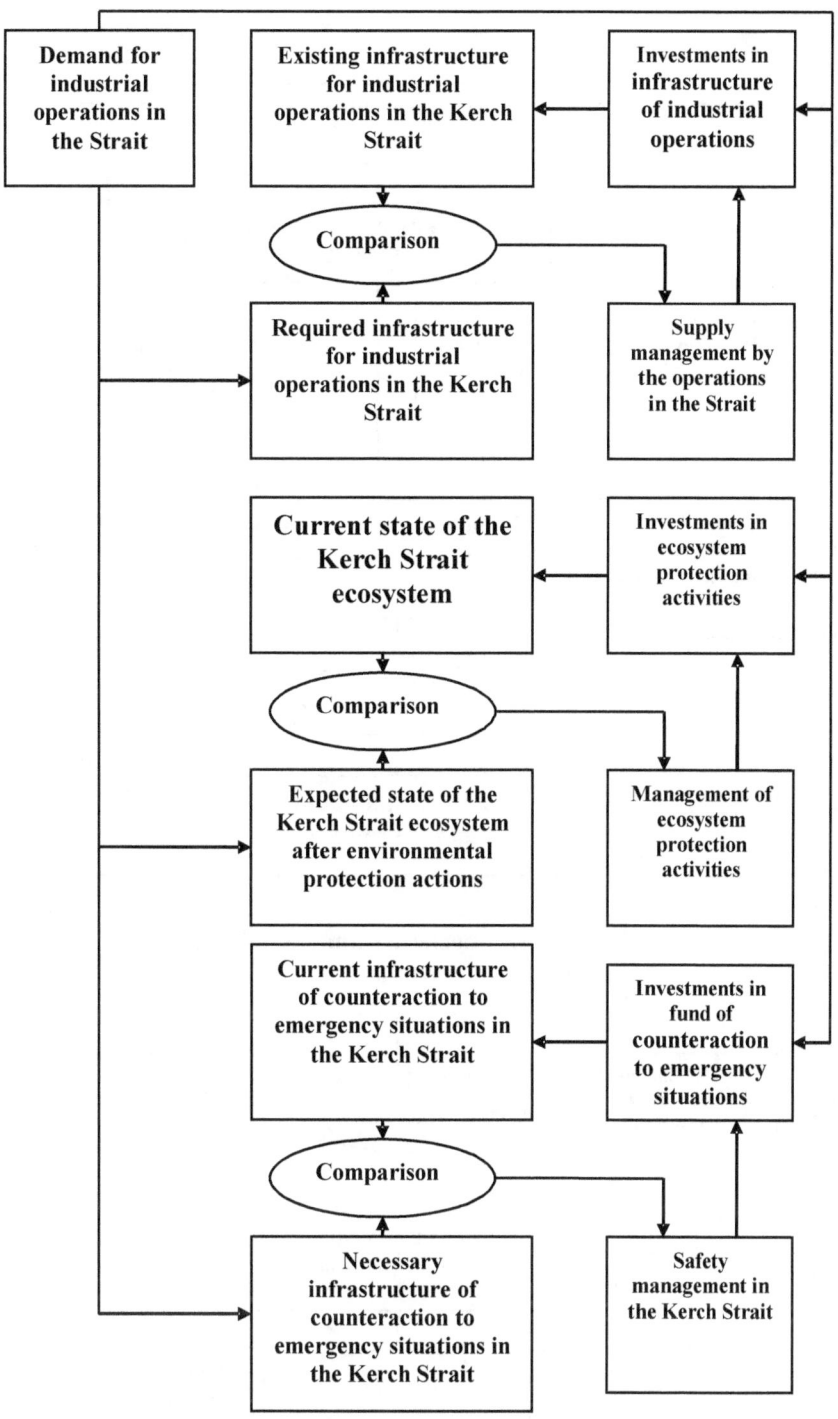

Fig. 7.3. Interaction between three types of infrastructures for ecological-economic balance of economic use of the Kerch Strait

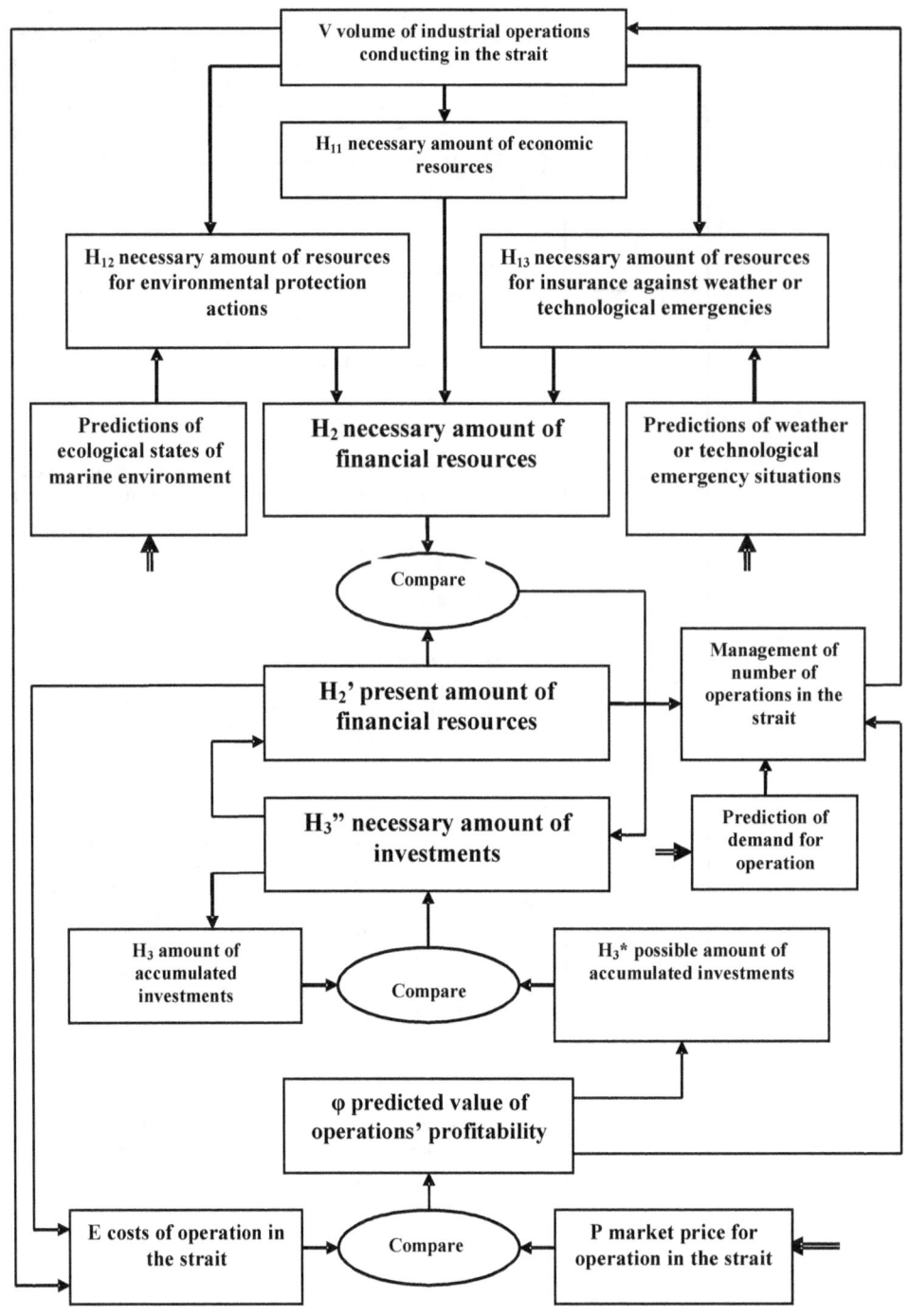

Fig. 7.4. Conceptual model of economic system of operations in the Strait

presents a conceptual model of the system of economic operations in the Strait, in which management of operations depends on their profitability.

To be able to work successfully, the production system is forced to buy certain amounts of each of the three types of resources: economic, ecological and emergency-preventive management resources. The main regulators of ecological- economic balance in the strait are prices for each of the three types of resources. Before introducing prices of these resources into the model, it is necessary to clarify the meaning of the concept of a unit of each of the three types of resources.

Let us agree that a unit of economic type of resources identifies material and financial expenditures invested by the Strait's industrial system into conducting a typical operation of providing of a cargo transshipment into a vessel of medium tonnage. In monetary terms, the cost per unit of economic resources is determined by the ratio between the expenses assigned for maintaining the infrastructure of industrial operations in the strait and maximum possible number of typical operations, executed in it during a unit of time, for example – during one day. In other words, the number of units of economic resources available at a given time for the production system of the strait characterizes its capacity. Further increase in numbers of industrial operations is only possible by means of increasing the capacity of the system's infrastructure, which is associated with assigning additional investment to acquire insufficient amounts of economic resources.

The ecological state of marine environment in the strait area is characterized with an integral index, for example – a biodiversity index. Biological productiveness of sea waters is a complex function of many parameters, among which the main role is given to food chains of marine species, oxygen, nutrients and other chemicals' concentrations that determine conditions for existence of marine organisms of the ecosystem. The pollution of the environment by oil products and other harmful substances that are usually associated with cargo transportation and transshipment at sea, can significantly change the state of the ecosystem and can cause risk of its' degradation.

We will proceed from the assumption that, similarly to the economic measures directed at preserving the strait's economic infrastructure a certain appropriate level, environment protection measures ensuring conservation of the strait's marine ecosystem must be implemented continuously.

The industrial system of the Strait should allocate special funds to implement conservation activities which would be sufficient to maintain ecological situation at an environmentally justified level. The degree of environmental degradation associated with a unit of industrial operations in the Strait, can serve as a measure of economic sanctions imposed on the production system of the strait. The size of these sanctions can be as follows: fines for ecological damage or environmental protection taxes can be considered as a price per unit of ecological resources purchased for performing a single industrial operation.

Another kind of obligatory costs to pay by the economic system is associated with the risk of emergency situations (*ES*) in the strait, which may be caused by natural or anthropogenic factors. To prevent, or at least to mitigate the negative consequences of the emergency situations in the strait, there is a need in a permanently working infrastructure, which would serve to monitor potentially dangerous natural processes and human impact on ecological-economic system of the strait. We will call it a system of emergency situations prevention. A share of the costs assigned for operation of this system should be covered by the production system of the strait, because its work is a source of risk of emergency situations.

Unusual weather conditions in the strait are one of the sources of risk of emergency situations. The most significant in this respect are abnormally high wind velocities and abnormally low temperatures. The extreme weather conditions which cause accidents of vessels are making the specifics of industrial operations in the strait. They are often accompanied with pollution of waters of the Kerch Strait with oil and other harmful substances having negative impact on the marine ecosystem.

We will assume that the risk of anthropogenic emergencies is proportional to the volume of industrial operations in the Strait, and the risk of abnormal weather situations is proportional to the values of abnormally high wind velocity and abnormally low temperatures. Then, to denote a unit of resource of anti-emergency actions, we will use a single insurance payment made by this system into a special fund of emergency situations' prevention. We'll consider that this fund is used for maintenance and development of the *ES* infrastructure .

Dynamics of demand to conduct industrial operations in the Strait is a source of external influence on the economic system. Traditionally, demand determines the capacity of a production infrastructure; it also determines the anticipated state of the ecosystem, by taking into account the planned environment protection activities and activities of the *ES* infrastructure. To assess possibility of the economic system to satisfy the demand in industrial operations, the required values of parameters of the system's state are compared with existing (current) values of these parameters. Drawing continuous comparisons will allow us to control the entire system, and to see if it is capable to meet the demand. At the same time, a proper level of the marine ecosystem's development must be maintained, and the level of security operations in the strait must be properly managed. Investments of the economic system into each of the three types of resources serve to achieve this goal.

In these conditions, a fundamental question occurs, whether the working capital of the production system is sufficient for acquisition of required amounts of resources. The required amount of financial resources, which is the summary investment assigned for purchasing of the three kinds of resources, depends on the volume of industrial operations in the strait. The volume of available working capital is compared with the funds which are required to fully meet the demand for operations in the strait. If working capital is not enough, the production system must be ready to assign additional investment for acquisition of insufficient resources.

However, additional investment means some increase in total expenses of production, and as a sequence – some increase in total costs of the operations. Therefore, constant and continuous monitoring of accumulated credit' amount (due to investments in the production system of the strait) must be provided. It is necessary to control that the economic system does not exceed the limit value of admissible production level. Simultaneously with this control, continuous forecasts of cost-effectiveness of industrial operations should be executed; each time, it should process the new values of accumulated credit. In addition to the demand for industrial operations in the strait, the profitability of operations depends on another one external factor: the market value of an operation, i.e. on its cost.

7.4. Dynamic Model of Marine Industry Operations in an Ecological-Economic System of the Kerch Strait Area

The integrated model of resources management in the Kerch Strait zone should respect the economic profitability of industrial operations (ships pilotage, cargo transshipment) with ensuring of adequate ecological state of the marine environment (biodiversity, pollution). As it was noted above, the strait region represents the typical example of a complex system of the coastal zone area, where the intensive economic activity is in conflict with the protection of ecological state of the marine environment. Therefore, the integrated management model of ecological-economic processes in the strait zone, should predict the required amount of environmental protection activity, providing conditions for economic use of its resources.

In this formulation of the problem, the industrial system of the strait allocates a part of its profit to environment protection actions and measures directed at prevention of emergency situations. It is forced to purchase certain amounts of the three types of resources: the economic, the ecological and the emergency-prevention type of resources. Thus, the prices of each of these three types of resources have become major regulators of ecological-economic balance of the economic use of the strait.

The problem of resources management of the Kerch Strait area is to provide such conditions when profitability of industrial operations will have high value, and simultaneously, the state of the marine environment will be maintained at a sufficient level, when the risk of adverse effects of potential emergencies will be

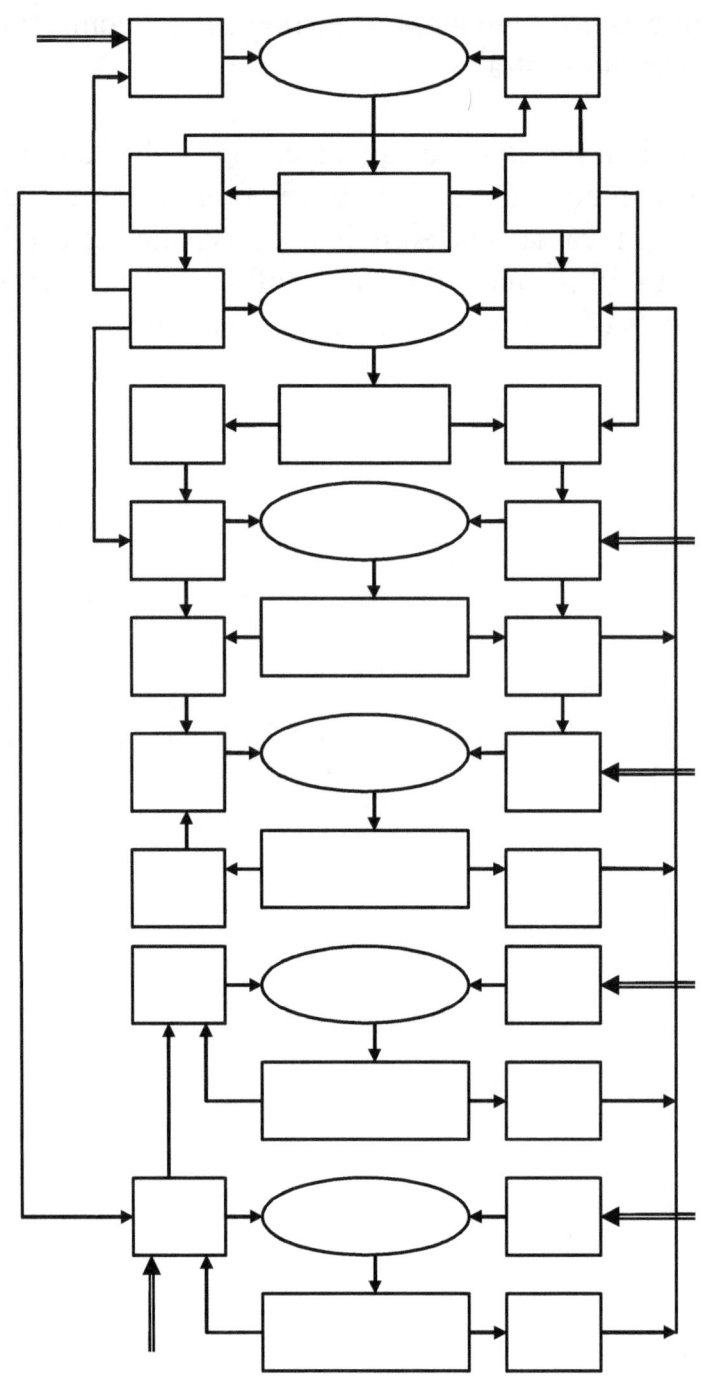

Fig. 7.5. Structure of the integrated model of ecological-economic system of the
Kerch Strait zone

minimized. A model, that satisfies these conditions, must consist of three main blocks: a block of conducting operations; a block of resources provision and a block of integrated assessment of the marine ecosystem's state. In addition to this, management agents controlling parameters of the ecosystem and managing intensity of industrial operations, based on ecological criteria of the marine environment, must be provided in it. The structure of the model is shown in Fig. 7.5.

Let us denote D as demand for industrial operations all over the region of the strait. Current demand is compared with available opportunities for operations H (the actions of comparison are drawn as ovals in Fig. 7.5), and in the case when the price of operation p exceeds their cost (prime price) e, the management agent $A[D, H]$ provides a command to perform operations and to increase (if necessary) existing capacity of industrial system V. Driving comparisons between prices and costs of operations and examining profitability of the economic system of the strait allow us to manage the entire system and at the same time, to maintain proper level of the marine ecosystem's state by controlling pollution concentrations in the marine environment of the strait.

The values of investments provided by the production system and directed at purchasing of the three main types of production resources (economic, ecological and emergency prevention resources) – all serve for the purpose of balancing ecological-economic situation in the strait area. The investments are proportional to the volume of industrial operations in the Strait, as well as to the quantities y_i of
the i-type of resources (i = 1, 2, 3), which are required for performing of a single industrial operation.

Values of investments are determined by management agent $A[p, e]$ which sets up the amounts of needed resources. Agent $A[H_2, H_2^*]$ compares amounts of financial resources of the production system: it compares its currently available working capital H_2 with the volume of H_2^* of the required capital to fully meet the demand. The value of H_2^* is formed under the influence of external factors, which include current prices for resources (external influences on the system are shown in Fig. 7.5 with bold arrows directed from outside).

Agent $A[H_3, H_3^*]$ determines the amount of loans, needed to purchase the missing amount of resources. It is assumed that the total amount of loans H_3 accumulated by the production system to this point in time should not exceed a certain maximum permissible value of H_3^*. Agent $A[H_3, H_3^*]$ monitors the implementation of this condition, taking into account the external influence, directed at H_3^*. This influence limits the economic activity of the system, establishing the maximum permissible value of investments in production. Interest on the accumulated credit θ affects the costs of industrial operation.

Managing the balance of ecological-economic resources is carried out by formation of prices for primary resources. Production cost depends on their values, and hence, the profitability of operations in the strait depends on them, too. As shown in Fig. 7.5, we defined a total cost of material and financial expenses of the strait's economic system related to performing a typical industrial operation (shipment and cargo transshipment for a medium-tonnage vessel) as a standard unit of economic resource. In monetary terms, the cost per unit of economic resources r_i is determined by the ratio of infrastructure maintenance cost which ensures performing of operations in the Strait, to the maximum possible number of typical operations carried out in it during a single time interval- for example, in a day.

The number of units of economic resources available at a given time in the strait's production system, characterizes its capacity. Further increase in numbers of industrial operations in the strait is possible only with an increase in capacity of the whole system, by means of making additional investments to acquire insufficient amounts of economic resources.

We made an assumption above that, similarly to the economic measures aimed at sustaining the economic infrastructure of the Strait, it is necessary to continuously implement environmental measures which would be aimed at preservation of the marine ecosystems of the Strait. The information about current index value of biodiversity in the model is compared with some reference value of the index. For example, average long-term norms Bd_m of these values can be used for this purpose in our model.

Excessive consumption of marine resources may disrupt the processes of natural recovery of biodiversity. Production system of the Strait should allocate special funds to implement conservation actions that would be sufficient to maintain biological status of the marine environment at some ecologically justified level. The functions of monitoring and assessment of required environment protection actions are performed by the agent $A[Bd, Bd_m]$.

The degree of ecological degradation associated with performance of a single industrial operation in the Strait, is a measure of economic sanctions that should be imposed on the production system of the Strait. The size of these sanctions – the amount of resource rent, or the environmental protection tax - can be seen as cost per unit of environmental resources, acquired to perform a single industrial operation.

Another type of the costs which may have to be paid by the production system is associated with the level of marine pollution in the strait, caused by performance of industrial operations in its area. In the model, we assumed that there exists direct dependence between the pollution level of marine environment, the total volume of industrial operations, and the amount of environment protection activities funded by the production system of the Strait.

High level of marine pollution in the strait caused by harmful substances (petroleum products, pesticides, heavy metals, etc.) is usually associated with abnormally high wind velocities and abnormally low air temperatures, which cause accidents of maritime vessels. To mitigate the negative effects of emergency situations in the Strait, as already noted above, constant monitoring of potentially dangerous natural processes and human impact on the ecosystem of the Strait is required. The environment protection activities, managed by system of emergency situations prevention, should be funded by the production system of the Strait.

Based on these assumptions, current values of contaminants concentration Pl are compared in the model with their average long-term norms Pl_m. As a result of this, the agent $A[Pl, Pl_m]$ determines necessary amount of environment protection activities. This way, cost per unit of resource r_3 is formed. It identifies environment protection tax, which transforms into environmental fines when concentration of pollutants (contaminants) in the marine environment becomes abnormally high.

Dynamic Management Model of the Kerch Strait' Production System. A formal model of the production system was constructed in accordance with previously developed informational technology of method of adaptive balance of causes (*ABC*-method), which was described in details in Chapter Two. Dynamic equations of the *ABC*-method have a standard structure which takes into account the scheme of causal relationships between variables of ecological-economic system, as shown in Fig. 7.5. For the convenience of modeling we will introduce dimensionless (reduced) values, varying in the range (0, 10). Keeping in mind the dependence between the demand for operations in the Strait and the costs of these operations, for the dimensionless variable $X_1 = 5\,D\,[M\,(D)]^{-1}$, representing the dynamics of demand, we will obtain the following equation of the *ABC*-model

$$\frac{dX_1}{dT} = X_1\left[1 - 2\left(X_1 + a_{12}X_2 - a_{1F_1}F_D\right)\right], \tag{7.16}$$

where X_2 is dimensionless price of industrial operation p; F_D is external influence imitating the demand dynamics. Coefficients a_{ij} in this and in the subsequent equations, take into account the degree of influence of some variables of the model on others. In the same way we represent the equation for the price of X_2 and the cost of operations X_2

$$\frac{dX_2}{dT} = X_2 \left[1 - 2 \left(X_2 + a_{23} X_3 \right) \right], \tag{7.17}$$

$$\frac{dX_3}{dT} = X_3 \left[1 - 2 \left(X_3 - \sum_{i=1}^{3} y_i r_i - Q \right) \right], \tag{7.18}$$

where r_i is the prices of resources, Q is the added cost.

The demand for industrial operations in the Strait is compared in the model with current total capacity of pilot vessels for cargoes transshipment. Let us denote the number of industrial operations, implementation of which will provide for existing capacity of the strait, as H. Then, the speed of operations can be represented by the following equation of the *ABC*-model

$$\frac{dH}{dT} = H \left[1 - 2 \left(H - V + S \right) \right], \tag{7.19}$$

$S =$ IF $(p < e; \ 0; \ R)$,

$R =$ IF $(D < H; \ D; \ H)$

$V =$ IF $(D < H; \ 0; \ M)$,

$M =$ IF $(D - H < M; \ D - H; \ M)$,

$M = \min (m_1; \ m_2; \ m_3)$,

$m_i = H_{1i} / y_i \quad (i = 1,2,3)$,

where V is the number of operations, prepared for implementation to the current moment in time, S is the number of executed operations, H_{1i} are the volumes of production, environment protection and proactive measures that industrial system of the Strait is capable to ensure with currently available resources, y_i are the quantities of each kind of resources, needed to conduct a single operation.

Because of continuously changing conditions of ecological-economic system's functioning, its ability to provide industrial operations and environment protection actions with the use of existing funds, it is also subject to changes. Therefore, they should be represented by the following equations

$$\frac{dH_{1i}}{dT} = H_{1i}\left[1 - 2\left(H_{1i} - V_{1i} + S_{1i}\right)\right], \quad (i = 1,2,3). \tag{7.20}$$

$$V_{1i} = IF\{D - H_{1i} < 0; 0; IF[y_i(D - H) < H_{1i}; 0; U_{1i}]\},$$

$$U_{1i} = IF\{y_i(D - H) - H_{1i} < \rho_i H_2 / r_i; y_i(D - H) - H_{1i};$$
$$IF[\rho_i(H_3^* - H_3) < 0; 0; U_{1i}^*]\}$$

$$S_{1i} = IF\{D - H < 0; 0; IF[y_i(D - H) < H_{1i}; y_i(D - H); H_{1i}]\},$$

in which the control function U_{1i}^* limits investments into providing industrial operations and environment protection activities, based on the conditions established for granting loans.

Management of working capital of the system with the account of investment, will result in the following equation

$$\frac{dH_2}{dT} = H_2\left[1 - 2\left(H_2 - pS + \sum_{i=1}^{3} S_2^i + S_3 + \chi H_2\right)\right], \tag{7.21}$$

$$S_2^i = IF[r_i y_i(D - H) - H_1^i < \rho_i H_2; r_i y_i(D - H) - H_1^i; \rho_i H_2],$$

$$S_3 = IF[\theta H_3 < H_2; \theta H_3; H_2],$$

$$\rho_i = \frac{r_i y_i}{r_1 y_1 + r_2 y_2 + r_3 y_3}, \quad (i = 1,2,3),$$

where χ is percentage of funds derived from the turnover, i.e. the net profit of the system.

Denote the volumes of purchased on credit resources as V_{11}, V_{12}, V_{13}. Then the equation for the accumulated credit takes the form

$$\frac{dH_3}{dt} = H_3\left[1 - 2\left(H_3 - \sum_{i=1}^{3} r_i V_{1i} + S_3\right)\right], \tag{7.22}$$

$$V_{1i} = IF[(D - H)y_i < H_{1i}; 0; F_i],$$

$$F_i = IF[r_i(y_i D - H_{1i}) < \rho_i H_2; y_i D - H_{1i}; F_i^*], \; (i = 1,2,3),$$

where functions F_i^* represent control agents that restrict the amount of resources purchased for credit.

Profitability of operations is determined by the ratio of income earned over a certain period of time, to total expenditures for this interval. As it was stated above, to assess economic profitability, it is convenient to use the formula

$$\varphi = \ln \frac{[pS + 1]}{[10 + \sum_{i=1}^{3} S_2^i + S_3 + \chi H_2]}, \qquad (7.23)$$

in which S_2^i are expenses of the production system on purchasing of all types of resources.

Integrated Management Model of Ecosystem in the Strait Area. The state of marine ecosystem in the area of the strait is integrally evaluated by two parameters: biodiversity index Bd and pollutants' concentration Pl, which, starting with a certain level, can adversely affect marine organisms of the ecosystem. We will use a weighted sum of the concentrations Br_i as biodiversity index; it denotes concentrations of the major marine species in the Strait zone, averaged over the water area of this zone

$$Bd = \sum_{i=1}^{N} g_i Br_i . \qquad (7.24)$$

Let us assume that we know the long-term variability value of biodiversity index, and let us denote its monthly averaged value as Bd_m. Current values of the index for each year undergo a deviation from the average long-term variability. As the main factors that shape these deviations, we can take variations in temperature of the upper layer of the sea T, wind speed modulus WF, and sea-surface light SR around the respective monthly averaged long-term values T_m, WF_m

and SR_m. The important role is also played by the pollution concentration Pl, when it exceeds the averaged long-term norm Pl_m.

In order to take into account the process of biodiversity index' restoration to a specified level by conducting environment protection actions, we must include a management agent $A(Bd, Bd_m)$ into the equation for the index Bd of the model. Therefore, the dynamic equation of biodiversity index takes the form

$$\frac{dBd}{dt} = Bd \left\{ 1 - 2 \begin{bmatrix} Bd + a_{Bd/Pl}Pl - A(Bd, Bd_m) + \\ a_{Bd/T}(T - T_m) - a_{Bd/WF}(WF - WF_m) - \\ a_{Bd/SR}(SR - SR_m)) \end{bmatrix} \right\};$$

(7.25)

$$A(Bd, Bd_m) = IF \left[Bd > Bd_m; 0; a_{Br/y_2}y_2 \left(1 - e^{-\alpha_{Bd}\tau}\right) \right];$$

Parameter $a_{Br/y_2}y_2$ determines how much the index of biological diversity influences on the funds that the production system provides for environment protection actions. Parameter α_{Bd} specifies the growth rate of the index.

The pollution level (averaged over the Strait area) can be integrally presented as a weighted sum of major pollutants' concentrations Pl_i which are typical of this area, and which form the averaged long-term course of the parameter Pl_m

$$Pl = \sum_{i=1}^{M} h_i Pl_i .$$

(7.26)

We will assume that the main source of pollution is the volume of operations S, which is opposed to natural self-cleaning of the marine environment $a_{Pl}Pl$ and to reduction of pollution due to environment protection activities simulated by the agent $A(Pl, Pl_m)$. Dynamics of pollution level is described by the equation

436

$$\frac{dPl}{dt} = Pl\left\{1 - 2\left[Pl - a_{Pl/S}S + a_{Pl}Pl + A(Pl, Pl_m)\right]\right\},\qquad(7.27)$$

$$A(Pl, Pl_m) = IF\left[Pl < Pl_m; 0; a_{Pl_m}Pl\left(1 - e^{-\alpha_{Pl_m}\tau}\right)\right].$$

We also assume that the environment protection activities were carried out when the pollution concentration exceeded the average long-term level.

7.5. Management of Environmental-Economic Balance of the Marine
Resources Consumption

In the ecological-economic system of the Kerch Strait area, profitability of industrial operations depends on the dynamics of profit obtained by the system, and on the expenditures related to implementation of operations and environment protection activities. Profits are determined by the demand for operations and by the market price of the operations themselves, including the value of added costs. To achieve ecological-economic balance of the system, the expenditures of the production system must be put in direct dependence with the state of marine environment. This means that, along with deterioration of ecological situation in the strait and with increase of the risk of adverse effects from natural or technological disaster, the costs of all types of resources consumed by the system should increase. In these conditions, the total amount of accumulated debt H_3 may make the authorities of the strait production system to limit the scope of operations, even in the conditions of high economic profitability of their performance.

Depending on ecological state of the marine ecosystem, we can consider two modes of formation of ecological-economic balance for resources consumption. In the first case, an increase in biodiversity index should be achieved by limitation of investments (loans) into industrial operations. In the second case, prices of the ecological type of resources, as well as the resource of emergency prevention

437

activities, should be increased proportionally to the decrease of biodiversity index values and to the growing pollution level.

The expenditures on industrial operations and the works performed to protect marine biodiversity, depend on the prices of the three main types of resources, consumed by the system: economic r_1, ecological r_2 and emergency-prevention resources r_3. The cost of economic resource is determined by production cost formation factors; the costs of two other types of resources are proportional to inclinations of biodiversity index Bd and to pollution level index Pl of their long-term mean annual variations.

The periods when costs of operations increase, can be associated with the periods of time with wind velocities' values WF are quite high compared to the average multi-year norm WF_m. To control the moments of high wind velocities, we will introduce a management agent $A[WF, WF_m]$ in the equation for the economic resources' price r_1

$$\frac{dr_1}{dt} = r_1 \left\{ -2 \left[r_1 - A[WF, WF_m] - a_{r_1} r_1^* \right] \right\}, \qquad (7.28)$$

$$A[WF, WF_m] = \mathrm{IF} \left[WF < WF_m; 0; a_{r_1/WF} (WF - WF_m) \right],$$

where r_1^* is a parameter that determines the cost of one operation taking place in normal conditions.

When forming the price of the ecological type of resources, we will take into account the conditions which we discussed above when we constructed the integrated model of the marine ecosystem. Biodiversity index Bd depends on the level of pollution of the marine environment Pl, and is mainly connected with the volumes of industrial operations in the Strait S. Therefore, the cost of a unit of ecological resource can be regarded as a resource rent charged on the production system for performing a unit of production operations. This value should depend on the value of the difference between ecological state of marine environment Bd and annual value Bd_a, averaged for many years. In addition to this, if the index Bd takes a critically low value, which means that the ecosystem is on the edge of degradation, the price of ecological resource must increase substantially. This function in the equation for r_2 must be performed by agent $A(Bd_m, Bd_a)$. If we accept these conditions, the equation for the price of ecological resource takes the form

$$\frac{dr_2}{dt} = r_2 \left\{ 1 - 2 \left[\frac{r_2 + a_{r_2/Bd}(Bd - Bd_m) -}{A(Bd_m, Bd_a) - a_{r_2} r_2^*} \right] \right\}, \tag{7.29}$$

$$A(Bd_m, Bd_a) = \text{IF}\left[Bd > Bd_a; 0; a_{r_2}(Bd_m - Bd_a)(1 - e^{-\alpha_{Bd_a}\tau}) \right],$$

where r_2^* is the parameter that determines the value of the resource rent.

The cost of the emergency prevention resource may be determined by a constant, continuous insurance payment, which forms a fund intended for prevention of negative consequences of emergencies. During implementation of environment protection actions, the value of this type of resource must grow along with increasing level of marine environment pollution. In times of emergencies associated with abnormally high level of marine pollution Pl_a, the price of the emergency prevention resource should increase significantly. To meet this condition, an agent $A(Pl_m; Pl_a)$ was included into the equation for r_3. Thus, the equation for value of the emergency prevention resource can be written in the following form

$$\frac{dr_3}{dt} = r_3 \{ 1 - 2[r_3 - a_{r_3/Pl}(Pl - Pl_m) - A(Pl_m; Pl_a) - a_{r_3} r_3^*] \}$$

$$\tag{7.30}$$

$$A(Pl_m; Pl_a) = \text{IF}\left[Pl < Pl_a; 0; a_{r_3/Pl}(Pl_a - Pl_m)(1 - e^{-\alpha_{Pl_a}\tau}) \right],$$

where r_3^* is the parameter determining amount of payment to the emergency prevention fund.

Thus, the model of resources' costs enables management of ecological-economic balance of the strait, because it controls operations costs' dynamics in connection with changes of ecological state of the marine environment. By predicting possible scenarios of weather processes and risks of anthropogenic emergencies, we can construct the scenarios of changes of prices for resources and consequently, we can assess cost-effectiveness of industrial operations. This also allows us to make decisions about the required environment protection actions and about management of processes taking place in the strait. In order to test the management principles embedded in the model, the following computational experiments were undertaken.

Management of Biodiversity Index Scenarios and Pollution Level by Planning Environment Protection Activities. To assess the impact of costs of the three resources types on the scenarios of economic processes, we carried out calculations on the model (7.16) – (7.30) for 500 days (the dimensionless time steps). In this and subsequent experiments, all the scenarios of processes were built in a dimensionless form with use of transformations similar to the formula (2.22). The dynamics of resources' costs, shown in Fig. 7.6, a, simulated deterioration of ecological situation in the strait during the period from day 200 to day 400, as the cost of resources r_2 and r_3 in this period significantly increased. A sharp peak in the cost of emergency prevention resource r_3 simulated contamination of the area resulting from an emergency situation. In this regard, the permissible levels of investments into industrial operations in the strait H_3^* were limited, as shown in Fig. 7.6, b.

Demand for operations D in this period was also reduced. As a result, execution of operations ceased to be profitable, and management agents in the equations (7.16) – (7.23) "ordered" to cease operations (see Fig. 7.8, a). After 450 days, when the prices for resources went down and allowable amounts of investments went up (see Fig. 7.6, a and b), operations were resumed again and became profitable (see Fig. 7.8, a and b). Thus, the suggested model adequately responded to the dynamics of the demand for operations, to the changes in prices for resources and to the control actions.

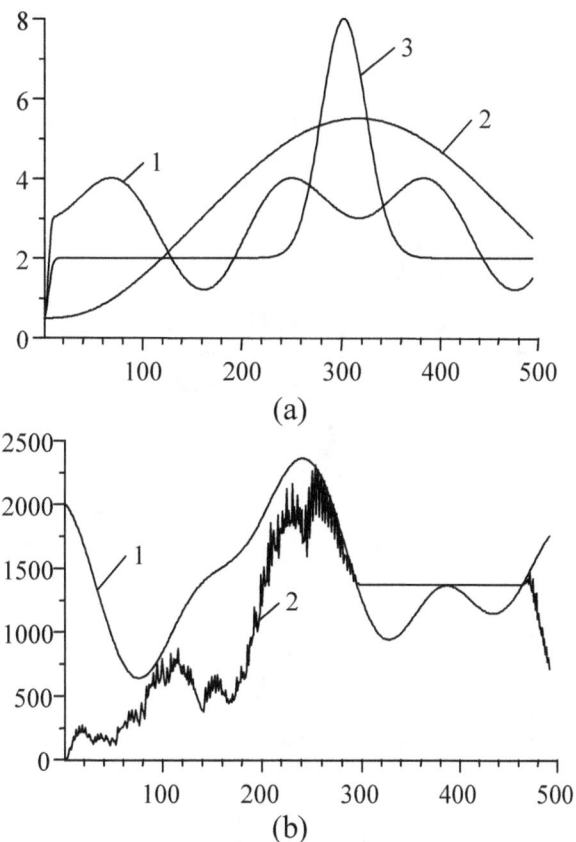

(a)

(b)

Fig. 7.6. Model predictions of scenarios of economic processes. (a) simulated

dynamics of resources' prices: *1* – economic r_1, *2* – ecological r_2,

3 – emergency prevention r_3; (b) limitation of system's crediting:

1 – control over the volume of accumulated credit H_3^*, *2* – dynamics

of accumulated credit H_3;

In the next series of experiments we simulated the role of environment protection actions in the Strait, which were performed by management agents included into equations (7.25) and (7.27). We took constant values of prices for resources: $r_1 = 3$, $r_2 = 5$, $r_3 = 7$.

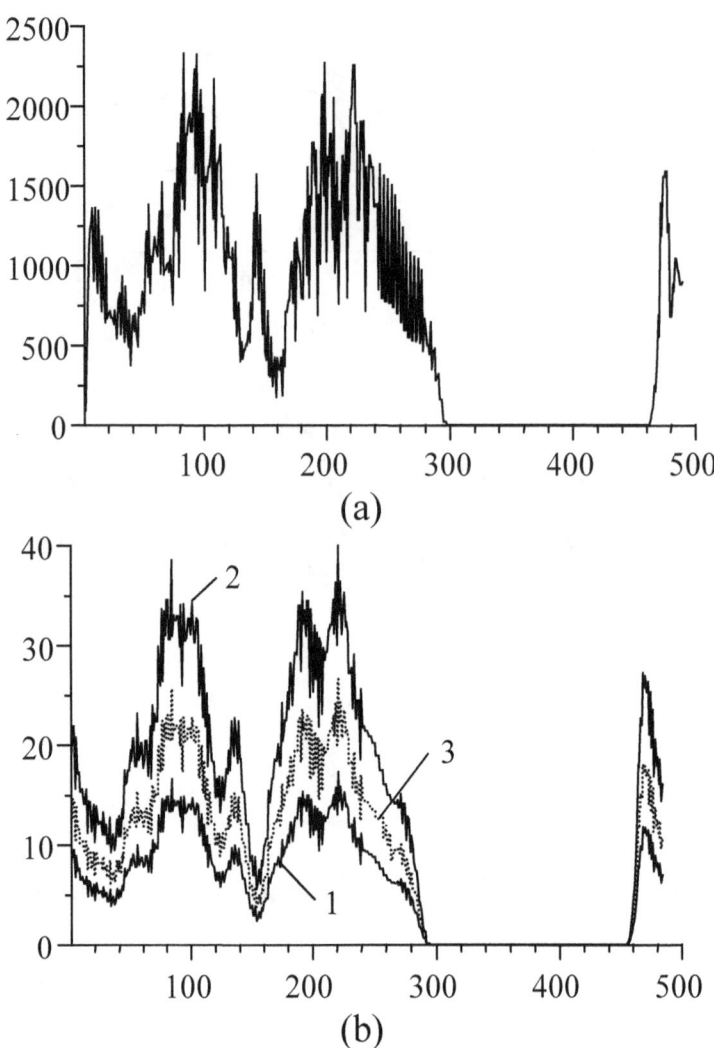

Fig. 7.7. Model predictions of scenarios of economic processes: (a) dynamics of working capital H_2; (b) dynamics of resources consumption:

1 – economic V_{11}, 2 – ecological V_{12}, 3 – counter emergency V_{13}.

Restriction on execution of operations was stopped because the allowable amount of investment was adopted by a constant and high $H_3^* = 2000$ for the same demand for the operation (see Fig. 7.8, *a*). Therefore, operations were performed during all the time of calculations.

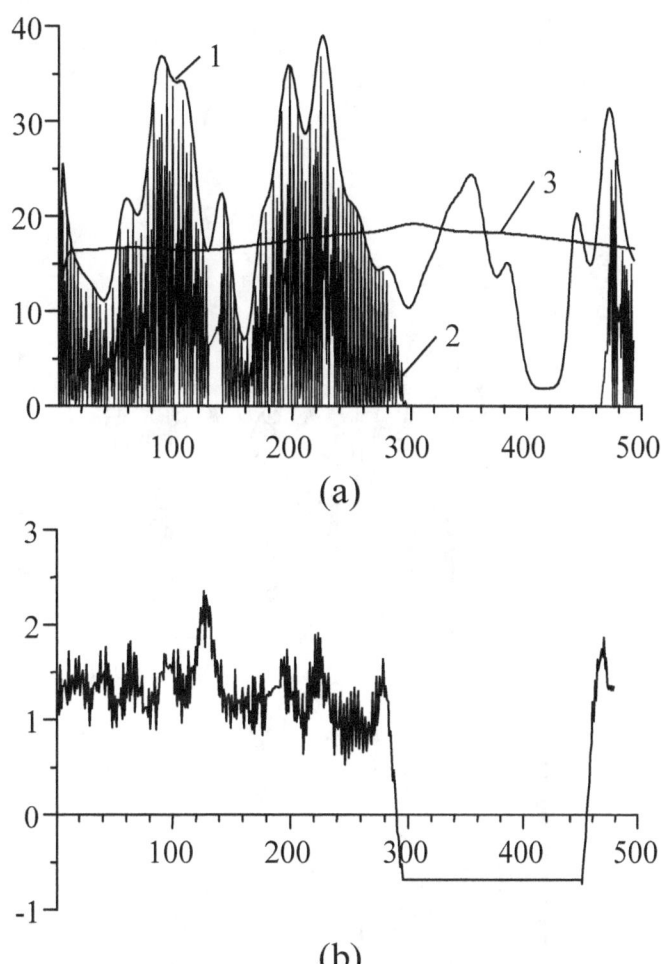

Fig. 7.8. Model predictions of scenarios of economic processes: (*a*) meeting the   demand for operations: *1* – simulated demand *D*, *2* – volumes of

        executed industrial operations *S*, *3* – simulated price of industrial operations p; (*b*) profitability of industrial operations φ.

The level of pollution in the Strait in the absence of environment protection activities (for example, neutralizing oil film on the surface of the sea), was also to remain, as shown in Fig. 7.9, *a* (curve *1*). As a result of environment protection activities, concentration of pollutants was reduced to a level represented by curve 3 in Fig. 7.9, *a*. In periods when concentrations of contaminants exceeded the value $Pl_m = 4$, the management agent continued to act, imitating additional environment protection activities which were required in the situation (see scenario 4 in Fig. 7.9, *a*).

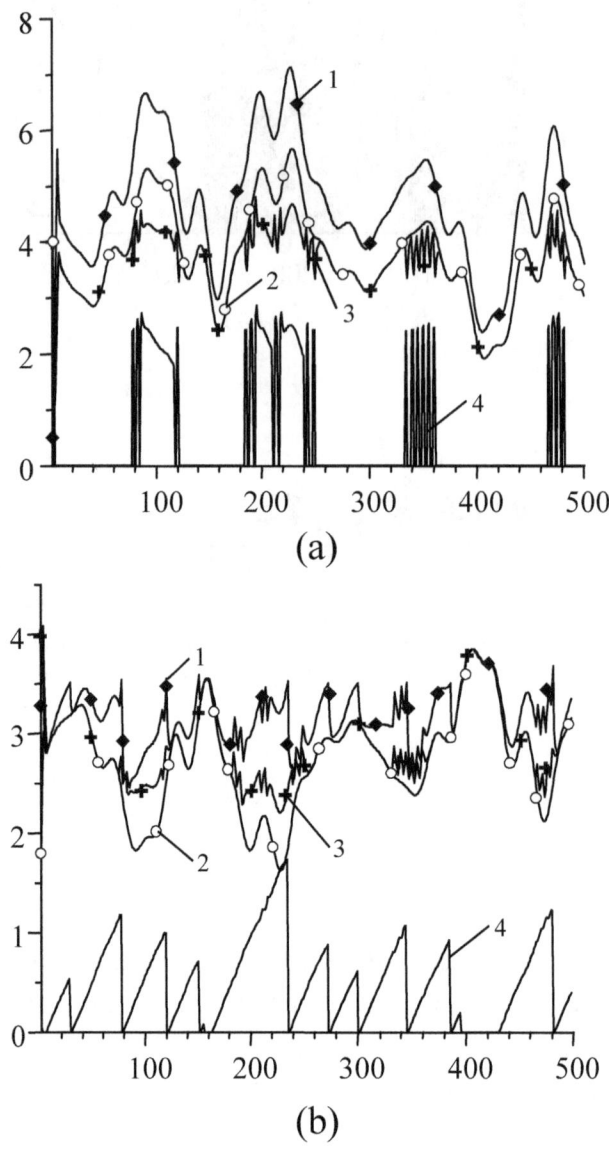

(a)

(b)

444

Fig. 7.9. Predicted reduction in pollution levels Pl and increase of biodiversity Bd through natural processes and environment protection activities (EPA): (a) scenarios of pollution levels Pl: 1 – initial, 2 – with an account of self-cleaning of the environment, 3 – with an account of the EPA ($Pl_m =$ 4), 4 – scenario of the agent $A(Pl, Pl_m)$ function based on the EPA;

(b) scenarios of biodiversity index Bd: 1 – with an account of the EPA on bio-resources ($Bd_m = 3,5$) and on pollution ($Pl_m = 4$), 2 – initial Bd

values, 3 – with an account of the EPA on pollution only. 4 – planning of the EPA on bio-resources by the agent $A(Bd, Bd_m)$

In order to take into account the factor of natural self-cleaning of the marine environment due to biochemical processes and turbulent mixing, the influence factor $a_{Pl}Pl$ in the equation (7.26) imitated decrease in concentration of pollutions. The rate of the decrease was chosen so that its level fell down by 50% in 15 days (curve 2 in Fig. 7.9, a).

Management agent in equation (7.27) gave an order to carry out environment protection actions in order to reduce concentration of contaminants when their level exceeded the value $Pl_m = 4$. The chosen value of the speed of decontamination parameter $\alpha_{Pl_m} = 0,4$ reflected the decrease of the level of Pl by 50% within 20 days. In constructing scenarios of biodiversity index using equation (7.25) in this series of experiments, the influence of weather conditions was not taken into account, because we only considered the role of pollution level, caused by the production activities in the strait.

Fig. 7.9, b shows scenario two of biodiversity index Bd – scenarios of biodiversity index with an account of the EPA on bio-resources ($Bd_m = 3,5$) and on pollution ($Pl_m = 4$). It could be compared with the value that was anticipated to be registered in the Strait in the absence of environment protection actions. As a result of decreasing level of contamination (described above), the scenario of biodiversity index rose in Fig. 7.9, *b* to the position of curve *3*.

Under the conditions of the experiment, management agent $A(Bd, Bd_m)$ in the equation (7.25) was supposed to simulate future environment protection activities for bio-resources – for example, restriction of fishery to improve *Bd* index values to the level of $Bd_m =$ 3,5. Under the chosen level of biodiversity, the rate of its recovery was 50% for 70 days. The resulting scenario of biodiversity index is represented by curve *1*, and the script of the control agent's actions are illustrated by curve *4* in Fig. 7.9, *b*.

Calculations confirmed the possibility to plan and manage volumes and durations of activities on prevention of environment pollution and protection of bio-resources by carrying out simulation experiments for different values of model parameters.

7.6. Weather Factors' Impact on Scenarios of Ecological-Economic Processes in the Kerch Strait Area

In the equations of integrated model of ecological-economic system, the influence of mean monthly values of weather factors was taken into account. Scenarios of wind speed, temperature of the upper sea layer and solar radiation at the surface of the sea with such averaging can be predicted rather accurately, for 5 – 10 days in advance. By comparing these scenarios with their known average long-term progress, we can estimate the influence of weather factors in a given year on the index value of biodiversity (7.24) and on the cost of economic resources (7.28).

Fig. 4 shows archival observational data for the area of the Kerch Strait: monthly mean air temperatures at the sea surface $T_{m\ air.}$, temperature of the upper layer of the sea T_m, the intensity of the light flux SR_m, module of the wind velocity at the sea surface WF_m [25, 42]. These data were used to simulate weather conditions for a particular year in the form of random deviations of the parameters from their averaged long-term norms.

In realization of computational experiment, we used the same initial conditions as the ones which were described in the previous section: constant prices of resources ($r_1 = 3$, $r_2 = 5$, $r_3 = 7$), a constant amount of allowable investments ($H_3^* = 2000$) with the same demand for the industrial operations (see Fig. 7.8a). However, in the equation (7.25), the effects of weather conditions were included (they are shown in Fig. 7.10 – 7.11). The results of calculations are shown in Fig. 7.12.

Since the operations in this experiment were not restricted by any conditions, the volume of operations (curve *1* in Fig. 7.12, *a*), the scenario of pollution concentration (curve *2* in Fig. 7.12, *a*) actually followed the scenario of demand function. The plan of environment protection activities directed at eliminatiton of pollution, was aimed at reduction of the pollution level to the average long-term annual level $Pl_m = 4$ (20% in 30 days). The resulting scenario of Pl (curve *3* in Fig. 7.9, *a*) demonstrates balance of two tendencies: the growth of pollutants' concentrations, caused by industrial activity, and the reduction of it due to environment protection activities.

Scenarios of biodiversity index, shown in Fig. 7.12, *b*, along with the influence of pollutants' concentration, reflect the impact of weather on the ecosystem of the Strait. Including the impact of weather factors led to some decrease in the level of biodiversity, as it follows from the comparison of curves *1* and *2* in Fig. 7.12, *b*. This is explained by the fact that the simulated variations in temperature and illumination (see Fig. 7.10, *b* and 7.11, *a*) are mostly exceeding the average long-term norms, and the coefficient of temperature influence $a_{Bd/T}$ on the index Bd in the equation (7.25) had a greater value than the coefficient of illumination's influence $a_{Bd/SR}$.

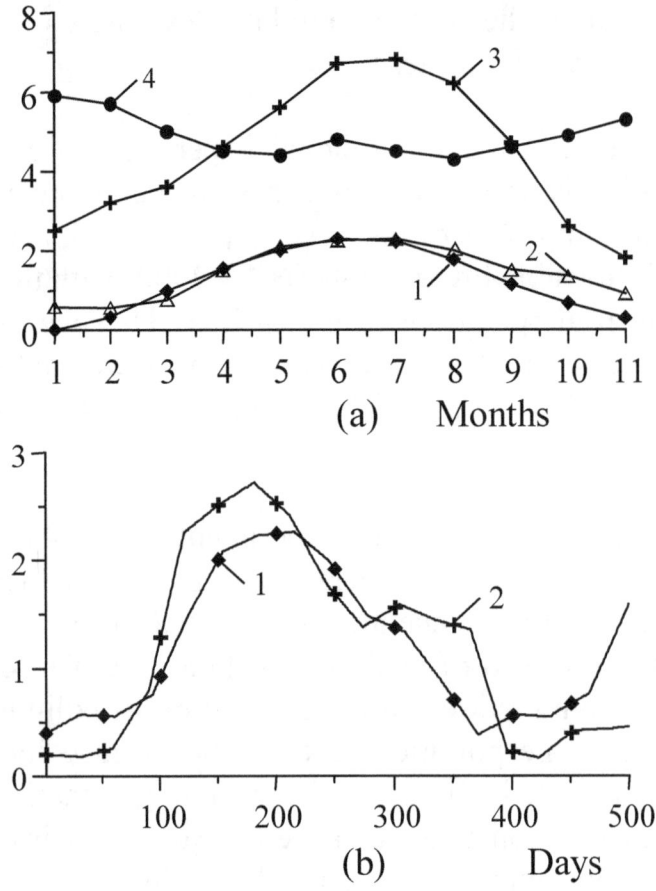

Fig. 7.10. Deviations of weather factors in the Kerch strait area from the

long-term averaged values: (*a*) long-term monthly mean values of weather factors from January to November: *1* – air temperature at the sea surface ($0,1 \cdot T_{m\ air}$ degrees C), *2* – temperature of the upper layer of the sea ($0,1 \cdot T_m$ degrees C), *3* – the intensity of light at sea surface SR_m

(relative units), *4* –module of wind speed at sea surface WF_m (m/sec);

(*b*) imitated deviations of monthly mean temperature of sea water ($0,1 \cdot T_{m\ air}$ degrees C): long-term norms (curves *1*), values for a

given year
(curve *2*).

448

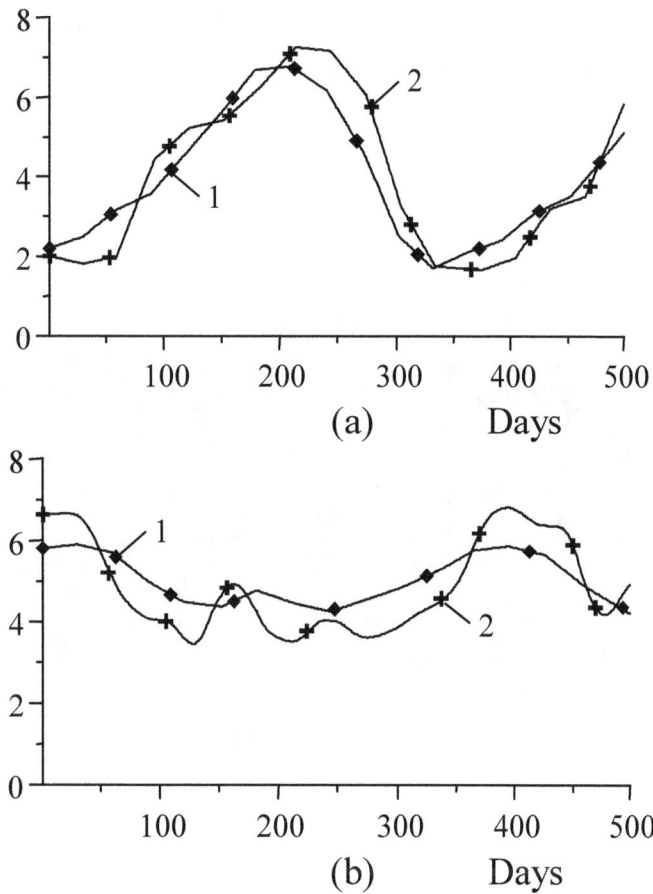

Fig. 7.11. Deviations of weather factors in the Kerch strait area from the long-   term norms, simulating their values for a given year: (*a*) 1 – averaged

values of intensity of light at sea surface SR_m (relative units),
2 – deviations of intensity of light; (*b*) 1 – averaged values of module of   wind speed at sea surface WF_m (m/sec), 2 – wind speed values for a   given year

Simulation of environment protection activities on biological resources in this experiment provided for a slow increase in the index

Bd up to the mean long-term level $Bd_m = 4$. Scenario of a management agent for the index is represented by curve *4* in Fig. 7.12, *b*, and the resulting scenario – by curve *3*. It is obvious that the speed of implementation of environment protection activities happened to be insufficient to increase the value of biodiversity index up to prescribed control level $Bd_m = 4$.

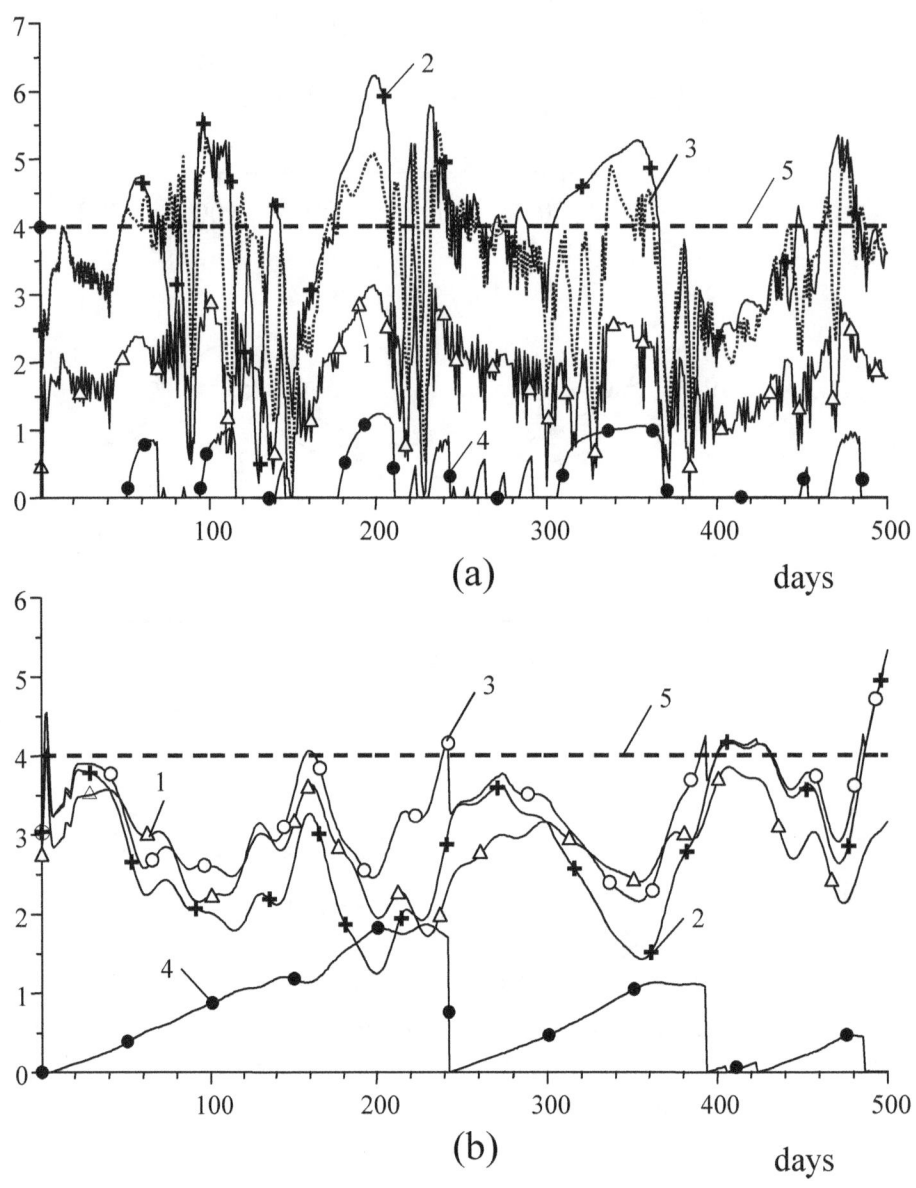

(a)

days

(b)

days

Fig. 7.9. Scenarios of processes in the ecosystem of the Strait, taking into account   various influence factors: (*a*) *1* – volumes of operations in the Strait *S*

(averaged for 4 days), *2* – scenarios of pollution level *Pl* before

implementation of environment protection activities (*EPA*) on pollution level, *3* – the same after *EPA*, *4* – script of the *EPA* agent $A(Pl, Pl_m)$ function, *5* – average long-term annual level of pollution Pl_m; (*b*) Scenarios of biodiversity *Bd* taking into account: *1* – pollution   level *Pl*, *2* – pollution level *Pl* and weather conditions *T*, *WF*, *SR*,

3 – *EPA* on biodiversity level. *4* –scenario of the *EPA* agent $A(Bd, Bd_m)$, *5* – average long-term annual index of biodiversity Bd_m.

Thus, accounting for the influence of weather conditions on the state of marine environment, may significantly change scenarios of biodiversity index and pollution level. As we will show later, this fact has a noticeable effect on profitability of the production system, which has to spend a part of its profit to support appropriate ecological state of marine environment.

7.7. Cooperative Planning of Marine Industrial Operations and Environment Protection Activity

Computational experiments conducted with the model of ecological-economic system, confirmed the possibility to plan volumes of environment protection activities, in which industrial operations in the strait do not lead to a drop in the index of biodiversity below a prescribed control level. The computational algorithm of the model allows for obtaining the scenarios of development processes under different external conditions changing in time. By setting at the input of the model expected demand's curve for industrial operations, predicted dynamics of weather factors and average prices for three major types of resources it is possible to obtain prognostic scenarios of costs and profitability of operations, as well as to assess volumes of required actions, supporting the value of biodiversity index at the proper level.

Let us suppose, for example, that anticipated dynamics of the demand for operations in the Strait develops as it is shown in Fig. 7.13, *a*. We will use the variations of the weather factors *T*, *WF*, *SR* (considered above), as predicted impacts on the marine ecosystem, and let us define average prices for resources, as follows: $r_1 = 3$, $r_2 = 5$, $r_3 = 7$. Let us also define a number of conditions that affect profitability of operations in the Strait. We will assume that, at each time step of calculations 40% of the net profit is taken from the working capital of the production system, but the system has the opportunity to obtain loans for purchasing resources, and redemption coefficient of the accumulated credit has the value $\theta = 0,2$.

One option to manage this system is to limit investments assigned for implementation of operations H_3 by the maximum admissible values H_3^*, which depend

on biodiversity index $H_3^* = a_{Bd}Bd$, either on pollution level $H_3^* = a_{Pl}(10 - Pl)$.

The second option is to curry out environment protection activities by introducing differential pricing of resources consumed by the production system. Let us consider the use of each of these options of management.

The first version of management is more lenient with respect to the economic system, and may not be effective in relation to the marine ecosystem. As an example, Fig. 7.13, *a* demonstrates the dynamics of volumes of performed industrial operations, trying to meet the demand for the operations when the control on the biodiversity index is specified as a function H_3^*, shown in Fig. 7.13, *b*.

In addition to the influence of pollution, the index of biodiversity was subject to the impact of weather factors, shown in Fig. 7.10 and 7.11. As follows from Fig. 7.13, *a*, some decrease in limitation of crediting of the economic system's operations coincides with those time periods when the volumes of industrial operations take the highest values. However, industrial activity in these periods had not been weakened due to the high demand, and the volumes of working capital even increased. Therefore, pollution concentration also increased (curve *3* in Fig. 7.14, *a*), while biodiversity index decreased (curve *1* in Fig. 7.14, *a*).

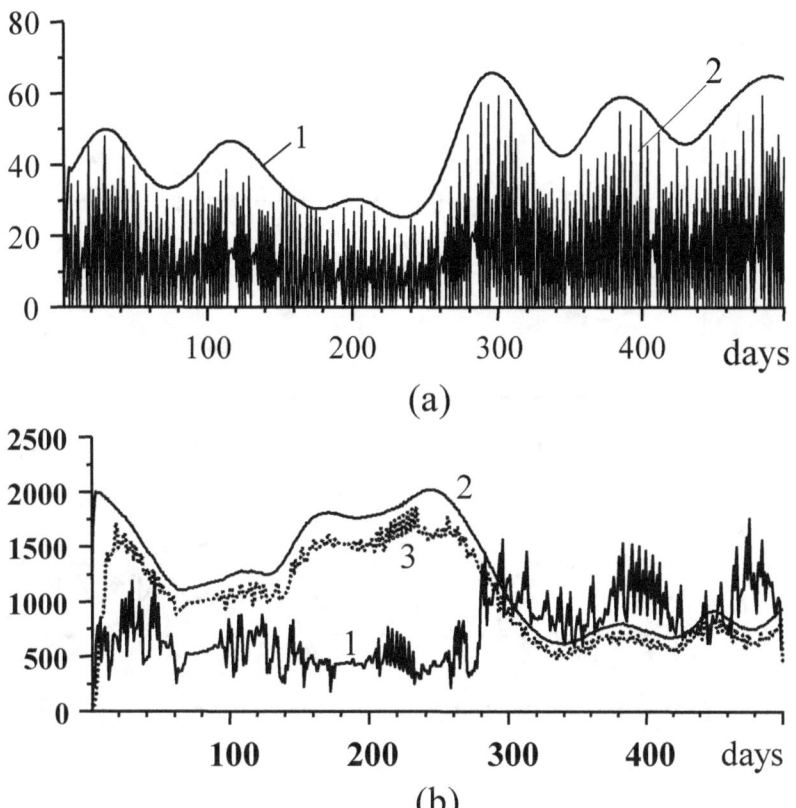

(a)

(b)

Fig. 7.13. Simulated management scenarios for industrial operations in the Kerch strait: (*a*) 1 – demand for number of operations D, 2 – daily

numbers of executed operations S in the absence of management;

(*b*) 1 – working capital H_2 at constant prices on resources, 2 – admissible level of accumulated credit H_3^* as a function of the biodiversity index,

3 – accumulated credit

The second version of management proved to be more effective in this situation. In order to return the ecosystem to its normal state by conducting environment protection activities, we introduced two conditions: that the level of pollution should not exceed the value $Pl_m = 7$, and biodiversity index should not be lower than $Bd_m = 5$ (see Fig. 7.14, *a*). In these conditions, agents of environmental actions in the model of the ecosystem switched on variable prices for resources purchased by the industrial system. In Fig. 7.14, *b* scenarios of prices for resources in relation to their initial constant values $r_{10} = 3$, $r_{20} = 5$, $r_{30} = 7$ are given.

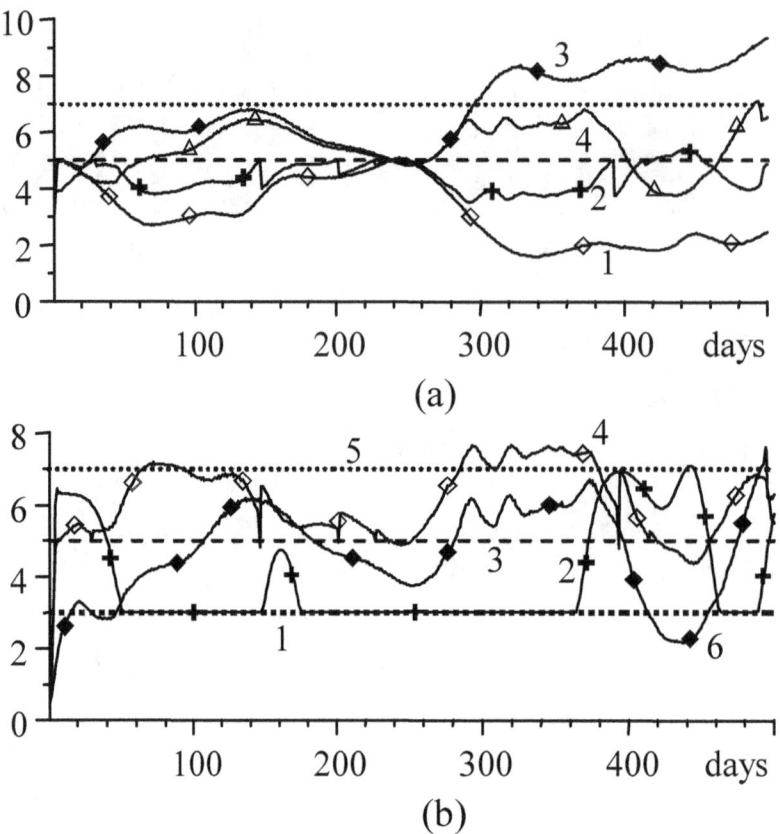

Fig. 7.11. Influence of prices for resources and environment protection activities   (EPA) on the development scenarios: (*a*) 1 – index of biodiversity *Bd*

before implementation of EPA, 2 – index of biodiversity *Bd* after EPA,

3 – pollution level *Pl* before EPA, 4 – pollution level *Pl* after EPA,   dashed line – lower admissible level of biodiversity index $Bd_a = 4$, dotted   line – the upper admissible pollution level $Pl_m = 7$; (*b*) 1 – initial price for   economic resource r_{10}, 2 – price for economic

454

resource r_1 in dependence   of weather conditions, 3 – initial price for ecological resource r_{20}, 4 – price for ecological resource r_2 in dependence of weather conditions and   EPA on biodiversity, 5 – initial price for emergency prevention resource r_{30}, 6 – price for emergency prevention resource r_3

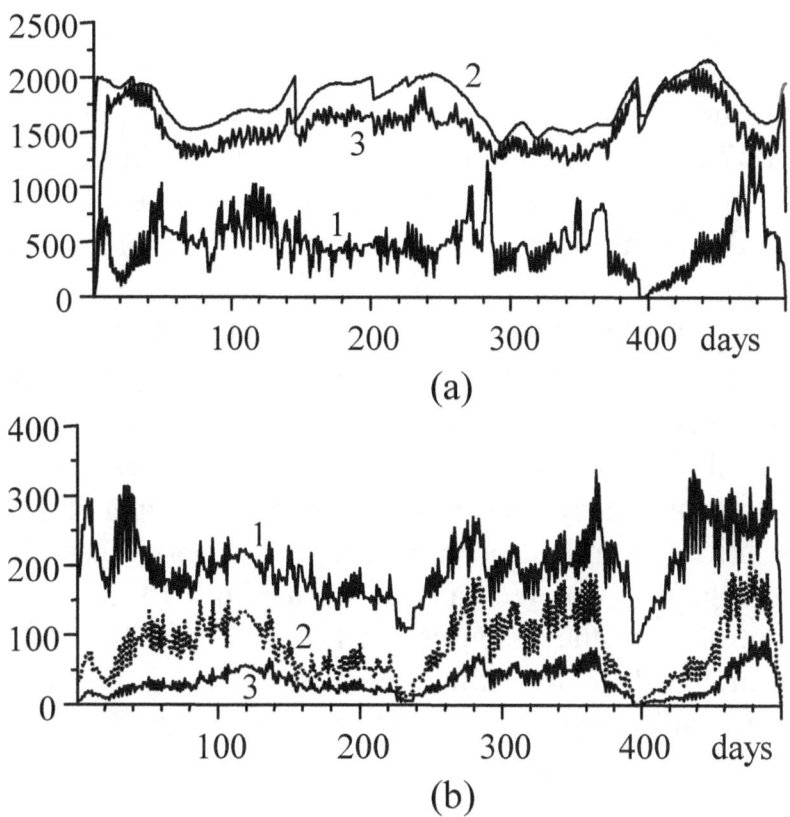

Fig. 7.15. Scenarios of working capital, credits and expenditures of the economic   system: (*a*) 1 – working capital H_2 (1), 2 – limit value of accumulated   credit H_3^* as a function of the biodiversity index *Bd*, 3 – current value of accumulated   credit H_3; (*b*) 1 – charges for economic resources, 2 –   charges   for   ecological resources, 3 – charges for counter emergency
    resources (averaged for 5 days)

Similarly to equation (7.28), a control agent was introduced. It took into account the influence of strong winds on execution of marine operations. The cost of economic resource r_1 (curve *4* in Fig. 7.14, *b*) experienced significant growth during those periods when mean monthly values of wind speed modulus exceeded their long-standing norms (see Fig. 7.11, *b*).

To build forecasts of the costs for ecological resources r_2, we set a supplementary condition: the state of the ecosystem was considered critical when the value of biodiversity index fell down lower than the value $Bd_a = 4$. During these periods of time, the influence of the agent adding value to the price of the resource was switched on in the equation (7.25), which ultimately led to large fluctuations of the curve *4* in Fig. 7.14, *b* in relation to the value $r_{20} = 5$, which corresponded to the average long-term norm of the index $Bd_m = 5$.

For the scenario of price r_3 on the counter emergency resource, we also established critical level of pollution $Pl_a = 8$, but the curve of pollution concentration (see curve *4* in Fig. 7.14, *a*) did not reach this critical level due to environment protection activities. Therefore, the price of the emergency prevention resource r_3 happened to be lower than the price which had been accepted as a norm $r_{30} = 7$ (see the line *5* in Fig. 7.14, *b*). We can see from Fig. 7.14, *a*, that resulting from control over the resources' prices, the scenario of pollution concentration fell below the specified level, and the curve of biodiversity index came close to the specified value $Bd_m = 5$.

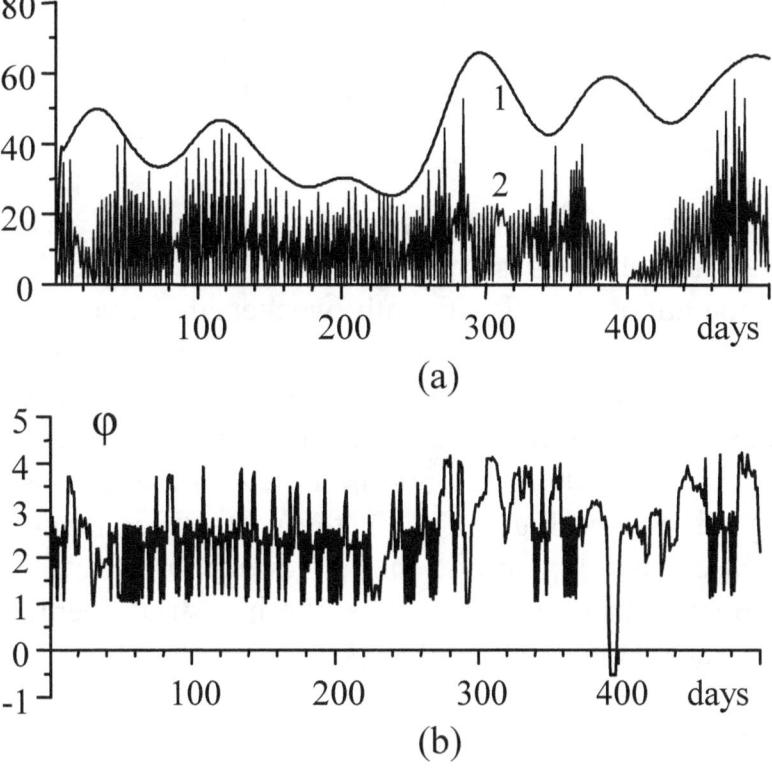

Fig. 7.16. Volumes of industrial operations in the Kerch Strait taking into

account environmental protection actions: (*a*) 1 – demand for operations *D*, 2 –volumes of operations *S*; (*b*) profitability of industrial operations with accounting of environmental protection actions

Fig. 7.15 and 7.16 represent scenarios of processes for the case when two abovementioned management options were implemented simultaneously. Since the main influence on biodiversity index was imposed by pollution concentration *Pl*, the restriction on investments into marine operations was established in accordance with the scenario $H_3^* = a_{Pl}(10 - Pl)$ (see Fig. 7.15, *a*). Having the credit repayment scheme used in this experiment, the economic system had sufficient working capital (see curve *1* in Fig. 7.15, *a*) which allowed it to avoid accumulating credits (curve *3* in Fig. 7.15, *a*) above the limits which were set in course of management (curve *2* in Fig. 7.15, *a*). Some exceptions were only made for a few days around step 400 of calculations, when investments for acquisition of ecological and emergency prevention' resources were insufficient (curves *2* and *3* in Fig. 7.15, *b*).

The dynamics of meeting demand for operations accounting for ecological limitation is shown in Fig. 7.16, *a*. The intensity of planned operations was significantly weakened in the second half of the forecast period, which was a consequence of resource pricing scenarios depicted in Fig. 7.15, *a*. Costs for ecological and for emergency prevention resources were increased from day 200 to day 400, and from day 400 to day 500 in a view of the forecasted period of strong winds (see Fig. 7.11, *b*); the cost of economic resources also increased sharply. Despite the withdrawal of a part of profit for the environment protection activities, the industrial operations in the Strait maintained their profitability (see Fig. 7.16, *b*).

The ultimate goal of modeling of the natural-economic systems of the coastal zone area is creation of informational technologies for decision-making support in the process of management of development of these zones. In the proposed model of ecological-economic system of the Kerch Strait, management of balance of ecological and economic processes takes place by means of limiting investment to acquire resources and establishing of differential prices on resources. Application of dynamic equations of the method of adaptive balance of causes in the model with including the management agents in it allowed for construction of the information technology of management by industrial operations in the Strait, which provided forecasts of development scenarios for various use of resources and changing environmental conditions.

The conducted computational experiments demonstrated that economic cost-effectiveness of industrial operations in the Strait depends essentially on the expected weather conditions and on ecological state of the marine environment. The scheme of management by the state of the Strait area's marine ecosystem on two integral parameters: biodiversity index and pollution level of marine environment, allowed us to plan intensity and duration of environment protection activities which could maintain the desired balance between ecological and economic conditions, for future beneficial use of the strait. Thus, the proposed model of management by resources of the Kerch Strait area offers an efficient way of simulation and management over complex natural-economic systems of coastal zone areas.

CONCLUSION

Integrated management of resources encompasses a vast range of problems of coastal zone area's (*CZA*) sustainable development. A central objective of management is adaptation of *CZA'* development goals to actually existing conditions of development, with understanding of social, environmental and economic constraints on the use of all types of resources. The comprehensive nature of this problem requires a systems methodology for its solution. In this work, the emphasis was made on practical utilization of systematic methodology for the task of management of complex natural and economic objects of the coastal zone.

Systems methodology determines sequence of events which lead to creation of informational technologies supporting management of sustainable development of *CZA* resources. Computerized models of development processes which provide forecasts of development scenarios of various types of resources make up the basis for these technologies. By following principal steps of systems modeling and control, we studied conceptual models of ecological-economic systems of environmental management in relation to three main types of coastal zone resources of the sea: biological resources, recreational resources and environmental resources, which are consumed along with performance of economic operations.

Systems methodology provides basis for constructing schemes of development management of the coastal zone. Possible development scenarios can be provided by dynamic models of development. The predicted scenarios allow us to choose options of management that are consistent with social, environmental and economic objectives of *CZA* in both, short-tern and long-term perspectives. Dynamic models have become indispensable tools for management, since they allow to predict outcome of planned actions. Evaluation of management efficiency, which is implemented by the monitoring system of resources and of ecological-economic state of *CZA*, allows us to carry out ongoing monitoring of actual results against anticipated results (target devices) of development. This information lays a basis for adaptation of both, development goals and the dynamic models themselves, which provide predictions of anticipated results.

Thus, a great significance is given to simple and reliable methods of constructing dynamic marine ecosystems' models and CZA ecological-economic models. In this work, we examined a number of integrated approaches to building models. Among them, the method of systems dynamics by Forrester [50], and the method of adaptive balance of causes (ABC-method) [181] have proved to be the most suitable for computer simulation of complex social, ecological and economic systems of *CZA*, where models of management become an integral part of resource development management. The method of adaptive balance of causes allows us to perform objective evaluation of coefficients in equations of dynamic models, where archived records of observations of simulated processes are used. This method can utilize patterns in equations of current monitoring of development processes, which makes the projected scenarios significantly closer to the processes taking place in reality. Therefore, we made a conclusion about the benefits of the *ABC*-method to create informational technology of CZA management.

Formalization of these models by the method of adaptive balance of causes (*ABC*-method) is a way to build prediction scenarios of ecological and economic processes in the coastal zone of the sea, which have a large number of control parameters. Application of informational technology ABC AGENT allowed us to use management agents in model equations to perform control of factors that limit development of systems. A principal advantage of such computer-based technology of management is its capacity to perform simulation experiments and, in a variety of resulting scenarios, to choose the most appropriate development scenarios which satisfy the goals of sustainable development

Among the immediate objectives of the ABC-method are- the construction of an informational technology of *CZA* management, based on integrated models of bio-geo-chemical processes in the coastal sea area, and on the dynamic input-output models, which are perfect tools for management of coastal area ecological and economic systems. In our opinion, the lack of systematic observational data describing complex processes of socio-economic development, which would be synchronized with observations of natural processes taking place in *CZA*s, should not be an obstacle to the creation of such models. As it was demonstrated in this book by the examples of marine ecosystems' models and coastal area ecological and economic systems, many important aspects of informational technology of management can and should be resolved by means of simulation experiments.

This partially explains the fact that the problems of collection, systematization and processing of observational data of development processes remained outside of our consideration in this research. In recent years, significant progress has been reached in obtaining and using remote sensing data of territories and water resources of *CZA*. Operating systems of diagnosis and prediction of coastal seas and oceans were also developed recently. Assimilation of satellite data in *CZA* dynamic models opens new possibilities for decision making in management of their resources. To realize these opportunities, the main efforts should be aimed at creating models of the processes of development-oriented rapid assimilation of observational data.

One of the promising areas of *CZA* resources research is creation of specialized GIS technologies, which allow experts to assess economic profitability of resources consumption in local parts of *CZA*s, taking into account social and environmental constraints. The concept of conditional resources potential of coastal waters and *CZA* coastal areas which we discussed in this research, can be used to serve this goal.

With adaptive balance models of environmental-economic systems, we can construct a dynamic inventory of natural and socio-economic resources reflecting the dynamics of potential profitability of their use. Realization of this idea involves development and application of integral, spatial and temporal patterns of terrestrial and marine resources.

REFERENCES

1. Abakumova, T.N. (1988). Distribution of nitrate and nitrite nitrogen in the Black Sea. *Processes of Formation and Annual Variability of Hydrophysical and Hydrochemical Fields of the Black Sea.* [In Russian] – Sevastopol.: MGI AN UkrSSR. 158 – 161.

2. Abrosov, N.S., and Bogolyubov, A.G. (1988). *Environmental and Genetic Patterns of Coexistence and Coevolution of Sspecies.* [In Russian]. M. Nauka. 333 pp.

3. Alekseev, D.V., Ivanov, V.A., Ivancha, E.V., Fomin, V.V., Cherkesov, L.V. (2008). The influence of cyclonic activity on the distribution of particulate matter in sea water in the north-western Black Sea. [In Russian]. *Meteorology and Hydrology.* **3**. 68 – 78.

4. Allan, C., and Stankey, G.H. (Eds.) (2009). Adaptive Environmental Management. A Practitioner's Guide. *CSIRO Publishing XVI.* Collingwood, Australia. 352 pp

5. Ashby, W.R. (1962). Principles of the self-organizing system, in: *Principles of Self-Organization*, von Foerster H. & Zopf G.(eds.), Pergamon, Oxford. 255 – 278

6. Ashmanov, S.A. (1980). *Mathematical Models and Methods in Economics.*
[In Russian]. M. MSU. 230 pp.

7. Bambach, R.K., Knoll, A.H., and Wang, S.C. (2004), "Origination, extinction, and mass depletions of marine diversity", *Paleobiology* 30 (4). 522 – 542

8. Baturin, V.A., Gurman, V.I., et al. (1981). *Models of Natural Resources Management.* [In Russian].M. Nauka. 360 pp.

9. Belokopytin, Y.S. (1993). *Energy Metabolism of Marine Fishes.* [In Russian]. Kiev. Naukova Dumka. 128 pp.

10.   Belyaev, V.I. (1987). *Modelling of Marine Ecosystems*. [In Russian]. Kiev. Naukova Dumka. 240 pp.

11.   Belyaev, V.I., and Timchenko, I.E. (1969). Displacement of measurements and accuracy of fields' reconstruction at automated collection and processing of hydrophysical information. [In Russian]. *Automation of Scientific Research of Seas and Oceans*. Part I. Sevastopol. MGI AN UkrSSR. 327 – 341.

12.   Belyaev, V.I., and Timchenko, I.E. (1972). On the application of objective and four dimensional analysis in oceanography. [In Russian]. *Morsk. Gidrofiz. Issledovaniya*. **2**.   80 – 92

13.   Bertalanffy, L. (1975). *Perspectives of General Systems Theory*, Braziller, 1975.

14.   Biodiversity. *Encyclopedia of Earth*. http://www.eoearth.org/article/Biodiversity

15.   Brown, L.R. (2001). *Eco-Economy: Building an Economy for the Earth*. Norton & Co., NY. Earth Policy Institute

16.   Bryson, A. and Ho Yu-Chi. (1972). *Applied Theory of Optimal Control*. [In Russian].M. Mir. 544 pp.

17.   *Bulletin of Russian Ministry for Emergency Situations*. (2007). http://www.mchs.gov.ru

18.   Charlson, R., Lovelock, J., Andreae, M., and Warren, S. (1987).Ocean phytoplankton, atmospheric sulfur, cloud albedo and climate. *Nature*, **326**   655-661

19.   Cherkashin, A.K. (1994). Mathematical methods for economic valuation of natural resources. [In Russian]. *Geography and Natural Resources*. № 1. 162 – 170

20.   Chernyakova, A.P. (1965). Typical wind fields of the Black Sea. *Collected Works of the Basin Hydro-meteorological Observatory of the Black and Azov Seas.* [In Russian]. Leningrad. Gidrometeoizdat. Vol. 3. 78 - 121

21.   Cicin-Saint, B., and Knecht, R.W. (1998). *Integrated Coastal and Ocean Management. Concepts and Practices.* Island Press. Washington, D.C. – Covelo, California. 518 pp.

22.   Clark, J. R. (1992). *Integrated Management of Coastal Zones.* FAO Fisheries Technical Paper No. 327. Rome: Food and Agriculture Organization of the United Nations.

23.   *Climate   of   Kerch.* http://www.meteoprog.ua/ru/climate/Kerch/

24.   Codispoti, L. A., Frederich, G. E., Murrey, J. W., and Sacamoto, S. M. (1991). Chemical variability in the Black Sea: implications of continuous vertical profiles that penetrated the oxic/anoxic interface. *Deep-Sea Res.* No. 38, Suppl. 2. 691 – 710

25.   Cokacar T., Ozsoy E. (1998). Comparative analyses and modeling for regional ecosystem of the Black Sea. *Ecosystem Modeling as a Management Tool for the Black Sea. NATO ASI Series 2. Environment.* Kluwer Academic Publishers. **2**. 323 – 357

26.   Common, M. and Stagl, S. (2005). *Ecological Economics: An Introduction.* New York: Cambridge University Press

27.   Costanza, R et al. (1998). The value of the world's ecosystem services and natural capital. *Ecological Economics.* 25 (1). 3 – 15

28.   Daly, H., and Farley, J. (2004). *Ecological Economics: Principles and Applications.* Washington: Island Press.

29.   Danilevsky, N.N., and Mayorov, A.A. (1979). Basic laws of population dynamics of the plankton-feeding Black Sea fish.

Anchovy. *Food Resources of the Black Sea.* [In Russian]. Moscow. Food Industry. 25 – 73

30.   Denman, K.L., and Platt, T. (1979). *Biological Forecast in the Sea. Simulation and Prediction of the Upper Ocean.* Leningrad. Gidrometeoizdat. 299 – 311

31.   Dobrovolsky, A.D., and Zalogin, B.S. (1965). *Seas of the USSR,* [In Russian].Moscow.

32.   Dolotov, V.V., Amosha, A.I., Ivanov, V.A., and Salomatina, P.N. (2006). The main concepts of the cadastral evaluation of beaches. *Ecological Safety of Near-shore and Shelf Zones and the Comprehensive Use of the Shelf Resources.* [In Russian]. Sevastopol. ECOSY-Hydrophysica. **14**. 28 – 36

33.   Dolotov, V.V., and Ivanov, V.A. (2007). *Increasing the Recreational Potential of the Ukraine: Crimean Beaches Cadastral Evaluation.* [In Russian]. Sevastopol. ECOSY-Hydrophysica. 192 pp.

34.   Ecosystem-based management: Markers for assessing progress. (2006). UNEP/GPA, The Hague. *United Nations Environment Program*

35.   Eremeev, V.N., and Ivanov, V.A. (2003). Oceanographic conditions and environmental problems of the Kerch Strait. [In Russian]. *Marine Ecological Journal.* № 3, v. II.
      27 – 39

36.   Eremeev, V.N., and Ivanov, V.A. (2005). *Natural Conditions in the Southern and Western Crimean Coast: Ecological Examination of Recreational Opportunities and Rehabilitation of the Population's Health* (Scientific handbook*).* [In Russian]. Sevastopol. ECOSY-Hydrophysica. 107 pp.

37.   Eremeev, V.N., Igumnova, E.M., Timchenko, I.E. (2002). Modeling of cause-effect relationships in marine ecosystems. [In Russian]. *Marine Ecological Journal.* № 1, v. I .

16 –32.

38.   Eremeev, V.N., Igumnova, E.M., and Timchenko, I.E. (2004). *Modeling of Ecological-Economic Systems*. [In Russian]. Sevastopol. ECOSY-Hydrophysica.   320 pp.

39.   Eremeev, V.N., Timchenko, I.E., et al. (2007). *Systems Modeling of Marine Ecological-Economic Processes*. [In Russian]. Sevastopol. ECOSY-Hydrophysica.   451 pp.

40.   Faber, M. (2008). How to be an ecological economist. *Ecological Economics* 66(1). 1 – 7

41.   Fasham, M., Ducklow, H., and McKelvie, S. (1990). A nitrogen-based model of plankton dynamics in the ocean mixed layer. *J. Mar. Res.* № 48. 591 – 639

42.   Fedorov, V.D., and Gilmanov, T.G. (1980). *Ecology*. [In Russian]. Moscow. MGU.   464 pp.

43.   Fedra, K. (2000). Environmental decision support systems: a conceptual framework and   application   examples. *Thése prèsentèe á la Facultè des sciences, de l'Université de Genéve. Imprimerie de l'Université de Genéve*, 368 pp.

44.   Fedra, K. (2004). *Coastal Zone Resource Management: Tools for aPparticipatory Planning and Decision Making Process.* In: Green, D.R. et al. [eds.]: Delivering Sustainable Coasts: Connecting Science and policy. Proceedings of Littoral 2004, September, Aberdeen, Scotland, UK. Volume 1. 281 – 286

45.   Fedra, K. (2002).GIS and simulation models for Water resources Management: A case study of the Kelantan River, Malaysia. *GIS Development.* **6/8**   39 – 43

46.   Felzenbaum, A.I. (1960). *Theoretical Basis and Methods of Calculation of Steady Ocean Currents.* [In Russian]. M. AS USSR, 1960. 127 pp.

47.   Fiddaman, T.S. (1997). *Feedback Complexity in Integrated Climate-Economy Models*, Massachusetts Institute of Technology Press. 300 pp.

48.   Field J.G., Hempel G., Summerhayes C.P. (2002). *Oceans 2020. Science, Trends, and the Challenge of Sustainability*. Island Press. Washington-Covelo-London. 365 p.

49.   Fomin, V.V., and Ivanov, V.A. (2007). Joint modeling of currents and wind waves in the Kerch Strait. [In Russian]. *Marine Hydrophysical Journal*. № 5. 3 – 20

50.   Forrester, J.W.(1968). *Principles of Systems*. Cambridge MA, Productivity Press. 1968.

51.   Forrester, J.W. (1973). *World Dynamics*. (2 ed.), Cambridge MA, Productivity Press.

52.   Forrester, J.W. (1991).*System Dynamics and the Lessons of 35 Years*. A chapter for "The Systemic Basis of Policy Making in the 1990s" edited by Kenyon B. De Greene. Boston. MIT Press. 38 pp.

53.   Freeman, A. (1993). The Measurement of Environmental and Resource Values: Theory and Methods. *Resources for the Future*. Washington DC

54.   Gaston, K.J., and Spicer, J.I. (2004). *Biodiversity. An introduction*. 2nd Edition. Blackwell.

55.   Ghil, M., and Malanotte – Rizzoli, P. (1999). Data assimilation in meteorology and oceanography. *Advances in Geophysics*. **33.** 141 – 266

56.   Gilbert, N., and Troitzsch, K. G. (2005). *Simulation for the Social Scientist.* Second edition. Milton Keynes: Open University Press

57.   Gozhik, P.F., and   Ivanov, V.A. (2008). Natural resources and their exploitation in offshore areas of the Azov-Black Sea basin. *Ecological Safety of Near-shore and Shelf Zones and the Comprehensive Use of the Shelf Resources.* [In Russian]. Sevastopol. ECOSY-Hydrophysica. **17**. 12 – 26

58.   Grese, V.N., Boguslavsky, S.G., et al. (1979). *Bases of Biological Productivity of the Black Sea.* [In Russian]. Kiev. Naukova Dumka. 391 pp.

59.   Gucu, A.C. (1997). *Role of Fishing in the Black Sea Ecosystem. Sensitivity to Change: Black Sea, Baltic Sea and North Sea.* Netherlands: Kluver Academic Publishers.   149 – 162.

60.   Hackett, S. (2006). *Environmental and Natural Resources Economics: Theory, Policy, and the Sustainable Society.* 3rd Edition. M.E. Sharpe, Publishers.

61.   Handbook for Measuring the Progress and Outcomes of Integrated Coastal and Ocean Management. (2006). *IOC Manuals and Guides, 46; ICAM Dossier, 2.* Paris, UNESCO

62.   Hanley, N., J. Shogren, and B. White (2007). *Environmental Economics in Theory and Practice,* Palgrave, London.

63.   Hare, M., and Deadman, P. (2004). Further towards a taxonomy of agent-based simulation models in environmental management. *Math. Comput. Simul.* Vol. 64(1). 25 – 40

64.   Harris, J. (2006). *Environmental and Natural Resource Economics: A Contemporary Approach.* Houghton Mifflin Company.

65.   Hayes, W.H., and Lynne, G.D. (2004). A centerpiece for ecological economics. *Ecological Economics.* 49. 287 – 301.

66.   Henocque, Y. and Denis, J. (editors) (2001). A methodological guide: steps and tools towards Integrated Coastal Area Management. *IOC Manuals and Guides* No 42. Paris. UNESCO

67.   Heylighen, F. (1991). Modelling Emergence. *World Futures.* **31**. 89 – 104

68.   Holling, C.S. (1978). Adaptive Environmental Assessment and Management. *Wiley IIASA International Series on Applied Systems Analysis.* New York. John Wiley & Sons.

69.   Horn, R. (1972). *Marine Chemistry.* [In Russian]. M. Nauka. 398 pp.

70.   Hydrometeorology and hydrochemistry of the seas of the USSR (1992). **4**. *Black Sea.* Vol. 2. *Hydrochemical Conditions and Oceanographic Fundamentals of Biological Productivity.* [In Russian].St. Petersburg. Gidrometeoizdat. 468 pp.

71.   Hydrometeorology and hydrochemistry of the seas of the USSR. (1991). **4**. *Black Sea.* Vol. 1. [In Russian]. *Hydrometeorological Conditions.* St. Petersburg. Gidrometeoizdat. 429 pp.

72.   *Hydrometeorological Charts of the Black and Azov Seas.* (1987). [In Russian]. Leningrad. GUNiO.

73.   Igumnova, E.M., and Timchenko, I.E. (2003). Modeling of adaptation processes in ecosystems. [In Russian]. *Marine Hydrophysical Journal.* № 1. 46 – 57.

74.   IISD definitions to sustainable development. (2004). *SD Gateway.htm. International Institute for Sustainable Development.* Home page: http//: www.iisd.com

75.   Ivanov, V.A., and Fomin, V.V. (2008). *Mathematical Modeling of Dynamic Processes in the Sea-land Zone.* [In Russian]. Sevastopol. ECOSY-Hydrophysica. 364 pp.

76.   Ivanov, V.A., and Ilyin, Y.P. (2009). The consequences of pollution contamination of the Kerch Strait and neighbouring areas in the storm on 10-12 November 2007. [In Russian]. *Dopovidi NASU.* v. 22, no. 1. 177 – 180.

77.   Ivanov, V.A., and Tuchkovenko, Y.S. (2006). *Applied Mathematical Water-quality Modeling of Shelf Marine Ecosystems.* [In Russian]. Sevastopol. ECOSY-Hydrophysica.   365 pp.

78.   Ivanov, V.A., and Tuchkovenko, Yu.S. (2009). *Applied Mathematical Water-Quality Modeling of Shelf Marine Ecosystems.* The Marine Hydrophysical Institute, National Academy of Sciences of Ukraine. Sevastopol. 311 pp.

79.   Ivanov, V.A., Cherkesov, L.V., and Shulga, T.Ya. (2009). The role of the Kerch Strait in the formation of steady flows and wave motions in the Sea of Azov. *Environmental Monitoring Systems.* [In Russian]. Sevastopol. ECOSY-Hydrophysica.   192 – 197.

80.   Ivanov, V.A., Pokazeev, K.V., and Schrader, A.A. (2005). *Fundamentals of Oceanography.*   [In Russian]. Sevastopol. ECOSY-Hydrophysica.   446 pp.

81.   Ivanov, V.A., Yastreb, V.P., and Khmara, T.V. (2008). Applications of the results of oceanographic research in marine resources management. *Ecological Safety of Near-shore and Shelf Zones and the Comprehensive Use of the Shelf Resources.* [In Russian]. Sevastopol. ECOSY-Hydrophysica. 17. 150 – 170

82.   Ivanov, V.A., Igumnova, E.M., Latun, V.S., and Timchenko, I.E. (2007). *Models of Coastal Zone Resources Management.* [In Russian]. Sevastopol. ECOSY-Hydrophysica. 258 pp.

83.   Ivanov, V.A., Lyubartseva, S.P., Mikhailova, E.N, and Shapiro, N.B. (2000). Seasonal variability of the ecological system of the north-western shelf from the results of a three-dimensional numerical simulation of the Black Sea ecosystem. *Ecological Safety of Near-shore and Shelf Zones and the Comprehensive Use of the Shelf Resources*. [In Russian]. Sevastopol. ECOSY-Hydrophysica. 251 - 260.

84.   Ivanov, V.A., Lyubartseva, S.P., Mikhailova, E.N, and Shapiro, N.B. (2000). Effect of wind conditions on the evolution of ecosystems near the mouth zone of the Danube in the summer *Ecological Safety of Near-shore and Shelf Zones and the Comprehensive Use of the Shelf Resources*. [In Russian]. Sevastopol. ECOSY-Hydrophysica. 261 - 268.

85.   Ivanov, V.A., Lyubartseva, S.P., Mikhailova, E.N, and Shapiro, N.B. (2003). Modern inter-discipline model of the Black Sea ecosystem. [In Russian]. *Marine Ecological Journal*. № 1. v. II. 3 – 25

86.   *Journal of Russian Ministry of Emergency Situations*. (2009). [In Russian]. http://www.mchs.gov.ru/

87.   Kalman, R., Falb, P., and Arbib, M. (1971). *Essays on Mathematical Systems Theory*. [In Russian]. M. Mir. 400 pp.

88.   Kalman, R.E. (1960). A New Approach to Linear Filtering and Prediction Problems.// *Journal of Basic Engen., Trans. of ASME*, March 1960. 35 – 45

89.   Kochergin, V.P., and Timchenko, I.E. (1987). *Monitoring of Hydrophysical Fields of the Ocean* . [In Russian]. Leningrad: Gidrometeoizdat,. 280 pp.

90.   Kolmogorov, A.N. (1941). Interpolation and extrapolation of stationary random sequences.   [In Russian]. *Izv. of AS USSR. Series Math.* **5**.

91.   Korotaev, G.K., and Eremeev, V.N. (2006). *Introduction to Operational Oceanography of the Black Sea*. [In Russian]. Sevastopol. ECOSY-Hydrophysica.   382 pp.

92.   Kotelnikov, V.A. (1933). *On the Carrying Capacity of the Ether*. [In Russian]. M., RKKA communication dep.   62 pp.

93.   Kovalev, A.V., Finenko, Z.Z., et al. (1993). *Black Sea Plankton*. [In Russian]. Kiev. Naukova Dumka.   280 pp.

94.   Land–Ocean interactions in the coastal zone. (2005). *IGBP Report 51 / IHDP Report 18*. IGBP Secretariat Box 50005 SE-104 05, Stockholm

95.   Latun, V.S. (2007). Ecological-mathematical model of the phytoplankton – zooplankton – anchovies – scad. *Environmental Monitoring Systems*. [In Russian]. Sevastopol. ECOSY-Hydrophysica. 127 – 134

96.   Lemeshev, M.J., and Shcherbina O.A. (1985). *Optimization of Recreational Activities*. [In Russian]. M. Economics. 160 pp.

97.   Leontiev, V.V. (1958). *Investigation of the Structure of the American Economy*. [In Russian]. M. Gosstatizdat.   150 pp.

98.   Levin, S. (1998). Ecosystems and the biosphere as complex adaptive systems. *Ecosystems* 1(5).   431 – 436

99.   Lorenz, E. N. (1963). Deterministic nonperiodic flow. *J. Atmos. Sci.* **20.** 130 – 141

100.   Lovelock, J. E. (1979). *Gaia: a New Look at Life on Earth*, OUP, Oxford. 124 pp.

101.   Lovelock, J.E. (1991). *Gaia: the Practical Science of Planetary Medicine*. Gaia Books Limited. London..

102.   Lyubartseva, S.P., Mikhailova, E.N, and Shapiro, N.B. (2000). Environmental three-dimensional numerical model of the Black Sea: seasonal evolution of the ecosystem's euphotic zone. [In Russian].
Marine Hydrophysical Journal. № 5. 55 – 80.

103.   McClain, R.J., and Lee, R.G. (1996) Adaptive Management: Promises and Pitfalls. *Environmental Management*, 20. 437 – 448

104.   McConnell, K., and Brue, S. (1972). *Economics: Principles, Problems and Policies*. [In Russian]. Moscow. Republic, 740 pp

105.   Margulis, L., and Sagan, D. (1995). *What is Life*. New York. Simon and Schuster. 280 pp.

106.   Martinez-Alier, J., Ropke, I. eds. (2008). *Recent Developments in Ecological Economics*. 2 vols. Cheltenham, UK,.

107.   *Mathematical Models in Biological Oceanography*. (1984). Editors: Platt, T., Mann, K.H., and Ulanovich, R.E. [In Russian]. Paris. Publ. UNESCO. 195 pp.

108.   Meadows, D.H., Meadows, D.L., Randers, J., and Behrens, W.W. (1972).*The Limits to Growth: A Report for the Club of Rome's Project on the Predicament of Mankind*. New York. Universe Books.

109.   Meadows, D.H., Meadows, D.L., and Randers, J. (1992). *Beyond the Limits: Confronting Global Collapse, Envisioning a Sustainable Future*. Post Mills VT. Chelsea Green.

110. Measuring Sustainable Development of the Coast (2004). *A Report to the EU ICZM Expert Group by the Working Group on Indicators and Data Led by the ETCTE. ETC/TE*, Brussels.

111. Mikhailova, E.N., Ivanov, V.A., Kubriakov, A.I., and Shapiro, N.B. (1998). Peculiarities of water circulation in the vicinity of the Zmeiny island under the influence of winds of different directions. [In Russian]. *Marine Hydrophysical Journal* № 4. 17 - 23

112. Moiseyev, N.N.(1981). *Mathematical Problems of System Analysis*. [In Russian]. M. Nauka. 487 pp.

113. Moiseev, N.N., Alexandrov, V.V., and Tarko, A.M. (1985). *Man and the Biosphere*. [In Russian]. M. Nauka. - 275 pp.

114. Monin, A.S., and Yaglom, A.M. (1965). *Statistical Hydromechanics*. [In Russian]. M. Nauka. 630 pp.

115. Morgan, R. (1999). Preferences and Priorities of Recreational Beach Users. *Journal of Coastal Research* **15(3)**. 653 – 667

116. Murray, C. and Marmorek, D. (2003). *Adaptive Management and Ecological Restoration*. Chapter 24, in: Freiderici, P. (ed.). 2003. Ecological Restoration of Southwestern Ponderosa Pine Forests. Island Press. Washington-Covelo-London. 417 – 428

117. Murray, J.D. (2003). *Mathematical Biology I: An Introduction*. Springer-Verlag.

118. Myshkis, A.D. (1964). *Lectures on High Mathematics*. [In Russian]. M. Nauka. 610 pp.

119. Nagata, K. Wayne. (2006). *Nonlinear Dynamics and Chaos: Mathematics 345 Lecture Notes*. Vancouver. University of B.C.

120.   Nelepo, B.A., and Timchenko, I.E. (1978). *System Principles of Analysis of Observations in Ocean.* [In Russian]. Kiev. Naukova Dumka. 222 pp.

121.   Newton, Isaac. (1999).*The Principia: Mathematical Principles of Natural Philosophy.* University of California Press. 974 pp.

122.   Nicolis, G., and Prigogine, I. (1979). *Self-organization in Non-equilibrium Systems.*   [In Russian]. M. Mir. 350 pp.

123.   Nihoul, J.C.J. (1998).   Modelling Sustainable Development as a Problem in Earth Science. *Math. Comput. Modeling.* Vol. 28. No 10. 1 – 6.

124.   Nordhaus, W.D. (1994). *Managing the Global Commons.* Cambridge, Mass. The MIT Press

125.   Nordhaus, W.D., and Yang, Z. (1995). *RICE: A Regional Dynamic General Equilibrium   Model   of Optimal Climate-Change Policy.* New Haven, Conn. Yale University and Massachusetts Institute of Technology.

126.   Norton, B.G., and Steinemann, A.C. (2001). Environmental Values and Adaptive Management. *Environmental Values.* 10. 473 – 506

127.   Norton, B.G., and Toman, M. (1997). Sustainability: Ecological and Economic Perspectives. *Land Economics.* 73(4). 553 – 568.

128.   Odum H.T. (1972). *Environment, Power and Society.* N.Y. John Willey & Sons,   331 pp.

129.   Oguz T., Malanotte-Rizzoli P., and Ducklow, H.W. (1996). *Towards Coupling Three Dimensional Eddy Resolving General Circulation and Biochemical Models in the Black Sea.*

Sensitivity to Change: Black Sea, Baltic Sea and North Sea. Kluwer Acad. Publ.
465 – 485

130.   Oguz,T., Dippner, J.W., and Kaymaz, Z. (2006). Climatic regulation of the Black Sea hydro-meteorological and ecological properties at interannual-to-decadal time scales. *Journal of Marine Systems*. 60 235 – 254

131.   Olsson, P., Folke, C., and Berkes, F. (2004). Adaptive co-management for building resilience in social-ecological systems. *Environmental Management* 34. 75 – 90

132.   Ostrom, E. (1990). *Governing the Commons*. Cambridge: Cambridge University Press.

133.   Oven, L.S., Gordina, A.D., Giragosov, V.E. et al. (1991). The current state of some exploited fish populations in the Black Sea. *The Black Sea Ecosystem Variability: Natural and Anthropogenic Factors*. [In Russian]. M. Nauka. 349 pp.

134.   Panin N. (1996). *Danube delta genesis, evolution, geological setting and sedimentology*.   Danube Delta-Black Sea System under Global Change Impact. Constantsa. IOR Publisher.
7 – 23.

135.   Parson, E.A., and Fisher-Vanden, K. (1995). Searching for Integrated Assessment: A Preliminary Investigation of Methods, Models, and Projects in the Integrated Assessment of Global Climatic Change. *Consortium for International Earth Science Information Network (CIESIN)*. University Center, Mich.

136.   Parsons, G.R., and Powell M. (2001). Measuring the Cost of Beach Retreat. *Coastal   Management* **29**. 91 – 103

137.   Pears, D., and Moran, D. (1994). *The Economic Value of Biodiversity*. London. Earthscan Publications. 172 pp.

138.   Petersen, D., and Middleton, D. (1962). Sampling and reconstruction of wave-number-limited functions in N-dimensional Euclidean spaces. *Information and Control*, **5**.   81 – 104

139.   Petrosyan, L.A., and Zakharov, V.V. (1997). *Mathematical models in ecology*.   [In Russian].   Izd.   St. Peters-burg State University.   380 pp.

140.   Plummer, R. and Armitage, D. (2007). A resilience-based framework for evaluating adaptive co-management: Linking ecology, economics and society in a complex world. *Ecological Economics* 61(1). 62 – 74

141.   Powersim solutions. (2002). *Proceedings of the 20th International Conference of the System Dynamics Society*. Palermo. 28 pp.

142.   Protecting the coastal and marine environment from impacts of land-based activities: A guide for national action. (2006). *Global Program of Action for the Protection of the Marine Environment from Land-based Activities*. United Nations Environment Program. Hague.

143.   Puhtyar, L.D., Stanichny S.V., and Timchenko, I.E. (2009). Optimal interpolation of sea surface remote sensing data. [In Russian]. *Marine Hydrophysical Journal*. № 4.
        34 – 50

144.   Raw material resources of the Black Sea. (1979). [In Russian]. Moscow: Food Industry.
        323 pp.

145.   Robinson A.R., Lermusiaux P.F.J. (2000). Overview of data assimilation. *Harvard Reports in Physical Interdisciplinary Ocean Science*. № 62. Harvard University. Cambridge, Massachusetts. 28 pp.

146.   Røpke, I. (2005). Trends in the development of ecological economics from the late 1980s to the early 2000s. Ecological Economics 55(2). 262 – 290

147.   Rotmans, J. (1989). IMAGE: *An Integrated Model to Assess the Greenhouse Effect. Amsterdam. The Netherlands*. Kluwer Academic Publishers

148.   Ryabtsev, Y. N., and Shapiro, N.B. (1995). Modeling of the formation and evolution of the cold inter-mediate layer in the Black Sea. [In Russian]. *Marine Hydrophysical Journal*. № 1. 51 – 68

149.   Ryabtsev, Y.N., and Shapiro, N.B. (1997). Modeling of the seasonal variability of the Black Sea. [In Russian]. *Marine Hydrophysical Journal*. № 1. 17 – 23

150.   Saaty, T. L. (2000). *The Fundamentals of Decision Making and Priority Theory with the Analytic   Hierarchy   Process*. RWS Publ. Vol. VI, AHP Series. 478 pp.

151.   Saaty, T.L. (1996). *Decision making with Dependence and Feedback*. RWS Publications. Pittsburg, PA.

152.   Salas, F., Marcos, C., Neto, P. J., Perez-Ruzafa, A. and Marques, J.C. (2006). User-friendly guide for using benthic ecological indicators in coastal and marine quality assessment. *Oceans & Coastal Management*. **49.** 308 – 331

153.   Samuelson, P.A. (1948). A Foundation of Economic Analysis // Cambridge, Mass.320 pp.

154.   Sarmiento, J., Slater R., Fasham, M., Ducklow, H., Toggweiler, J., and Evans, G. (1993). A seasonal three-dimensional ecosystem model of nitrogen cycling in the north Atlantic euphotic zone. *Global biogeochemical cycles*. № 7. 417 – 450

155.   Senge, P.M. (1990). *The Fifth Discipline: The Art and Practice of the learning Organization.*   New   York. Doubleday/Currency. 424 pp.

156.   Shapiro, N.B. (1998). Formation of circulation in quasi-picnical model of the Black Sea, taking into account stochastic wind stress. [In Russian]. *Marine Hydrophysical Journal.* № 6. 26 – 40

157.   Sherwood D. (2004). *Global warming and the Gaia theory – a systems approach.* Papers presented at the 2004 International Conference of the System Dynamic Society. Oxford. UK. URL: www. systemdynamic.org

158.   Shilov, G.E. (1956). *Introduction to the theory of linear spaces.* [In Russian]. M. GITIZ,. 303 pp.

159.   Shulman, G.E. (1972). *Physiological-biochemical features of the annual cycles of fish.* [In Russian]. Moscow. Food Industry. 368 pp.

160.   Simonovic, S. P. (2002). Assessment of Water Resources Through System Dynamics Simulation: From Global Issues to Regional Solutions. *Proceedings of the 36th Hawaii International Conference on System Sciences.* (HICSS'03) IEEE

161.   Soros, G. (1999). *Crisis of the world capitalism. Open society is in danger.* [In Russian]. M. NFRA-M. 262 pp.

162.   Spash, C. L. (1999) The development of environmental thinking in economics. Environmental Values 8(4). 413 – 435

163.   Sprott, J. C. (2003). *Chaos and Time-Series Analysis* .Oxford University Press

164.   Steinemann, A. (2001). Improving Alternatives for Environmental Impact Assessment. *Environmental Impact Assessment Review.* 21. 3 – 12.

165.   Stelmakh, I. V., Yunev, O. A., Finenko, Z. Z., Vedernikov V. I., et al. (1998). Peculiarities   of   seasonal variability of primary production in the Black Sea. *Ecosystem Modelling as a   Management Tool for the Black Sea*. 47. 93 – 104

166.   Sterman, J.D. (1999). *Business Dynamics: Systems Thinking and Modeling for a Complex   World*.   Irwin/McGraw-Hill.

167.   Stern Review (2007). The Economics of Climate Change: The Stern Review. Cambridge.University Press. Cambridge. UK

168.   Tesfatsion, L. (2002). Agent-based computational economics: Growing economies from the bottom up, Artificial Life. 8. 55 – 82

169.   Timchenko, I.E. (1973). Prediction of hydrophysical processes based on the Kalman filter. [In Russian]. *MGI AN UkrSSR Publications*, Sevastopol. № 1 (60).

170.   Timchenko, I.E. (1984). *Stochastic Modeling of Ocean Dynamics*. Harwood Acad. Publ.   Chur-London-Paris-New-York. 320 pp.

171.   Timchenko, I.E. (1988). *System Methods in Ocean Hydrophysics*. [In Russian].   Kiev. Naukova Dumka. 180 pp.

172.   Timchenko, I. E., and Igumnova, E. M. (2000). Space and time variability of the parameters of ecological-economical systems. *Phys. Oceanogr.*, Vol. 11, No. 2, 153 – 166

173.   Timchenko, I.E. and Igumnova, E.M. (2000). Systems Reanalysis. *Proceedings of the Second WCRP International Conference on Reanalysis (Reading, UK, August 1999)*. World Meteorological Organization Publications: WMO/TD-No. 985, January.

174. Timchenko, I.E., and Igumnova, E.M. (2003). Prediction of the natural processes by the adaptive balance of causes method. [In Russian]. *Marine Hydrophysical Journal*.
№ 6. 30 – 41

175. Timchenko, I.E., and Igumnova, E.M. (2005). *Modeling of Sustainable Development of Territories*. [In Russian]. Simferopol. Publication of Tavric National University. 132 pp.

176. Timchenko, I.E., and Igumnova, E. M. (2005). Integrated Management of Economic-Ecological Sea-Land Systems. *Physical Oceanography*. Vol. 15. № 4. October 2005.
247 – 263

177. Timchenko, I.E., and Igumnova, E.M. (2006). System analysis of environmental resources properties. *Ecological Safety of Near-shore and Shelf Zones and the Comprehensive Use of the Shelf Resources*. [In Russian]. Sevastopol. ECOSY-Hydrophysica. 60 – 85

178. Timchenko, I.E., and Igumnova, E.M. (2009). Assimilation of observational data and adaptive prediction of natural processes. [In Russian]. *Marine Hydrophysical Journal*. № 6. 47 – 70

179. Timchenko, I.E., Igumnova, E.M., and Solodova, S.M. (2001). *Natural Resources Management. Simulation technology ABC AGENT*. [In Russian]. Preprint. Sevastopol. ECOSY-Hydrophysica, 2001. 95 pp.

180. Timchenko, I.E., Igumnova, E. M., and Solodova, S. M. (2007). Adaptive balance of global development processes. *Physical Oceanography*. Vol.17, No 3, May. 2007. Http://www.springerlink.com/content/k1t6240361774225

181. Timchenko, I.E., Igumnova, E.M., and Timchenko, I.I. (2000). *System Management and the ABC-technology for Sustainable*

Development. [In Russian]. Sevastopol. ECOSY-Hydrophysica. 225 pp.

182.   Timchenko, I.E., Igumnova, E.M., and Timchenko, I.I. (2002). Multi-Agent Simulation of Sustainable Development Scenarios. *Proceedings of the 20th International Conference of the System Dynamics Society.* Palermo. 2002. 139 – 140

183.   Timchenko, I.I., Igumnova, E.M., and Timchenko, I.E. (2004). *Education and Sustainable Development. Systems Methodology.* [In Russian]. Sevastopol. ECOSY-Hydrophysica. 522 pp.

184.   Timchenko, I.E., Igumnova, E.M., Solodova, S.M., and Timchenko, I.I. (2007). Systems management with integrated model of ecological state of the environment. *Ecological Safety of Near-shore and Shelf Zones and the Comprehensive Use of the Shelf Resources.* [In Russian]. Sevastopol. ECOSY-Hydrophysica. **14**. 156 – 160

185.   Timchenko, I.E., Igumnova, E.M., and Naboykina, A.V. (2009). Systems modeling of nature management processes. *Collect. Papers of Int. Conf. "Lomonosov-2009" at the Black Sea Branch of Moscow State Univ.* Sevastopol.

186.   Timchenko, I.E., Yarin, V.D., Vasechkina, E.F., and Igumnova, E.M. (1996). *System Analysis of Marine Environment.* [In Russian]. Sevastopol. Izd. MHI NASU.
        225 pp.

187.   Towards a European Integrated Coastal Zone Management (ICZM) (1999). Strategy: General Principles and Policy Options. Luxembourg: *Office for Official Publications of the European Communities*

188.   Uemov, A.I. (1978). *The Systems Approach and General Systems Theory.* [In Russian]. Moscow. Mysl. 271 pp.

189.   Valiela, I. (1984). *Marine Ecological Processes.* N.-Y. – Berlin – Heidelberg – Tokyo.   Springer-Verlag. 547 pp.

190.   Vedernikov, V. I., and Demidov, A. B. (1993). Primary production and chlorophyll in the deep regions of the Black Sea .*Oceanology.* **33.** № 5. 639 – 647

191.   Vinogradov, M. E. (1992). Long-term variability of the pelagic community structure in the   open part of the Black Sea. *Problems of the Black Sea.* Sevastopol. MHI NASU Publ   . 19 – 33

192.   Ventana Systems. (2002). *Proceedings of the 20th International Conference of the System Dynamics Society.* Palermo. 2002. 29 – 30

193.   Vernadsky, V.I. (1931). On conditions for the emergence of life on the Earth. [In Russian]. *Izv. of AS   SSR. Geology and Geochemistry.* **2.** 633 – 653.

194.   Vinogradov, M.E., Sapozhnikov, V.V., and Shushkina, E.A. (1992). *Black Sea Ecosystem.*   [In Russian].M. Nauka,.   112 pp.

195.   Vinogradov, KA, Rozengurt, M. S., and Tolmazin, D.M. (1966). *Atlas of Hydrological Characteristics of the North-western Black Sea (in fisheries purposes).* [In Russian].Kiev.   Naukova Dumka. 96 pp

196.   UNEP (2007). Guidelines for Conducting Economic Valuation   of   Coastal   Ecosystem   Goods   and   Services, *UNEP/GEF/SCS Technical Publication* No. 8.

197.   Walmsley, D. and Arbour, J. (2005). Application of the IOC Handbook for Measuring the Progress and Outcomes of Integrated Coastal and Ocean Management. The Eastern Scotian

Shelf Integrated Management Area: A Canadian Test Case. *Report to the Intergovernmental Oceanographic Commission*

198. Waltham, MA (1974). *Pegasus Communications*. 637 pp.

199. Watson, A.J., and Lovelock, J.E. (1983). Biological homeostasis of the global environment: the parable of Daisyworld. *Tellus* 35B:284

200. Watzold, F., Drechsler, M., and Armstrong, C.W. (2004). Ecological-Economic Modeling for Biodiversity Management: Potential, Pitfalls, and Prospects. *Conservation Biology*. Volume 20, No. 4, 1034 – 1041

201. Weinstein, M. P. et al. (2007). Managing coastal resources in the 21st century. *Front Ecol. Environ.* **5(1).** 43 – 48

202. Wooldridge, M. (2002) *An Introduction to Multi-Agent Systems*. John Wiley and Sons Limited. Chichester

203. World Commission on Environment and Development. (1987). *Our Common Future, World Commission on Environment and Development*. Oxford. University Press, Oxford, UK.

204. Yaglom, A.M. (1981). *Correlation Theory of Stationary Random Functions*. [In Russian]. L. Gidrometeoizdat,. - 280 pp.